Lecture Notes in Computer Science   9867

Commenced Publication in 1973
Founding and Former Series Editors:
Gerhard Goos, Juris Hartmanis, and Jan van Leeuwen

More information about this series at http://www.springer.com/series/7409

Josep Domingo-Ferrer · Mirjana Pejić-Bach (Eds.)

Privacy in Statistical Databases

UNESCO Chair in Data Privacy
International Conference, PSD 2016
Dubrovnik, Croatia, September 14–16, 2016
Proceedings

 Springer

Editors
Josep Domingo-Ferrer
Universitat Rovira i Virgili
Tarragona
Spain

Mirjana Pejić-Bach
University of Zagreb
Zagreb
Croatia

ISSN 0302-9743   ISSN 1611-3349   (electronic)
Lecture Notes in Computer Science
ISBN 978-3-319-45380-4   ISBN 978-3-319-45381-1   (eBook)
DOI 10.1007/978-3-319-45381-1

Library of Congress Control Number: 2016948609

LNCS Sublibrary: SL3 – Information Systems and Applications, incl. Internet/Web, and HCI

Printed on acid-free paper

This Springer imprint is published by Springer Nature
The registered company is Springer International Publishing AG Switzerland

Preface

Privacy in statistical databases is a discipline whose purpose it is to provide solutions to the tension between the social, political, economic, and corporate demand for accurate information, and the legal and ethical obligation to protect the privacy of the various parties involved. Those parties are the subjects, sometimes also known as respondents (the individuals and enterprises to which the data refer), the data controllers (those organizations collecting, curating, and to some extent sharing or releasing the data), and the users (the ones querying the database or the search engine, who would like their queries to stay confidential). Beyond law and ethics, there are also practical reasons for data controllers to invest in subject privacy: if individual subjects feel their privacy is guaranteed, they are likely to provide more accurate responses. Data controller privacy is primarily motivated by practical considerations: if an enterprise collects data at its own expense and responsibility, it may wish to minimize leakage of those data to other enterprises (even to those with whom joint data exploitation is planned). Finally, user privacy results in increased user satisfaction, even if it may curtail the ability of the data controller to profile users.

There are at least two traditions in statistical database privacy, both of which started in the 1970s: the first one stems from official statistics, where the discipline is also known as statistical disclosure control (SDC) or statistical disclosure limitation (SDL), and the second one originates from computer science and database technology. In official statistics, the basic concern is subject privacy. In computer science, the initial motivation was also subject privacy but, from 2000 onwards, growing attention has been devoted to controller privacy (privacy-preserving data mining) and user privacy (private information retrieval). In the last few years, the interest and the achievements of computer scientists in the topic have substantially increased, as reflected in the contents of this volume. At the same time, the generalization of big data is challenging privacy technologies in many ways: this volume also contains recent research aimed at tackling some of these challenges.

"Privacy in Statistical Databases 2016" (PSD 2016) was held under the sponsorship of the UNESCO Chair in Data Privacy, which has provided a stable umbrella for the PSD biennial conference series since 2008. Previous PSD conferences were PSD 2014, held in Eivissa; PSD 2012, held in Palermo; PSD 2010, held in Corfu; PSD 2008, held in Istanbul; PSD 2006, the final conference of the Eurostat-funded CENEX-SDC project, held in Rome; and PSD 2004, the final conference of the European FP5 CASC project, held in Barcelona.

Proceedings of PSD 2014, PSD 2012, PSD 2010, PSD 2008, PSD 2006, and PSD 2004 were published by Springer in LNCS 8744, LNCS 7556, LNCS 6344, LNCS 5262, LNCS 4302, and LNCS 3050, respectively.

The seven PSD conferences held so far are a follow-up of a series of high-quality technical conferences on SDC that started eighteen years ago with "Statistical Data Protection-SDP'98", held in Lisbon in 1998 and with proceedings published by

OPOCE, and continued with the AMRADS project SDC Workshop, held in Luxemburg in 2001 and with proceedings published by Springer in LNCS 2316.

The PSD 2016 Program Committee accepted for publication in this volume 19 papers out of 35 submissions. Furthermore, 5 of the above submissions were reviewed for short presentation at the conference and inclusion in the companion CD proceedings. Papers came from 14 different countries and four different continents. Each submitted paper received at least two reviews. The revised versions of the 19 accepted papers in this volume are a fine blend of contributions from official statistics and computer science.

Covered topics include tabular data protection, microdata and big data masking, protection using privacy models, synthetic data, disclosure risk assessment, remote and cloud access, and co-utile anonymization.

We are indebted to many people. First, to the Organization Committee for making the conference possible and especially to Jesús A. Manjón, who helped prepare these proceedings, and Goran Lesaja, who helped in the local arrangements. In evaluating the papers we were assisted by the Program Committee and by Yu-Xiang Wang as an external reviewer.

We also wish to thank all the authors of submitted papers and we apologize for possible omissions.

Finally, we dedicate this volume to the memory of Dr Lawrence Cox, who was a Program Committee member of all past editions of the PSD conference.

July 2016

Josep Domingo-Ferrer
Mirjana Pejić-Bach

Organization

Program Committee

Jane Bambauer	University of Arizona, USA
Bettina Berendt	Katholieke Universiteit Leuven, Belgium
Elisa Bertino	CERIAS, Purdue University, USA
Aleksandra Bujnowska	EUROSTAT, European Union
Jordi Castro	Polytechnical University of Catalonia, Catalonia
Lawrence Cox	National Institute of Statistical Sciences, USA
Josep Domingo-Ferrer	Universitat Rovira i Virgili, Catalonia
Jörg Drechsler	IAB, Germany
Mark Elliot	Manchester University, UK
Stephen Fienberg	Carnegie Mellon University, USA
Sarah Giessing	Destatis, Germany
Sara Hajian	Eurecat Technology Center, Catalonia
Julia Lane	New York University, USA
Bradley Malin	Vanderbilt University, USA
Oliver Mason	National University of Ireland-Maynooth, Ireland
Laura McKenna	Census Bureau, USA
Gerome Miklau	University of Massachusetts-Amherst, USA
Krishnamurty Muralidhar	The University of Oklahoma, USA
Anna Oganian	National Center for Health Statistics, USA
Christine O'Keefe	CSIRO, Australia
Jerry Reiter	Duke University, USA
Yosef Rinott	Hebrew University, Israel
Juan José Salazar	University of La Laguna, Spain
Pierangela Samarati	University of Milan, Italy
David Sánchez	Universitat Rovira i Virgili, Catalonia
Eric Schulte-Nordholt	Statistics Netherlands
Natalie Shlomo	University of Manchester, UK
Aleksandra Slavković	Penn State University, USA
Jordi Soria-Comas	Universitat Rovira i Virgili, Catalonia
Tamir Tassa	The Open University, Israel
Vicenç Torra	Skövde University, Sweden
Vassilios Verykios	Hellenic Open University, Greece
William E. Winkler	Census Bureau, USA
Peter-Paul de Wolf	Statistics Netherlands

Program Chair

Josep Domingo-Ferrer UNESCO Chair in Data Privacy,
Universitat Rovira i Virgili, Catalonia

General Chair

Mirjana Pejić-Bach Faculty of Business & Economics,
University of Zagreb, Croatia

Organization Committee

Vlasta Brunsko Centre for Advanced Academic Studies,
University of Zagreb, Croatia
Ksenija Dumicic Faculty of Business & Economics,
University of Zagreb, Croatia
Joaquín García-Alfaro Télécom SudParis, France
Goran Lesaja Georgia Southern University, USA
Jesús A. Manjón Universitat Rovira i Virgili, Catalonia
Tamar Molina Universitat Rovira i Virgili, Catalonia
Sara Ricci Universitat Rovira i Virgili, Catalonia

Contents

Tabular Data Protection

Revisiting Interval Protection, a.k.a. Partial Cell Suppression, for Tabular Data

Jordi Castro[1(✉)] and Anna Via[2]

[1] Department of Statistics and Operations Research,
Universitat Politècnica de Catalunya,
Jordi Girona 1-3, 08034 Barcelona, Catalonia, Spain
jordi.castro@upc.edu
[2] School of Mathematics and Statistics, Universitat Politècnica de Catalunya,
Pau Gargallo 5, 08028 Barcelona, Catalonia, Spain
annaa35@gmail.com

Abstract. Interval protection or partial cell suppression was introduced in "M. Fischetti, J.-J. Salazar, Partial cell suppression: A new methodology for statistical disclosure control, *Statistics and Computing*, 13, 13–21, 2003" as a "linearization" of the difficult cell suppression problem. Interval protection replaces some cells by intervals containing the original cell value, unlike in cell suppression where the values are suppressed. Although the resulting optimization problem is still huge—as in cell suppression, it is linear, thus allowing the application of efficient procedures. In this work we present preliminary results with a prototype implementation of Benders decomposition for interval protection. Although the above seminal publication about partial cell suppression applied a similar methodology, our approach differs in two aspects: (i) the boundaries of the intervals are completely independent in our implementation, whereas the one of 2003 solved a simpler variant where boundaries must satisfy a certain ratio; (ii) our prototype is applied to a set of seven general and hierarchical tables, whereas only three two-dimensional tables were solved with the implementation of 2003.

Keywords: Statistical disclosure control · Tabular data · Interval protection · Cell suppression · Linear optimization · Large-scale optimization

1 Introduction

Post-tabular data protection methods are based on modifying or suppressing some of the table cells, yet satisfying the table additivity (that is, the sum of the "inner" cells has to be equal to the marginal cell) and preserving the original value of a subset of cells (e.g., some subtotal or total cells). This is the main

Supported by grant MTM2015-65362-R of the Spanish Ministry of Economy and Competitiveness.

J. Domingo-Ferrer and M. Pejić-Bach (Eds.): PSD 2016, LNCS 9867, pp. 3–14, 2016.
DOI: 10.1007/978-3-319-45381-1_1

difference compared to pre-tabular methods, which at the same time cannot guarantee table additivity and the original value of a subset of cells. Among post-tabular data protection methods we find cell suppression [4,9] and controlled tabular adjustment [1,3], both formulating difficult mixed integer linear optimization problems. More details can be found in the monograph [12] and the survey [5].

Interval protection or partial cell suppression was introduced in [10] as a linearization of the difficult cell suppression problem. Unlike in cell suppression, interval protection replaces some cell values by intervals containing the true value. From those intervals, no attacker can be able to recompute the true value within some predefined lower and upper protection levels. One of the great advantages of interval suppression against alternative approaches is that the resulting optimization problem is convex and continuous, which means that theoretically it can be efficiently solved in polynomial time by, for instance, interior-point methods [13]. Therefore, theoretically, this approach is valid for big tables from the big-data era.

However, attempting to solve the resulting "monolithic" linear optimization model by some state-of-the-art solver is almost impossible for huge tables: we will either exhaust the RAM memory of the computer, or we will require a large CPU time. Alternative approaches to be tried include a Benders decomposition of this huge linear optimization problem. In this work we present preliminary results with a prototype implementation of Benders decomposition. A similar approach was used in the seminal publication [10] about partial cell suppression. However, this work differs in two substantial aspects: (i) our implementation considers two independent boundaries for each cell interval, whereas those two boundaries were forced to satisfy a ratio in the code of [10] (that is, actually only one boundary was considered in the 2003 code, thus solving a simpler variant of the problem); (ii) we applied our prototype to a set of seven general and hierarchical tables, where results for only three two-dimensional tables were reported in [10]. As we will see, our "not-too efficient and tuned" classical Benders decomposition prototype still outperforms state-of-the-art solvers in these complex tables.

The paper is organized as follows. Section 2 describes the general interval protection method. Section 3 outlines the Benders solution approach. The particular form of Benders for interval protection is shown in Sect. 4, which is illustrated by a small example in Subsect. 4.1. Finally, Sect. 5 reports computational results with some general and hierarchical tables.

2   The General Interval Protection Problem Formulation

We are given a table (i.e., a set of cells $a_i, i \in \mathcal{N} = \{1, \ldots, n\}$), satisfying m linear relations $Aa = b$, $A \in \mathbb{R}^{m \times n}, b \in \mathbb{R}^m$. Any set of values x satisfying $Ax = b, l \leq x \leq u$, is a valid table, $l \in \mathbb{R}^n$, $u \in \mathbb{R}^n$ being known a priori lower and upper bounds for cell values. For positive tables we have $l_i = 0$, $u_i = +\infty$, $i = 1, \ldots, n$, but the procedure here outlined is also valid for general tables.

For instance, we may consider the cells provide information about some attribute for several individual states (e.g., member states of European Union), as well as the highest-level of aggregated information (e.g., at European Union level). The set of multi-state cells, or cells providing this highest-level of aggregated information could be the ones to be replaced by intervals, and they will be denoted as $\mathcal{H} \subseteq \mathcal{N}$.

Let $\mathcal{F}, \mathcal{S}, \mathcal{M}$ be a partition of $\mathcal{N}$, i.e., $\mathcal{N} = \mathcal{F} \cup \mathcal{S} \cup \mathcal{M}$, and $\mathcal{F} \cap \mathcal{S} = \mathcal{F} \cap \mathcal{M} = \mathcal{S} \cap \mathcal{M} = \emptyset$. $\mathcal{S}$ is the set of sensitive cells to be protected, with upper and lower protection levels upl_s and lpl_s for each cell $s \in \mathcal{S}$. $\mathcal{F}$ is the set of cells whose values are known (e.g., they have been previously published by individual states). $\mathcal{M}$ is the set of non-sensitive and non previously published cells. To simplify the formulation of the forthcoming optimization problems, we can assume that for $f \in \mathcal{F}$ we have $l_f = u_f = a_f$, and then cells from $\mathcal{F}$ can be considered elements of $\mathcal{M}$, that is, $\mathcal{M} \leftarrow \mathcal{M} \cup \mathcal{F}$ and $\mathcal{F} \leftarrow \emptyset$. Following our example, we have that, in general, cells in $\mathcal{S}$ provide information at state level, but in some cases multi-state cells may also be sensitive; thus we may have $\mathcal{S} \cap \mathcal{H} \neq \emptyset$. In a similar way, since multi-state cells may not have been previously published we may also have $\mathcal{M} \cap \mathcal{H} \neq \emptyset$. To make the formulation more general our only assumption will be that $\mathcal{H} \subseteq \mathcal{N}$. When $\mathcal{H} = \mathcal{N}$ we just have the standard "interval protection" or "partial cell suppression" introduced in [10].

Our purpose is to publish the set of smallest intervals $[lb_h, ub_h]$—where $l_h \leq lb_h$ and $ub_h \leq u_h$— for each cell $h \in \mathcal{H}$ instead of the real value $a_h \in [lb_h, ub_h]$, such that, from these intervals, no attacker can determine that $a_s \in (a_s - lpl_s, a_s + upl_s)$ for all sensitive cells $s \in \mathcal{S}$. This means that

$$\underline{a_s} \leq a_s - lpl_s \quad \text{and} \quad \overline{a_s} \geq a_s + upl_s, \tag{1}$$

$\underline{a_s}$ and $\overline{a_s}$ being defined as

$$
\begin{array}{ll}
\underline{a_s} = \min x_s & \overline{a_s} = \max x_s \\
\quad \text{s.to } Ax = b & \quad \text{s.to } Ax = b \\
\quad l_i \leq x_i \leq u_i \quad i \in \mathcal{N} \setminus \mathcal{H} \quad \text{and} & \quad l_i \leq x_i \leq u_i \quad i \in \mathcal{N} \setminus \mathcal{H} \\
\quad lb_i \leq x_i \leq ub_i \ i \in \mathcal{H} & \quad lb_i \leq x_i \leq ub_i \ i \in \mathcal{H}
\end{array}
\tag{2}
$$

Clearly, for cells $i \in \mathcal{H} \cap \mathcal{S}$, (1) and (2) imply that $lb_i \leq a_i - lpl_i$ and $ub_i \geq a_i + upl_i$.

The previous problem can be formulated as a large-scale linear optimization problem. For each primary cell $s \in \mathcal{S}$, two auxiliary vectors $x^{l,s} \in \mathbb{R}^n$ and $x^{u,s} \in \mathbb{R}^n$ are introduced to impose, respectively, the lower and upper protection requirement of (1). The problem formulation is as follows:

$$\min \sum_{i \in \mathcal{H}} w_i(ub_i - lb_i)$$

s.to

$$
\left.
\begin{array}{l}
\quad Ax^{l,s} = b \\
l_i \le \ x_i^{l,s} \ \le u_i \qquad i \in \mathcal{N} \setminus \mathcal{H} \\
lb_i \le \ x_i^{l,s} \ \le ub_i \qquad i \in \mathcal{H} \\
\qquad x_s^{l,s} \le a_s - lpl_s \\
\\
\quad Ax^{u,s} = b \\
l_i \le \ x_i^{u,s} \ \le u_i \qquad i \in \mathcal{N} \setminus \mathcal{H} \\
lb_i \le \ x_i^{u,s} \ \le ub_i \qquad i \in \mathcal{H} \\
\qquad x_s^{u,s} \ge a_s + upl_s
\end{array}
\right\} \quad \forall \, s \in \mathcal{S}
\tag{3}
$$

$$
\begin{array}{l}
l_i \le lb_i \le a_i \quad i \in \mathcal{H} \\
a_i \le ub_i \le u_i \quad i \in \mathcal{H}
\end{array}
$$

where w_i is a weight for the information loss associated with cell a_i.

Problem (3) is very large (easily in the order of millions of variables and constraints), but it is linear (no binary, no integer variables), and thus theoretically it can be efficiently solved in polynomial time by general or by specialized interior-point algorithms [7, 13].

3   Outline of Benders Decomposition

Benders decomposition [2] was suggested for problems with two types of variables, one of them considered as "complicating variables". In MILP models complicating variables are the binary/integer ones; in continuous problems, the complicating variables are usually associated to linking variables between groups of constraints (i.e., variables lb and ub in (3)). Consider the following primal problem (P) with two groups of variables (x, y)

$$
(P) \quad
\begin{array}{ll}
\min & c^{\top}x + d^{\top}y \\
\text{s. to} & A_1 x + A_2 y = b \\
& x \ge 0 \\
& y \in Y,
\end{array}
$$

where y are the complicating variables, $c, x \in \mathbb{R}^{n_1}$, $d, y \in \mathbb{R}^{n_2}$, $A_1 \in \mathbb{R}^{m \times n_1}$ and $A_2 \in \mathbb{R}^{m \times n_2}$. Fixing some $y \in Y$, we obtain:

$$
(Q) \quad
\begin{array}{ll}
\min & c^{\top}x \\
\text{s. to} & A_1 x = b - A_2 y \\
& x \ge 0.
\end{array}
$$

The dual of (Q) is:

$$\begin{aligned} &\max && u^\top(b - A_2y) \\ (Q_D) \quad &\text{s. to} && A_1^\top u \le c \\ & && u \in \mathbb{R}^m. \end{aligned}$$

It is known that if (Q_D) has a solution then (Q) has a solution too, and both objective functions coincide; if (Q_D) is unbounded, then (Q) is infeasible. Let assume that (Q_D) is never infeasible (in the interval protection problem this is always the case). If, as notation convention, we consider that the objective of (Q) is $+\infty$ when it is infeasible, then (P) can be written as

$$(P') \quad \begin{aligned} &\min && \{d^\top y + \max\{u^\top(b - A_2y)|A_1^\top u \le c, u \in \mathbb{R}^m\}\} \\ &\text{s. to} && y \in Y. \end{aligned}$$

Let $U = \{u|A_1^\top u \le c, u \in \mathbb{R}^m\}$ be the convex feasible set of (Q_D). By Minkowski representation we know that every point $u \in U$ may be represented as a convex combination of the vertices $u^1, \ldots, u^s$ and extreme rays $v^1, \ldots, v^t$ of the convex polytope U. Therefore any $u \in U$ may be written as

$$\begin{aligned} u &= \sum_{i=1}^s \lambda_i u^i + \sum_{j=1}^t \mu_j v^j \\ &\sum_{i=1}^s \lambda_i = 1 \\ \lambda_i &\ge 0 \quad i = 1, \ldots, s \\ \mu_j &\ge 0 \quad j = 1, \ldots, t. \end{aligned}$$

If ${v^j}^\top(b - A_2y) > 0$ for some $j \in \{1, \ldots, t\}$ then (Q_D) is unbounded, and thus (Q) is infeasible. We then impose

$${v^j}^\top(b - A_2y) \le 0 \quad j = 1, \ldots, t.$$

The optimal solution of (Q_D) is then known to be in a vertex of U, and (P') may be rewritten as

$$(P'') \quad \begin{aligned} &\min && d^\top y + \max_{i=1,\ldots,s}({u^i}^\top(b - A_2y)) \\ &\text{s. to} && {v^j}^\top(b - A_2y) \le 0 \quad j = 1, \ldots, t \\ & && y \in Y. \end{aligned}$$

Introducing variable θ, (P'') is equivalent to the Benders problem (BP):

$$(BP) \quad \begin{aligned} &\min && \theta \\ &\text{s. to} && \theta \ge d^\top y + {u^i}^\top(b - A_2y) \quad i = 1, \ldots, s \\ & && {v^j}^\top(b - A_2y) \le 0 \quad j = 1, \ldots, t \\ & && y \in Y. \end{aligned}$$

Problem (BP) is impractical since s and t can be very large, and in addition the vertices and extreme rays are unknown. Instead, the method considers a relaxation (BP_r) with a subset of the vertices and extreme rays. The relaxed Benders problem (or master problem) is thus:

$$
(BP_r) \quad
\begin{array}{ll}
\min & \theta \\
\text{s. to} & \theta \geq d^\top y + {u^i}^\top (b - A_2 y) \quad i \in I \subseteq \{1, \ldots, s\} \\
& {v^j}^\top (b - A_2 y) \leq 0 \quad\quad\quad j \in J \subseteq \{1, \ldots, t\} \\
& y \in Y.
\end{array}
$$

Initially $I = J = \emptyset$, and new vertices and extreme rays provided by the subproblem (Q_D) are added to the master problem, until the optimal solution is found. In summary, the steps of the Benders algorithm are:

Benders algorithm

0. Initially $I = \emptyset$ and $J = \emptyset$. Let (θ_r^*, y_r^*) be the solution of current master problem (BP_r), and (θ^*, y^*) the optimal solution of (BP).
1. Solve master problem (BP_r) obtaining θ_r^* and y_r^*. At first iteration, $\theta_r^* = -\infty$ and y_r is any feasible point in Y.
2. Solve subproblem (Q_D) using $y = y_r^*$. There are two cases:
   (a) (Q_D) has finite optimal solution in vertex u^{i0}.
      – If $\theta_r^* = d^\top y_r^* + {u^{i0}}^\top (b - A_2 y_r^*)$ then **STOP**. Optimal solution is $y^* = y_r^*$ with cost $\theta^* = \theta_r^*$.
      – If $\theta_r^* < d^\top y_r^* + {u^{i0}}^\top (b - A_2 y_r^*)$ then this solution violates constraint of (BP) $\theta > d^\top y + {u^{i0}}^\top (b - A_2 y)$. Add this new constraint to (BP_r): $I \leftarrow I \cup \{i_0\}$.
   (b) (Q_D) is unbounded along segment $u^{i0} + \lambda v^{j0}$ (u^{i0} is current vertex, v^{j0} is extreme ray). Then this solution violates constraint of (BP) ${v^{j0}}^\top (b - A_2 w) \leq 0$. Add this new constraint to (BP_r): $J \leftarrow J \cup \{j_0\}$; vertex may also be added: $I \leftarrow I \cup \{i_0\}$.
3. Go to step 1 above.

Convergence is guaranteed since at each iteration one or two constraints are added to (BP_r), no constraints are repeated, and the maximum number of constraints is $s + t$.

4   Benders Decomposition for the Interval Protection Problem

Problem (3) has two groups of variables: $x^{l,s} \in \mathbb{R}^n$, $x^{u,s} \in \mathbb{R}^n$; and $lb \in \mathbb{R}^{|\mathcal{H}|}$, $ub \in \mathbb{R}^{|\mathcal{H}|}$, which can be seen as the complicating variables, since if they are fixed, the resulting problem in variables $x^{l,s}$ and $x^{u,s}$ is separable, as shown below. Indeed, projecting out the $x^{l,s}$, $x^{u,s}$ variables, (3) can be written as

$$\min \sum_{i \in \mathcal{H}} w_i(ub_i - lb_i) + Q(ub, lb)$$
$$\text{s.to } l_i \le lb_i \le a_i \quad i \in \mathcal{H}$$
$$a_i \le ub_i \le u_i \quad i \in \mathcal{H}$$

(4)

where

$$Q(ub, lb) = \min \sum_{s \in \mathcal{S}} (0_n^\top x^{l,s} + 0_n^\top x^{u,s}) = 0$$

s.to

$$\left. \begin{array}{ll} Ax^{l,s} = b & \\ l_i \le x_i^{l,s} \le u_i & i \in \mathcal{N} \setminus \mathcal{H} \\ lb_i \le x_i^{l,s} \le ub_i & i \in \mathcal{H} \\ x_s^{l,s} \le a_s - lpl_s & \\ \\ Ax^{u,s} = b & \\ l_i \le x_i^{u,s} \le u_i & i \in \mathcal{N} \setminus \mathcal{H} \\ lb_i \le x_i^{u,s} \le ub_i & i \in \mathcal{H} \\ x_s^{u,s} \ge a_s + upl_s & \end{array} \right\} \quad \forall \, s \in \mathcal{S},$$

(5)

$0_n \in \mathbb{R}^n$ denoting the zero vector. Problem (5) is separable in the $x^{l,s}$, $x^{u,s}$ variables for each $s \in \mathcal{S}$ so it can be replaced by the solution of $2|\mathcal{S}|$ smaller problems of the form

$$Q^{l,s}(ub, lb) = \min 0_n^\top x^{l,s} = 0$$
$$\text{s.to} \quad Ax^{l,s} = b$$
$$l_i \le x_i^{l,s} \le u_i \quad i \in \mathcal{N} \setminus \mathcal{H}$$
$$lb_i \le x_i^{l,s} \le ub_i \quad i \in \mathcal{H}$$
$$x_s^{l,s} \le a_s - lpl_s \, ,$$

(6)

for the lower protection of sensitive cell $s \in \mathcal{S}$, and

$$Q^{u,s}(ub, lb) = \min 0_n^\top x^{u,s} = 0$$
$$\text{s.to} \quad Ax^{u,s} = b$$
$$l_i \le x_i^{u,s} \le u_i \quad i \in \mathcal{N} \setminus \mathcal{H}$$
$$lb_i \le x_i^{u,s} \le ub_i \quad i \in \mathcal{H}$$
$$x_s^{u,s} \ge a_s + upl_s \, .$$

(7)

for the upper protection of sensitive cell $s \in \mathcal{S}$. Note that (5)–(7) are just feasibility problems with a constant (dummy) objective function.

Problems (6) and (7) are our Benders subproblems. Due to its constant objective function, (6) and (7) are feasibility problems. Therefore Benders algorithm will only include extreme rays of the dual formulations of (6) and (7) to guarantee the feasibility of the values of lb and ub provided by the master problem.

Denoting the j-th extreme ray of the dual formulation of (6) as $v_{l,s}^j = (v_j^{\lambda^{l,s}}, v_j^{\mu_l^{l,s}}, v_j^{\mu_u^{l,s}}, v_j^{\nu^{l,s}})$, where $\lambda^{l,s}$, $\mu_l^{l,s}$, $\mu_u^{l,s}$ and $\nu^{l,s}$ refer to the indices of the

Lagrange multipliers of the constraints of (6), it can be shown that the feasibility cut to be added to the master problem would be

$$0 \geq \sum_{i=1}^{m} v_{j,i}^{\lambda^{l,s}} b_i + \sum_{i \in \mathcal{N} \setminus \mathcal{H}} (-v_{j,i}^{\mu_u^{l,s}} u_i + v_{j,i}^{\mu_l^{l,s}} l_i) + \sum_{i \in \mathcal{H}} (-v_{j,i}^{\mu_u^{l,s}} ub_i + v_{j,i}^{\mu_l^{l,s}} lb_i) - (a_s - lpl_s) v_j^{\nu^{l,s}}$$
$$= g_{l,s}^{j}(ub, lb).$$

(8)

The extreme rays of the dual of (7) have an analogous form $v_{u,s}^{j} = (v_j^{\lambda^{u,s}}, v_j^{\mu_l^{u,s}}, v_j^{\mu_u^{u,s}}, v_j^{\nu^{u,s}})$ and so does the feasibility cut to be added to the master problem:

$$0 \geq \sum_{i=1}^{m} v_{j,i}^{\lambda^{u,s}} b_i + \sum_{i \in \mathcal{N} \setminus \mathcal{H}} (-v_{j,i}^{\mu_u^{u,s}} u_i + v_{j,i}^{\mu_l^{u,s}} l_i) + \sum_{i \in \mathcal{H}} (-v_{j,i}^{\mu_u^{u,s}} ub_i + v_{j,i}^{\mu_l^{u,s}} lb_i) + (a_s - lpl_s) v_j^{\nu^{u,s}}$$
$$= g_{u,s}^{j}(ub, lb).$$

(9)

Denoting as $\mathcal{I}_{l,s}$ and $\mathcal{I}_{u,s}$ the set of indices of feasibility cuts obtained from $Q^{l,s}$ and $Q^{u,s}$, the master problem is:

$$\begin{aligned}
\min \ & \sum_{i \in \mathcal{H}} w_i(ub_i - lb_i) \\
\text{s.to} \ & g_j^{l,s}(ub, lb) \leq 0 \quad j \in \mathcal{I}_{l,s} \\
& g_j^{u,s}(ub, lb) \leq 0 \quad j \in \mathcal{I}_{u,s} \\
& l_i \leq lb_i \leq a_i \qquad i \in \mathcal{H} \\
& a_i \leq ub_i \leq u_i \qquad i \in \mathcal{H}.
\end{aligned}$$

(10)

The Benders decomposition algorithm will then solve (10) for the master problem and the duals of (6) and (7) for the subproblems.

4.1   Illustrative Example

Consider the following simple table

10	15	25
20	17	37

of $n = 6$ cells and $m = 2$ linear constraints associated to row totals

$$a_1 + a_2 - a_3 = 0$$
$$a_4 + a_5 - a_6 = 0$$

(we don't consider column totals to simplify the example), where $\mathcal{H} = \mathcal{N} = \{1, \ldots, 6\}$, and a_1 and a_5 as the two sensitive cells, whose parameters are given by

s	a_s	lpl_s	upl_s
1	10	5	5 .
5	17	7	4

Note that this example, in principle, can not be solved with the original implementation of [10] since the ratios between upper and lower protection levels are not the same for all sensitive cells.

We next show the application of Benders algorithm to the previous table:

1. **Initialization.**

   The number of cuts for the lb and the ub variables is set to 0, this means $\mathcal{I}_{l,s} = \mathcal{I}_{u,s} = \emptyset$. The first master problem to be solved is thus

   $$\min \sum_{i=1}^{6}(ub_i - lb_i)$$
   $$\text{s.to } l_i \leq lb_i \leq a_i \quad i = 1, \ldots, 6$$
   $$a_i \leq ub_i \leq u_i \quad i = 1, \ldots, 6,$$

   obtaining some initial values for lb, ub.

2. **Iterating Through Benders' Algorithm.**

   Cut generation is based on (8)–(9), details are omitted to simplify the exposition.

   – **Iteration 1.** The two Benders cuts obtained for cell 1 are $lb_1 \leq 5$ and $ub_1 \geq 15$. The two Benders cuts obtained for cell 5 are $lb_5 \leq 10$ and $ub_1 \geq 21$. Note these are obvious cuts associated to the protection levels of sensitive cells, that could have been added from the beginning in an efficient implementation, thus avoiding this first Benders iteration.

   – **Iteration 2.** The current master subproblem

   $$\min \sum_{i=1}^{6}(ub_i - lb_i)$$
   $$\text{s.to } l_i \leq lb_i \leq a_i \quad i \in 1, \ldots, 6$$
   $$a_i \leq ub_i \leq u_i \quad i \in 1, \ldots, 6$$
   $$lb_1 \leq 5, \qquad ub_1 \geq 15$$
   $$lb_5 \leq 10, \qquad ub_5 \geq 21$$

   has solution $lb = [5, 15, 25, 20, 10, 37]$ and $ub = [15, 15, 25, 20, 21, 37]$. Using this solution the two Benders cuts obtained for cell 1 are $lb_3 - ub_2 \leq 58$ and $lb_2 - ub_2 \geq 15$. The two cuts obtained for cell 5 are $lb_6 - ub_4 \leq 21$ and $ub_6 - lb_4 \geq 39$.

   – **Iteration 3.** The current master problem is

   $$\min \sum_{i=1}^{6}(ub_i - lb_i)$$
   $$\text{s.to } l_i \leq lb_i \leq a_i \quad i \in 1, \ldots, 6$$
   $$a_i \leq ub_i \leq u_i \quad i \in 1, \ldots, 6$$
   $$lb_1 \leq 5, \qquad ub_1 \geq 15$$
   $$lb_5 \leq 10, \qquad ub_5 \geq 21$$
   $$lb_3 - ub_2 \leq 58, \quad lb_2 - ub_2 \geq 15$$
   $$lb_6 - ub_4 \leq 21, \quad ub_6 - lb_4 \geq 39,$$

   with solution $lb = [5, 15, 20, 16, 10, 30]$ and $ub = [15, 15, 30, 20, 21, 37]$. Benders subproblems happen to be feasible with these values, thus we have an optimal solution of objective $\sum_{i=1}^{6}(ub_i - lb_i) = 42$. Since this table is small, the original model was solved using some off-the-shelf optimization solver, obtaining the same optimal objective function.

3. **Auditing.** Although this step is not needed with interval protection, to be sure that this solution satisfies that no attacker can determine that $a_s \in (a_s - lpl_s, a_s + upl_s)$ for $s \in \{1, 5\}$, the problems (2) were solved, obtaining $\underline{a_1} = 5$, $\overline{a_1} = 15$, $\underline{a_5} = 10$ and $\overline{a_5} = 21$. Therefore, it can be asserted that it is safe to publish this solution.

4. **Publication of the table.** The final safe table to be published would be

$$\begin{array}{|ccc|}
\hline
[5, 15] & 15 & [20, 30] \\
[16, 20] & [10, 21] & [30, 37] \\
\hline
\end{array}.$$

5   Computational Results

We developed a prototype implementation of the Benders algorithm for interval protection using the AMPL modeling language [11] and Cplex for the master and subproblems. We solved seven instances, whose dimensions are given in Table 1. Columns n, $|\mathcal{S}|$ and m provide, respectively, the number of cells, sensitive cells and table linear equations. Table "targus" is a general table, while the remaining six tables are 1H2D tables (i.e., two-dimensional hierarchical tables with one hierarchical variable) obtained with a generator used in the literature [1,8].

Table 1. Instance dimensions and results with Benders decomposition

Table	n	$|\mathcal{S}|$	m	CPU	it_B	it_S	obj
Targus	162	13	63	5.17	31	8872	2142265.7
Table 1	121	10	55	3.41	26	7167	136924
Table 2	1680	158	299	410.53	43	1104884	43715149
Table 3	600	53	170	26.38	43	131834	3624906
Table 4	756	68	243	50.92	33	144963	9134139
Table 5	168	14	62	3.95	19	5959	303844
Table 6	1584	143	485	966.28	70	1729767	21302104

The results obtained with the Benders decomposition are provided in the last columns of Table 1. Columns "CPU", "it_B", "it_S" and "obj" provide respectively the total CPU time, number of Benders iterations, overall number of simplex iterations, and the final optimal objective function obtained.

Table 2 provides results for the solution of the monolithic model (3) using Cplex default linear algorithm (dual simplex). Column "n.var" reports the number of variables of the resulting linear optimization problem. The meaning of remaining columns is the same as in Table 1. Three executions, clearly marked, were aborted because the CPU time was excessive compared with the solution

Table 2. Results using Cplex for monolithic model

Table	CPU	it$_S$	n.var	obj
Targus	36.0515	16532	4212	2142265.7
Table 1	3.43548	7452	2420	136924
Table 2	2944.87^a	—	530880	16056608400
Table 3	522.875^a	—	63600	260592812
Table 4	11085.6	436895	102816	9134139
Table 5	10.6764	17325	4704	303844
Table 6	7816.61^a	—	453024	4404161015

^{a} Aborted due to excessive CPU time

by Benders; in those cases column "obj" provides the value of the objective function when the algorithm was stopped. From these tables it is clear that the solution of the monolithic model is impractical and that an standard implementation of Benders can be more efficient for some classes of problems (namely, 1H2D tables).

6   Conclusions

Partial cell suppression or interval protection can be an alternative method for tabular data protection. Unlike other approaches, this method results in a huge but continuous optimization problem, which can be effectively solved by linear optimization algorithms. One of them is Benders decomposition: a prototype code was able to solve some nontrivial tables more efficiently than state-of-the-art solvers applied to the monolithic model. It is expected that a more sophisticated implementation of Benders algorithm would be able to solve even larger and more complex tables. An additional and promising line of research would be to consider highly efficient specialized interior-point methods for block-angular problems [6,7]. This is part of the further work to be done.

References

1. Baena, D., Castro, J., González, J.A.: Fix-and-relax approaches for controlled tabular adjustment. Comput. Oper. Res. **58**, 41–52 (2015)
2. Benders, J.F.: Partitioning procedures for solving mixed-variables programming problems. Comput. Manag. Sci. **2**, 3–19 (2005). English translation of the original paper appeared in Numerische Mathematik **4**, 238–252 (1962)
3. Castro, J.: Minimum-distance controlled perturbation methods for large-scale tabular data protection. Eur. J. Oper. Res. **171**, 39–52 (2006)
4. Castro, J.: A shortest paths heuristic for statistical disclosure control in positive tables. INFORMS J. Comput. **19**, 520–533 (2007)
5. Castro, J.: Recent advances in optimization techniques for statistical tabular data protection. Eur. J. Oper. Res. **216**, 257–269 (2012)

6. Castro, J.: Interior-point solver for convex separable block-angular problems. Optim. Methods Softw. **31**, 88–109 (2016)
7. Castro, J., Cuesta, J.: Quadratic regularizations in an interior-point method for primal block-angular problems. Math. Program. **130**, 415–445 (2011)
8. Castro, J., Frangioni, A., Gentile, C.: Perspective reformulations of the CTA problem with L_2 distances. Oper. Res. **62**, 891–909 (2014)
9. Fischetti, M., Salazar, J.J.: Solving the cell suppression problem on tabular data with linear constraints. Manag. Sci. **47**, 1008–1026 (2001)
10. Fischetti, M., Salazar, J.J.: Partial cell suppression: a new methodology for statistical disclosure control. Stat. Comput. **13**, 13–21 (2003)
11. Fourer, R., Gay, D.M., Kernighan, D.W.: AMPL: A Modeling Language for Mathematical Programming, 2nd edn. Thomson Brooks/Cole, Pacific Grove (2003)
12. Hundepool, A., Domingo-Ferrer, J., Franconi, L., Giessing, S., Schulte-Nordholt, E., Spicer, K., de Wolf, P.P.: Statistical Disclosure Control. Wiley, Chichester (2012)
13. Wright, S.J.: Primal-Dual Interior-Point Methods. SIAM, Philadelphia (1997)

Precision Threshold and Noise: An Alternative Framework of Sensitivity Measures

Darren Gray[(✉)]

Statistics Canada, Ottawa, Canada
darren.gray@canada.ca

Abstract. At many national statistical organizations, linear sensitivity measures such as the prior-posterior and dominance rules provide the basis for assessing statistical disclosure risk in tabular magnitude data. However, these measures are not always well-suited for issues present in survey data such as negative values, respondent waivers and sampling weights. In order to address this gap, this paper introduces the Precision Threshold and Noise framework, defining a new class of sensitivity measures. These measures expand upon existing theory by relaxing certain restrictions, providing a powerful, flexible and functional tool for national statistical organizations in the assessment of disclosure risk.

Keywords: Statistical disclosure control · Linear sensitivity rules · Prior-posterior rule · pq rule · PTN sensitivity · Precision threshold · Noise

1 Introduction

Most, if not all National Statistical Organizations (NSOs) are required by law to protect the confidentiality of respondents and ensure that the information they provide is protected against statistical disclosure. For tables of magnitude data totals, established sensitivity rules such as the prior-posterior and dominance rules (also referred to as the pq and nk rules) are frequently used to assess disclosure risk. The status of a cell (with respect to these rules) can be assessed using a linear sensitivity measure of the form

$$S = \sum_r \alpha_r x_r \qquad (1)$$

for a non-negative non-ascending finite input variable x_r (usually respondent contributions) and non-ascending finite coefficients α_r (determined by the choice of sensitivity rule). The cell is considered sensitive (i.e., at risk of disclosure) if $S > 0$ and safe otherwise.[1]

---

[1] Many NSOs have developed software to assess disclosure risk in tabular data; for examples please see [3,8]. For a detailed description of the prior posterior and dominance rules, we refer the reader to [4]; Chap. 4 gives an in-depth description of the rules, with examples. The expression of these rules as linear measures is given in [1] and [7, Chap. 6].

© Springer International Publishing Switzerland 2016
J. Domingo-Ferrer and M. Pejić-Bach (Eds.): PSD 2016, LNCS 9867, pp. 15–27, 2016.
DOI: 10.1007/978-3-319-45381-1_2

While powerful in their own right, these rules (and in general any sensitivity measure of the form above) were never designed to assess disclosure risk in the context of common survey issues such as negative values, respondent waivers and sampling weights. As an alternative, we introduce the Precision Threshold and Noise (PTN) framework of sensitivity measures. These measures require three input variables per respondent, which we collectively refer to as PTN variables: Precision Threshold (PT), Noise (N) and Self-Noise (SN). These variables are constructed to reflect the magnitude of protection required, and ambiguity provided, by a respondent contribution. The use of three input variables, instead of the single input variable present in linear sensitivity measures, allows for increased flexibility when dealing with survey data.

Along with these variables, the PTN sensitivity measures require two integer parameters, $n_t \geq 1$ and $n_s \geq 0$, to account for the variety of disclosure attack scenarios (or intruder scenarios) against which an NSO may wish to defend; the resulting measure is denoted $S_{n_s}^{n_t}$. In Sect. 2 we introduce S_1^1, the single target, single attacker sensitivity measure, and give a detailed definition of the PTN variables. Section 3 provides a demonstration of S_1^1 calculations, and explores other possible applications. A more detailed explanation of the parameters n_t and n_s is given in Sect. 4, along with some results on $S_{n_s}^{n_t}$ for arbitrary n_t, n_s.

2 PTN Pair Sensitivity

Within the PTN framework, S_1^1 is used to assess the risk of disclosure in a single target, single attacker scenario, and is referred to as PTN pair sensitivity. For a cell with two or more respondents, we assume that potential attackers have some knowledge of the size of the other contributions, in the form of upper and lower bounds. The concern is that this knowledge, combined with the publication of the cell total (and potentially other information) by the NSO may allow the attacker to estimate another respondent's contribution to within an unacceptably precise degree.

In this respect, S_1^1 can be considered a generalization of the prior-posterior rule. The prior-posterior rule (henceforth referred to as the pq rule) assumes that both the amount of protection required by, and attacker's prior knowledge of, a respondent contribution are proportional to the value of that contribution. In the PTN framework, we remove this restriction; we also allow for the possibility that attackers may not know the exact value of their own contribution to a cell total.

2.1   The Single Target, Single Attacker Premise

Let $T = \sum_r x_r$ represent the sum of respondent contributions $\{x_r\}$. We formulate a disclosure attack scenario whereby respondent s (the "suspect" or "attacker"; we use the two terms interchangeably) acting alone attempts to estimate the contribution x_t of respondent t (the "target") via the publication of total T. The suspect can derive bounds on x_t depending on their knowledge of the remainder $\sum_{r \neq t} x_r$, which includes their own contribution. Let $LB_s(\sum_{r \neq t} x_r)$

and $UB_s(\sum_{r\neq t} x_r)$ denote lower and upper bounds on this sum from the point of view of respondent s; they can then derive the following bounds on the target contribution:

$$T - UB_s\left(\sum_{r\neq t} x_r\right) \leq x_t \leq T - LB_s\left(\sum_{r\neq t} x_r\right) \tag{2}$$

Precision threshold is defined by the assumption that contribution x_t must be protected to within an interval $[x_t - \underline{PT}(t), x_t + \overline{PT}(t)]$ for some lower precision threshold $\underline{PT}(t) \geq 0$ and upper precision threshold $\overline{PT}(t) \geq 0$. The attack scenario formulated above is considered successful if this interval is not fully contained within the bounds defined in (2), in which case we refer to the target-suspect pair (t, s) as sensitive. A cell is considered sensitive if it contains any sensitive pairs, and safe otherwise.

2.2   Assumption: Suspect-Independent, Additive Bounds

To determine cell status (sensitive or safe) using (2) one must in theory determine $LB_s(\sum_{r\neq t} x_r)$ and $UB_s(\sum_{r\neq t} x_r)$ for every possible respondent pair (t, s). The problem is simplified if we make two assumptions:

1. For every respondent, there exist suspect-independent bounds $LB(r)$ and $UB(r)$ such that $LB_s(x_r) = LB(x_r)$ and $UB_s(x_r) = UB(x_r)$ for $r \neq s$.
2. Upper and lower bounds are additive over respondent sets.

Using the first assumption, we define lower noise $\underline{N}(r) = x_r - LB(x_r)$ and upper noise $\overline{N}(r) = UB(x_r) - x_r$. Let $LB_r(x_r)$ and $UB_r(x_r)$ denote bounds on respondent r's contribution from their own point of view, and define lower and upper self-noise as $\underline{SN}(r) = x_r - LB_r(x_r)$ and $\overline{SN}(r) = UB_r(x_r) - x_r$ respectively.

In many cases, it is reasonable to assume that respondents know their own contribution to a cell total exactly, in which case $LB_r(x_r) = UB_r(x_r) = x_r$ and both self-noise variables are zero; in this case we say the respondent is **self-aware**. However, we also wish to allow for scenarios where this might not hold, e.g., when T represents a weighted total and respondent r does not know the sampling weight assigned to them.

The second assumption allows us to rewrite (2) in terms of the upper and lower PTN variables; an equivalent definition of pair and cell sensitivity is then given below.

Definition 1. *For target/suspect pair (t, s) we respectively define PTN upper and lower pair sensitivity as follows:*

$$\overline{S}(t, s) = \overline{PT}(t) - \underline{SN}(s) - \sum_{r\neq s,t} \underline{N}(r)$$

$$\underline{S}(t, s) = \underline{PT}(t) - \overline{SN}(s) - \sum_{r\neq s,t} \overline{N}(r) \tag{3}$$

We say the pair (t, s) is sensitive if either $\overline{S}(t, s)$ or $\underline{S}(t, s)$ is positive and safe otherwise. Upper and lower pair sensitivity for the cell is defined as the maximum sensitivity taken over all possible distinct pairs:

$$\overline{S}_1^1 = max\left\{\overline{S}(t, s) \mid t \neq s\right\}$$
$$\underline{S}_1^1 = max\left\{\underline{S}(t, s) \mid t \neq s\right\} \tag{4}$$

Similarly, a cell is sensitive if $\overline{S}_1^1 > 0$ or $\underline{S}_1^1 > 0$ and safe otherwise.

Readers familiar with linear sensitivity forms of the pq and $p\%$ rules (see Eqs. 3.8 and 3.4 of [1]) may notice the similarity of those measures with the expressions above. There are some important differences. First, those rules do not allow for the possibility of non-zero self-noise associated with the attacker. Second, they make use of the fact that a worst-case disclosure attack occurs when the second-largest contributor attempts to estimate the largest contribution. In the PTN framework, this is not necessarily true; we show how to determine the worst-case scenario in the next section.

2.3 Maximal Pairs

Both upper and lower pair sensitivity take the form

$$S(t, s) = PT(t) - SN(s) - \sum_{r \neq s, t} N(r), \tag{5}$$

which we refer to as the general form. The general form for cell sensitivity can be similarly written as $S_1^1 = max\left\{S(t, s) \mid t \neq s\right\}$. For simplicity we will use these general forms for most discussion, and all proofs; any results on the general form apply to both upper and lower sensitivity as well. When $\underline{PT}(r) = \overline{PT}(r)$, $\underline{N}(r) = \overline{N}(r)$ and $\underline{SN}(r) = \overline{SN}(r)$ for each respondent we say that sensitivity is **symmetrical**; in this case the general form above can be used to describe both upper and lower sensitivity measures.

We define pair (t, s) as **maximal** if $S_1^1 = S(t, s)$, i.e., if the pair maximizes sensitivity within a cell. There is a clear motivation for finding maximal pairs: if both the upper and lower maximal pairs are safe, then the cell is safe as well. If either of the two are sensitive, then the cell is also sensitive.

Clearly, one can find maximal pairs (they are not necessarily unique) by simply calculating pair sensitivity over every possible pair. For n respondents, this represents $n(n - 1)$ calculations (one for each distinct pair). This is not necessary, as we demonstrate below. To begin, we define target function f_t and suspect function f_s on respondent set $\{r\}$ as follows:

- Target function $f_t(r) = PT(r) + N(r)$
- Suspect function $f_s(r) = N(r) - SN(r)$

Re-arranging (5) such that the sum does not depend on (t, s) and substituting f_t and f_s gives

$$S(t, s) = f_t(t) + f_s(s) - \sum_r N(r), \tag{6}$$

which we refer to as **maximal form**. It is then clear that pair (t, s) is maximal if and only if $f_t(t) + f_s(s) = max\{f_t(i) + f_s(j) \mid i \neq j\}$.

We can find maximal pairs by ordering the respondents with respect to f_t and f_s. Let $\tau = \tau_1, \tau_2, \ldots$ and $\sigma = \sigma_1, \sigma_2, \ldots$ be ordered respondent indexes such that f_t and f_s are non-ascending, i.e., $f_t(\tau_1) \geq f_t(\tau_2) \geq \cdots$ and $f_s(\sigma_1) \geq f_s(\sigma_2) \geq \cdots$. We refer to τ and σ as **target** and **suspect orderings** respectively, noting they are not necessarily unique.

Theorem 1. *If $\tau_1 \neq \sigma_1$ (i.e., they do not refer to the same respondent) then (τ_1, σ_1) is a maximal pair. Otherwise, at least one of (τ_1, σ_2) or (τ_2, σ_1) is maximal.*[2]

The important result of this theorem is that it limits the number of steps required to find a maximal pair. Once respondents τ_1, τ_2, σ_1 and σ_2 are identified (with possible overlap), the number of calculations to determine cell sensitivity is at most two, not $n(n - 1)$. By comparison, the pq rule requires only one calculation (once the top two respondents have been identified); calculating PTN pair sensitivity is at most twice as computationally demanding.

2.4 Relationship to the pq and $p\%$ Rules

The pq rule (for non-negative contributions) can be summarized as follows: given parameters $0 < p < q \leq 1$, the value of each contribution must be protected to within $p*100\%$ from disclosure attacks by other respondents. All respondents are self-aware, and can estimate the value of other contributions to within $q * 100\%$.

This fits the definition of a single target, single attacker scenario. The pq rule can be naturally expressed within the PTN framework using a symmetrical S_1^1 measure, and setting $PT(r) = px_r$, $N(r) = qx_r$ and $SN(r) = 0$ for all respondents. To show S_1^1 produces the same result as the pq rule under these conditions, we present the following theorem:

Theorem 2. *Suppose all respondents are self-aware. If there exists a respondent ordering $\eta = \eta_1, \eta_2, \ldots$ such that both PT and N are non-ascending, then (η_1, η_2) is maximal.*

Assuming $\{r\}$ is an ordered index such that the contributions x_r are non-ascending and applying Theorem 2 to our PTN interpretation of the pq rule, we determine that $(1, 2)$ must be a maximal pair. Then

$$S_1^1 = px_1 - \sum_{r \geq 3} qx_r,$$

which is exactly the pq rule as presented in [1], multiplied by a factor of q. (This factor does not affect cell status.)

A common variation on the pq rule is the $p\%$ rule, which assumes the only prior knowledge available to attackers about other respondent contributions is

---

[2] All theorem proofs appear in the Appendix.

that they are non-negative. Mathematically, the $p\%$ rule is equivalent to the pq rule with $q = 1$. Within the PTN framework, the $p\%$ rule can be expressed as an upper pair sensitivity measure $\overline{S}_1^1$ with $\overline{PT}(r) = px_r$, $\underline{N}(r) = x_r$ and $\underline{SN}(r) = 0$.

3   Pair Sensitivity Application

Having defined PTN pair sensitivity, we now demonstrate its effectiveness in treating common survey data issues such as negative values, waivers, and weights. For a good overview of the topic we refer readers to [6]; Tambay and Fillion provide proposals for dealing with these issues within G-Confid, the cell suppression software developed and used by Statistics Canada. Solutions are also proposed in [4] in a section titled *Sensitivity rules for special cases*, pp. 148–152.

In general, these solutions suggest some manipulation of the pq and/or $p\%$ rule; this may include altering the input dataset, or altering the rule in some way to obtain the desired result. We will show that many of these solutions can be replicated simply be choosing appropriate PTN variables.

3.1 S_1^1 Demonstration: Distribution Counts

To begin, we present a unique scenario that highlights the versatility of the PTN framework. Suppose we are given the following set of revenue data: $\{5000, 1100, 750, 500, 300\}$. Applying the $p\%$ rule with $p = 0.1$ to this dataset would produce a negative sensitivity value; the cell total would be considered safe for release. Should this result still apply if the total revenue for the cell is accompanied by the distribution counts displayed in Table 1? Clearly not; Table 1 provides non-zero lower bounds for all but the smallest respondent, contradicting the $p\%$ rule assumption that attackers only know respondent contributions to be non-negative.

Table 1. Revenue distribution and total revenue

Revenue range	Number of enterprises
$[0, 500)$	1
$[500, 1000)$	2
$[1000, 5000)$	1
$[5000, 10000)$	1
Total revenue:	\$7,650

The PTN framework can be used to apply the spirit of the $p\%$ rule in this scenario. We begin with the unmodified $\overline{S}_1^1$ interpretation of the $p\%$ rule given at the end of Sect. 2.4. To reflect the additional information available to potential attackers (i.e., the non-zero lower bounds), we set $\underline{N}(r) = x_r - LB(x_r)$ for each respondent, where $LB(x_r)$ is the lower bound of the revenue range containing x_r.

As the intervals $[x_r, (1+p)x_r]$ are fully contained within each contribution's respective revenue range, we leave $\overline{PT}(r)$ unchanged.

To apply Theorem 1, we calculate f_t and f_s for each respondent and rank them according to these values (allowing ties). These calculations, along with each respondent's contribution and relevant PTN variables, are found in Table 2. Applying the theorem, we determine that respondent pair $(01, 05)$ must be a maximal, giving

$$\overline{S}_1^1 = \overline{S}_1^1(01, 05) = f_t(01) + f_s(05) - \sum_r \underline{N}(r) = 150$$

and indicating that the cell is sensitive.

Table 2. Calculation of S_1^1

Respondent index	Contribution	Upper PT	Lower N	f_t	f_s	f_t rank	f_s rank
01	5000	500	0	500	0	1	4
02	1100	110	100	210	100	4	3
03	750	75	250	325	250	3	2
04	500	50	0	50	0	5	4
05	350	30	350	380	300	2	1

In addition to illustrating the versatility of the PTN framework, this example also demonstrates how Theorem 1 can be applied to quickly and efficiently find maximal pairs.

3.2   Negative Data

While the PTN variables are non-negative by definition, no such restriction is placed on the actual contributions x_r, making PTN sensitivity measures suitable for dealing with negative data. With respect to the pq rule, a potential solution consists of applying a symmetrical S_1^1 rule with $PT(r) = p|x_r|$, $N(r) = q|x_r|$ and $SN(r) = 0$ for each respondent. This is appropriate if we assume that each contribution must be protected to within $p * 100\%$ of its magnitude, and that potential attackers know the value of each contribution to within $q * 100\%$. Theorem 2 once again applies, this time ordering the set of respondents in terms of non-ascending magnitudes $\{|x_r|\}$. Then cell sensitivity S_1^1 is equal to

$$p|x_1| - \sum_{r \geq 3} q|x_r|,$$

which is exactly the pq rule applied to the absolute values. This is identical to a result obtained by Daalmans and de Waal in [2], who also provide a generalization of the pq rule allowing for negative contributions.

The assumptions about PT and N above may not make sense in all contexts. Tambay and Fillion bring up this exact point ([6, Sect. 4.3]), stating that the use of absolute values "may be acceptable if one thinks of the absolute value for a respondent as indicative of the level of protection that it needs as well as of the level of protective noise that it can offer to others" but that this is not always the case: for example, "if the variable of interest is profits then the fact that a respondent with 6 millions in revenues has generated profits of only 32,000 makes the latter figure inadequate as an indicator of the amount of protection required or provided". In this instance, they discuss the use of a proxy variable that incorporates revenue and profit into the pq rule calculations; the same result can be achieved within the PTN framework by incorporating this information into the construction of PT and N.

3.3   Respondent Waivers

In [6], Tambay and Fillion define a waiver as "an agreement where the respondent (enterprise) gives consent to a statistical agency to release their individual information". With respect to sensitivity calculations, they suggest replacing x_r by zero if respondent r provides a waiver. This naturally implies that the contribution neither requires nor provides protection; within the PTN framework this is equivalent to setting all PTN variables to zero, which provides the same result.

This method implicitly treats x_r as public knowledge; if this is not true, the method ignores a source of noise and potentially overestimates sensitivity. With respect to the pq and $p\%$ rules, an alternative is obtained by altering the PTN variables described in Sect. 2.4 in the presence of waivers: for respondents who sign a waiver, we set precision threshold to zero, but leave noise unchanged. To determine cell sensitivity, we make use of the suspect and target orderings (σ and τ) introduced in Theorem 1. In this context σ_1 and σ_2 represent the two largest contributors. If σ_1 has not signed a waiver, then it is easy to show that $\tau_1 = \sigma_1$ and (τ_1, σ_2) is maximal. On the other hand, suppose $\tau_1 \neq \sigma_1$; in this case (τ_1, σ_1) is maximal. If τ_1 has signed a waiver, then $S(\tau_1, \sigma_1) \leq 0$ and the cell is safe. Conversely, if the cell is sensitive, then τ_1 must *not* have signed a waiver; in fact they must be the largest contributor not to have done so.

In other words, if the cell is sensitive, the maximal target-suspect pair consists of the largest contributor without a waiver (τ_1) and the largest remaining contributor (σ_1 or σ_2). With respect to the $p\%$ rule, this is identical to the treatment of waivers proposed on page 148 of [4].

The following result shows that we do not need to identify τ_1 to determine cell status; we need only identify the two largest contributors.

Theorem 3. *Suppose all respondents are self-aware and that $PT(r) \leq N(r)$ for all respondents. Choose ordering η such that N is non-ascending, i.e., $N(\eta_1) \geq N(\eta_2) \geq \dots$. If the cell is sensitive, then one of (η_1, η_2) or (η_2, η_1) is maximal.*

If $\{x_r\}$ are indexed in non-ascending order, the theorem above shows that we only need to calculate $S(1,2)$ or $S(2,1)$ to determine whether or not a cell is sensitive, as all other target-suspect pairs are safe.

3.4 Sampling Weights

The treatment of sampling weights is, in the author's opinion, the most complex and interesting application of PTN sensitivity. As this paper is simply an introduction to PTN sensitivity, we explore a simple scenario: a PTN framework interpretation of the $p\%$ rule assuming all unweighted contributions are non-negative, and all weights are at least one. We also consider two possibilities: attackers know the weights exactly, or only know that they are greater than or equal to one.

Cell total T now consists of weighted contributions $x_r = w_r y_r$ for respondent weights w_r and unweighted contributions y_r. As $LB(y_r) = 0$ for all respondents (according to the $p\%$ rule assumptions), it is reasonable that $LB(w_r y_r)$ should be zero as well, even if w_r is known. This gives $\underline{N}(r) = w_r y_r$. Self-noise is a different matter: it would be equal to zero if the weights are known, but $(w_r - 1)y_r$ if respondents only know that the weights are greater or equal to one.

Choosing appropriate precision thresholds can be more difficult. We begin by assuming the unweighted values y_r must be protected to within $p * 100\,\%$. If respondent weights are known exactly, then we suggest setting $\overline{PT}(r) = p * w_r y_r$. Alternatively, if they are not known, $\overline{PT}(r) = p * y_r - (w_r - 1)y_r$ is not a bad choice; it accounts for the fact that the weighted portion of $w_r y_r$ provides some natural protection.

Both scenarios (weights known vs. unknown) can be shown to satisfy the conditions of Theorem 2. When weights are known, the resulting cell sensitivity S_1^1 is equivalent to the $p\%$ rule applied to x_r. When weights are unknown, S_1^1 is equivalent to the $p\%$ rule applied to y_r and reduced by $\sum_r (w_r - 1)x_r$. The latter coincides with a sensitivity measure proposed by O'Malley and Ernst in [5].

Tambay and Fillion point out in [6] that this measure can have a potentially undesirable outcome: cells with a single respondent are declared safe if the weight of the respondent is at least $1 + p$. They suggest that protection levels remain constant at $p * y_r$ for $w_r < 3$, and are set to zero otherwise (with a bridging function to avoid any discontinuity around $w_r = 3$). The elegance of PTN sensitivity is that such concerns can be easily addressed simply by altering the PTN variables.

4   Arbitrary n_t and n_s

We briefly discuss the more general form of PTN sensitivity, allowing for arbitrary $n_t \geq 1$ and $n_s \geq 0$. Let T be a set of n_t respondents, and let $PT(T) \geq 0$ indicate the amount of desired protection for the group's aggregate contribution $\sum_{t \in T} x_t$. Let S be a set of n_s respondents that does not intersect T, and let

$SN(S) \geq 0$ indicate the amount of self-noise associated with their combined contribution to the total.

Suppose group S (the "suspect" group) wishes to estimate the aggregate contribution of group T (the "target" group). Expanding on the assumptions of Sect. 2.2, we will assume that PT and SN are also suspect-independent and additive over respondent sets, i.e., there exist $PT(r)$ and $SN(r)$ for all respondents such that $PT(T) = \sum_{t \in T} PT(t)$ for all possible sets T and $SN(S) = \sum_{s \in S} SN(s)$ for all possible sets S. Then we define set pair sensitivity as follows:

$$S(T, S) = \sum_{t \in T} PT(t) - \sum_{s \in S} SN(s) - \sum_{r \notin T \cup S} N(r) \tag{7}$$

Suppose we wished to ensure that *every* possible aggregated total of n_t contributions was protected against *every* combination of n_s colluding respondents. (When $n_s = 0$, the targeted contributions are only protected against external attacks.) We accomplish this by defining $S_{n_s}^{n_t}$ as the maximum $S(T, S)$ taken over all non-intersecting sets T, S of size n_t and n_s respectively. We say the set pair (T, S) is maximal if $S_{n_s}^{n_t} = S(T, S)$.

With this definition we can interpret all linear sensitivity measures (satisfying some conditions on the coefficients α_r) within the PTN framework; we provide details in the appendix. In particular the nk rule as described in Eq. 3.6 of [1] can be represented by choosing parameters $n_t = n$, $n_s = 0$ and setting $\overline{PT}(r) = ((100 - k)/k)x_r$, $\underline{N}(r) = x_r$ and $\underline{SN}(r) = 0$ for non-negative contributions x_r.

We do not present a general algorithm for finding maximal set pairs with respect to $S_{n_s}^{n_t}$ in this paper. However, we do present an interesting result comparing cell sensitivity as we allow n_t and n_s to vary:

Theorem 4. *For a cell with at least $n_t + n_s + 1$ respondents, suppose the PTN variables are fixed and that $SN(r) \leq N(r)$ for all respondents. Then the following relationships hold:*

$$S_{n_s}^{n_t} \leq S_{n_s+1}^{n_t} \leq S_{n_s}^{n_t+1} \tag{8}$$

In particular, we note two corollaries: that $S_0^1 \leq S_1^1$ and $S_{n_s}^1 \leq S_0^{n_t}$ whenever $n_s \leq n_t - 1$. This demonstrates often-cited properties of the pq and nk rules: protecting individual respondents from internal attackers protects them from external attackers as well, and if a group of n_t respondents is protected from an external attack, every individual respondent in that group is protected from attacks by $n_t - 1$ (or fewer) colluding respondents.

5   Conclusion

We hope to have convinced the reader that the PTN framework offers a versatile tool in the context of statistical disclosure control. In particular, it offers potential solutions in the treatment of common survey data issues, and as we showed in Sect. 3, many of the solutions currently proposed in the statistical disclosure community can be implemented within this framework via the construction of appropriate PTN variables. As treatments rely solely on the choice of PTN variables, implementing and testing new methods is simplified, and accessible to users who may have little to no experience with linear sensitivity measures.

Acknowledgments. The author is very grateful to Peter Wright, Jean-Marc Fillion, Jean-Louis Tambay and Mark Stinner for their thoughtful feedback on this paper and the PTN framework in general. Additionally, the author thanks Peter Wright and Karla Fox for supporting the author's interest in this field of research.

Appendix

Proof of Theorem 1

Proof. We start with the first statement, assuming $\tau_1 \neq \sigma_1$. As $f_t(\tau_1) \geq f_t(t)$ for any t and $f_s(\sigma_1) \geq f_s(s)$ for any s, it should be clear from (6) that

$$S(\tau_1, \sigma_1) \geq S(t, s)$$

for any pair (t, s), proving the first part of the theorem.

For the second part, we begin with the condition that $\tau_1 = \sigma_1$. Now, suppose (τ_1, σ_2) is not maximal. Then there exists maximal (τ_i, σ_j) where $(i, j) \neq (1, 2)$ such that $f_t(\tau_i) + f_s(\sigma_j) > f_t(\tau_1) + f_s(\sigma_2)$. As $f_t(\tau_1) \geq f_t(\tau_i)$ by definition, it follows that $f_s(\sigma_j) > f_s(\sigma_2)$ and we can conclude that $j = 1$. Then $(\tau_i, \sigma_j) = (\tau_i, \sigma_1)$ for some $i \neq 1$. But we know that $f_t(\tau_2) \geq f_t(\tau_i)$ and so $S(\tau_2, \sigma_1) \geq S(\tau_i, \sigma_1)$ for any $i \neq 1$. This shows that if (τ_1, σ_2) is not maximal, (τ_2, σ_1) must be, completing the proof.   □

Proof of Theorem 2

Proof. When all respondents are self-aware, $f_t = PT + N$ and $f_s = N$, and consequently any ordering that results in non-ascending PT, N also results in non-ascending f_t, f_s. Setting $\tau = \sigma = \eta$ and applying Theorem 1, we conclude that one of (η_1, η_2) or (η_2, η_1) is maximal. From (6) we can see that

$$S(\eta_1, \eta_2) - S(\eta_2, \eta_1) = PT(\eta_1) - PT(\eta_2) \geq 0$$

showing $S(\eta_1, \eta_2) \geq S(\eta_2, \eta_1)$ and (η_1, η_2) is maximal.   □

Proof of Theorem 3

Proof. The proof is self-evident for cells with two or fewer respondents, so we will assume there are at least three. Applying Theorem 1 and noting $f_s = N$ we can conclude that there exists a maximal pair of the form (η_i, η_j) for $j \leq 2$. As this pair is maximal it can be used to calculated cell sensitivity:

$$S_1^1 = S(\eta_i, \eta_j) = PT(\eta_i) - \sum_{r \neq i,j} N(\eta_r)$$

As $j \leq 2$, if $i \geq 3$ then exactly one of $N(\eta_1)$ or $N(\eta_2)$ is included in the summation above. Both of these are $\geq N(\eta_i)$ by ordering η, which is $\geq PT(\eta_i)$ by assumption. This means $S_1^1 < 0$ and the cell is safe. Conversely, if the cell is sensitive, there must exist a maximal pair of the form (η_i, η_j) with both $i, j \leq 2$, completing the proof. $\square$

Interpreting Arbitrary Linear Sensitivity Measures in $S_{n_s}^{n_t}$ Form

All linear sensitivity measures of the form $\sum_r \alpha_r x_r$ can be expressed in PTN form, provided they satisfy the following conditions:

– Finite number of non-negative coefficients
– All positive coefficients have the same value, say α_+
– All negative coefficients have the same value, say α_-.

Assuming these conditions are met, an equivalent PTN sensitivity measure can be defined as follows:

– Set n_t equal to the number of positive coefficients
– Set n_s equal to the number of coefficients equal to zero
– Set $PT(r) = \alpha_+ x_r$ for all r
– Set $N(r) = |\alpha_-| x_r$ and $SN(r) = 0$ for all r

We show that the resulting PTN cell sensitivity measure is equivalent to $\sum_r \alpha_r x_r$ by first writing (7) as follows:

$$S(T, S) = \sum_{t \in T} (PT(t) + N(t)) + \sum_{s \in S} (N(s) - SN(s)) - \sum_r N(r) \qquad (9)$$

Substituting in the appropriate PTN values gives

$$S(T, S) = \sum_{t \in T} (\alpha_+ + |\alpha_-|) x_t + \sum_{s \in S} |\alpha_-| x_s - \sum_r |\alpha_-| x_r. \qquad (10)$$

It is easy to see that T and S should be selected from the largest $n_t + n_s$ respondents to maximize $S(T, S)$. If they are already indexed in non-ascending order, then sensitivity is maximized when $T = \{1, \ldots, n_t\}$ and $S = \{n_t + 1, \ldots, n_t + n_s\}$. Then cell sensitivity is given by

$$S_{n_s}^{n_t} = \sum_{r=1}^{n_t} \alpha_+ x_r - \sum_{r > n_t + n_s} |\alpha_-| x_r \qquad (11)$$

which is exactly $\sum_r \alpha_r x_r$.

Proof of Theorem 4

We begin with a simple lemma:

Lemma 1. *Let T and S be non-intersecting sets of respondents. Let k be a respondent in neither, and assume $SN(k) \leq N(k)$. Then*

$$S(T, S) \leq S(T, S \cup k) \leq S(T \cup k, S). \tag{12}$$

Proof. We write (7) in maximal form, substituting in the target and suspect functions:

$$S(T, S) = \sum_{t \in T} f_t(t) + \sum_{s \in S} f_s(s) - \sum_r N(r) \tag{13}$$

Then $S(T, S \cup k) - S(T, S) = f_s(k)$. As $SN(k) \leq N(k)$ by assumption (we expect this to be true anyway, as a respondent should never know less about their own contribution than the general public), $f_s \geq 0$ proves the first inequality. The second inequality holds because $f_t \geq f_s$ for all respondents, including k.   □

With this lemma, the proof of Theorem 4 is almost trivial:

Proof. Let (T, S) be maximal with respect to $S_{n_s}^{n_t}$. We know there exists at least one respondent $k \notin T \cup S$, and by Lemma 1, $S(T, S) \leq S(T, S \cup k)$, proving that $S_{n_s}^{n_t} \leq S_{n_s+1}^{n_t}$.

For the second inequality, we note that any set pair that is maximal with respect to $S_{n_s+1}^{n_t}$ can be written in the form $(T, S \cup k)$ for some T of size n_t, S of size n_s and single respondent k. Once again applying Lemma 1 we see that $S(T, S \cup k) \leq S(T \cup k, S)$ and consequently $S_{n_s+1}^{n_t} \leq S_{n_s}^{n_t+1}$.   □

References

1. Cox, L.H.: Disclosure risk for tabular economic data. In: Doyle, P., Lane, J., Theeuwes, J., Zayatz, L. (eds.) Confidentiality, Disclosure and Data Access, Chap. 8. North-Holland, Amsterdam (2001)
2. Daalmans, J., de Waal, T.: An improved formulation of the disclosure auditing problem for secondary cell suppression. Trans. Data Priv. **3**(3), 217–251 (2010)
3. Hundepool, A., van de Wetering, A., Ramaswamy, R., de Wolf, P., Giessing, S., Fischetti, M., Salazar-Gonzalez, J., Castro, J., Lowthian, P.: τ-argus users manual. Version 3.5. Essnet-project (2011)
4. Hundepool, A., Domingo-Ferrer, J., Franconi, L., Giessing, S., Nordholt, E.S., Spicer, K., De Wolf, P.P.: Statistical Disclosure Control. John Wiley & Sons, Hoboken (2012)
5. O'Malley, M., Ernst, L.: Practical considerations in applying the pq-rule for primary disclosure suppressions. http://www.bls.gov/osmr/abstract/st/st070080.htm
6. Tambay, J.L., Fillion, J.M.: Strategies for processing tabular data using the g-confid cell suppression software. In: Joint Statistical Meetings, Montréal, Canada, pp. 3–8 (2013)
7. Willenborg, L., De Waal, T.: Elements of Statistical Disclosure Control. Lecture Notes in Statistics, vol. 155. Springer, New York (2001)
8. Wright, P.: G-Confid: Turning the tables on disclosure risk. Joint UNECE/Eurostat work session on statistical data confidentiality. http://www.unece.org/stats/documents/2013.10.confidentiality.html

Empirical Analysis of Sensitivity Rules: Cells with Frequency Exceeding 10 that Should Be Suppressed Based on Descriptive Statistics

Kiyomi Shirakawa[1(✉)], Yutaka Abe[2], and Shinsuke Ito[3]

[1] National Statistics Center, Hitotsubashi University,
2-1 Naka, Kunitachi-shi, Tokyo 186-8603, Japan
kshirakawa@ier.hit-u.ac.jp
[2] Hitotsubashi University, 2-1 Naka, Kunitachi-shi, Tokyo 186-8603, Japan
y-abe@ier.hit-u.ac.jp
[3] Chuo University, 742-1 Higashinakano, Hachioji-shi, Tokyo 192-0393, Japan
ssitoh@tamacc.chuo-u.ac.jp

Abstract. In official statistics, it is a serious problem to be able to estimate the original data from the numerical values of result tables. To limit such problems, cell suppression is frequently applied when creating result tables, and when researchers create summary tables that include levels of descriptive statistics for remote access, statistical disclosure control must be applied. This research therefore focuses on higher-order moments in descriptive statistics to perform empirical analysis of the safety of statistical levels. The results from standard deviation (variance), skewness, and kurtosis confirm that cells with frequency of 10 or higher are unsafe.

Keywords: Intruder · Combination pattern · Unsafe · Higher moment

1 Introduction

In official statistics, it is a serious problem if the original data can be estimated from the numerical values of result tables. To keep this from occurring, cell suppression is frequently applied when creating result tables. Cell suppression is the standard method for statistical disclosure control (SDC) of individual data.

This standard can be divided into cases where survey planners will publicize only result tables and those that permit the use of individual data. The latter case can be further subdivided into rules of thumb and principles-based models.

The April 2009 revision of the Japanese Statistics Act introduced remote access for new secondary uses of individual data, starting from 2016. In remote access, SDC must be applied when researchers create summary tables or regression models that include descriptive statistics. For example, applications where minimum frequencies are 3 or higher have been considered, but in that case there is no consideration of SDC for higher frequencies.

© Springer International Publishing Switzerland 2016
J. Domingo-Ferrer and M. Pejić-Bach (Eds.): PSD 2016, LNCS 9867, pp. 28–40, 2016.
DOI: 10.1007/978-3-319-45381-1_3

This study therefore focuses on higher-moment descriptive statistics (variance, skewness, and kurtosis) for cells with frequencies of 10 or higher, and thereby demonstrates that such cells are unsafe.

The remainder of this paper is organized as follows. Section 2 describes related studies. In Sect. 3, specific combination patterns of frequencies of 10 or higher are created, and in Sect. 4, the original data combination is estimated through the standard deviation, skewness, and kurtosis of the combination patterns. Section 5 describes the conclusions of this paper from the findings presented and closes with topics for future study.

2 SDC and Sensitivity Rules for Individual Data

Official statistical data are disseminated in various forms, based on the required confidentiality and the needs of users. Official statistics can be used in the form of result tables and microdata, but microdata for official statistics in particular has been provided in a number of different forms, including anonymized microdata, individual data, tailor-made tabulation, and on-demand services (remote execution).

Techniques for creating anonymized microdata can be roughly divided into perturbative and non-perturbative methods (Willenborg and de Waal 2001). Non-perturbative methods include global and local recoding, record and attribute suppression, and top- or bottom-coding. Perturbative methods include additive and multiplicative noise, data swapping[1], rounding, micro-aggregation, and the Post Randomization Method (PRAM) (Domingo-Ferrer and Torra 2001; Willenborg and de Waal 2001; Duncan et al. 2011). Most of the anonymized microdata currently released in Japan, such as the Employment Status Survey and the National Survey of Family Income and Expenditures, have been created using non-perturbative methods such as top- or bottom-coding, recoding, and data deletion. Anonymized microdata such as the Population Census have been prepared using not only non-perturbative methods, but also perturbative methods such as swapping. Departments for the creation of statistical data in other countries have also been known to apply perturbative methods for anonymization when preparing microdata for official statistics. For example, the U.S. Census Bureau applied additive noise, swapping, and rounding when creating the Public Use Microdata Samples for the 2000 U.S. Census (Zayatz 2007), and the United Kingdom applied PRAM to the Samples of Anonymised Records for its 2001 census (De Kort and Wathan 2009).

The Research Data Center allows researchers to access microdata at on-site facilities and through remote access. European countries apply a standard called the Five Safes model as a framework for access to microdata (Ritchie 2008; Desai et al. 2016). This model is based on the concepts of safe projects, safe people, safe data, safe settings, and safe outputs. In detail, "safe projects" refers to access to microdata for only appropriate projects. "Safe people" refers to appropriate use by researchers who

---

[1] For empirical studies of data swapping in Japan, see, for example, Takemura (2002) and Ito and Hoshino (2013, 2014).

can be trusted to follow usage procedures. "Safe data" refers to the data itself, which should not disclose individual information. "Safe settings" refers to the technical management measures related to microdata access for preventing unauthorized alteration. "Safe outputs" means that individual information should not be included in the results of statistical analysis. By meeting these standards, "safe use" should be possible.

Focusing here on "safe outputs," verification of confidentiality in the final products of analysis makes it possible to publish the analysis results. In such cases, personnel use not rule-based approaches that determine suppression processing such as primary and secondary disclosures for cell frequency, but rather principles-based approaches that aim at anonymity through the cooperation of microdata providers and users (Ritchie and Welpton 2015).

As a principles-based approach, rule-of-thumb models have been applied to reduce confidentiality errors and errors related to efficiency (Brandt et al. 2010). There are four overall rules in these models:

1. In results tables, all cells must have a frequency of at least 10.
2. In all models, there must be at least 10 degrees of freedom.
3. In all results tables, the total frequency of a given cell within its containing row and column must not exceed 90 %.
4. In all results tables, the total frequency of a given cell in the table overall must not exceed 50 %.

European countries have enacted screening criteria based on the above standards that allow for removal of analysis results using microdata.

In the principles-based approach, permission for publication of analysis results generally requires verification by a responsible party. This party and researchers must therefore undergo SDC training.

After verification of analysis results, if the final product is determined to be "safe statistics," the results are provided to researchers. According to Brandt et al. (2010), these analysis results can be classified as "safe" or "unsafe." As Table 1 shows, the results for correlations and regression analyses are generally considered to be safe, with the exception of residuals. In contrast, summary tables and representative values such as means and percentiles, indices, and ratios are considered unsafe statistics. Brandt et al. (2010) also consider higher moments related to the distribution as safe statistics.

However, even when creating summary tables that fulfill the standards for "safe statistics" according to the rule-of-thumb model, it is possible to apply sensitivity rules for primary disclosure. Regarding cells contained in summary tables, this is a standard for determination of risk of identifying individuals, with representative sensitivity rules being minimum frequency (threshold) rules, (n, k)-dominance rules, and p % rules (Duncan et al. 2011; Hundepool et al. 2012). Loeve (2001) discussed a general formulation of such sensitivity rules. Sensitivity rules are also applied in τ-ARGUS (Giessing 2004), which allows users to set parameters to automatically apply anonymization of cells based on sensitivity rules such as minimum frequency (threshold) rules, dominance rules, and p % rules. In recent years, Bring and Wang (2014) revealed limitations regarding dominance rules and p % rules, and newly proposed the interval rule.

Table 1. Rules for output checking

Type of statistics	Type of output	Classification
Descriptive statistics	Frequency tables	Unsafe
	Magnitude tables	Unsafe
	Maxima, minima, and percentiles (incl. median)	Unsafe
	Mode	Safe
	Means, indices, ratios, indicators	Unsafe
	Concentration ratios	Safe
	Higher moments of distributions (incl. variance, covariance, kurtosis, skewness)	Safe
	Graphs: pictorial representations of actual data	Unsafe
Correlation and regression analysis	Linear regression coefficients	Safe
	Nonlinear regression coefficients	Safe
	Estimation residuals	Unsafe
	Summary and test statistics from estimates (R^2, X^2, etc.)	Safe
	Correlation coefficients	Safe

Source: Brandt et al. (2010).

3   Approach

This section focuses on utilizing of higher-order moments in descriptive statistics to perform empirical analysis of the safety of statistical levels.

3.1   Creation of Combination Patterns by Frequency

Conditions for combination pattern creation
This section considers appropriate methods for verifying confidentiality rules using statistical levels. The method creates all combination patterns according to frequency. Note that because infinitely many combinations would result from real-numbered values, the following conditions are established:

1. The total is 100.
2. An integer number of frequencies n is created.
3. Values are sorted in descending order.

Logic for combination pattern creation
From the conditions for combination pattern creation, the total is 100. Since each frequency n is an integer in descending order from X_1 to X_n, the creation logic is as follows.

In the case of frequency $n = 20$ …

$$X_1 + X_2 + \cdots + X_{19} + X_{20} = 100$$

$$X_1 \geq X_2 \geq \cdots \geq X_{19} \geq X_{20}$$

Table 2. n-th maximum and minimum value ($n = 20$)

	X_1	X_2	X_3	X_4	X_5	X_6	X_7	X_8	X_9	X_{10}
Mini.	5	0	0	0	0	0	0	0	0	0
Max.	100	50	33	25	20	16	14	12	11	10
	X_{11}	X_{12}	X_{13}	X_{14}	X_{15}	X_{16}	X_{17}	X_{18}	X_{19}	X_{20}
Mini.	0	0	0	0	0	0	0	0	0	0
Max.	9	8	7	7	6	6	5	5	5	5

The minimum value 5 for X_1 in Table 2 is the total divided by the frequency, rounded up to the next integer value, and the minimum values for X_2 to X_{20} are 0. On the other hand, the maximum value 100 for X_1, and the others are the total divided by each number of subscript. When the frequency is 10, the minimum value for X_1 is therefore 10, and the maximum and minimum values for X_2 to X_{10} are the same. When the frequency is 50, the minimum value for X_1 is 2.

According to this logic, there are 6,292,069 combination patterns for a frequency of 10. There are furthermore 97,132,873 combination patterns for a frequency $n = 20$, and 189,477,547 combination patterns for a frequency $n = 50$.

3.2 Standard Deviation by Combination Pattern

Maximum and minimum values

This research focuses on higher-moment descriptive statistics, but before discussing that we would like to consider the relation between standard deviations and maximum values. Table 3 shows the number of combination patterns for frequencies of 2 through 10, 20, and 50, and furthermore lists their mean and values for the maximum and minimum values and range for their standard deviation.

Regarding the maximum value for standard deviation for all frequencies in Table 3, X_1 is 100 for frequency n, and 0 for other values. Note that by calculating these combination patterns that include 0, we can include combinations where all values are real numbers. For the range of standard deviation, decimal values are approximated as nearly zero, and so are decided by maximum values. Furthermore, the larger the value of frequency n, the narrower the range of the standard deviation becomes.

Handling values of 0

One point to note about estimating combination patterns based on standard deviations is that there are different conditions for the cases where values of 0 are and are not included.

In Table 3, the maximum standard deviation for frequency $n = 2$ is 70.7 in the cases of $X_1 = 100$ and $X_2 = 0$. In the case of frequency 3 with $X_3 = 0$, we get the pattern $[100, 0, 0]$ with standard deviation 57.7, and the standard deviation for frequency 4 and $X_4 = 0$ is 50.0.

Table 3. Number of patterns and range of standard deviation by frequency

Freq.	SD			Mean	Patterns
	Max.	Mini.	Range		
2	70.7	0.0	70.7	50.0	51
3	57.7	0.6	57.2	33.3	884
4	50.0	0.0	50.0	25.0	8,037
5	44.7	0.0	44.7	20.0	46,262
6	40.8	0.5	40.3	16.7	189,509
7	37.8	0.5	37.3	14.3	596,763
8	35.4	0.5	34.8	12.5	1,527,675
9	33.3	0.3	33.0	11.1	3,314,203
10	31.6	0.0	31.6	10.0	6,292,069
20	22.4	0.0	22.4	5.0	97,132,873
50	14.1	0	14.1	2	189,477,547

Note: This table describes the case where values
for the standard deviation (SD) and minimum
calculations are integers. Where decimal values are
present, approximation is as close as possible to 0

The patterns where the standard deviation is minimized for frequency $n = 2$ are $X_1 = 50$ and $X_2 = 50$. Adding 0 to this pattern, for frequency 3 we get $[50, 50, 0]$, and for frequency 4 we get $[50, 50, 0, 0]$. Figure 1 shows changes in standard deviation and maximum value (X_1) for frequencies 2 (X_1, X_2), 3 $(X_1, X_2, 0)$, and 4 $(X_1, X_2, 0, 0)$.

If values of 0 are not included, all have frequency 2 (50, 50). However, when values of 0 are included in frequency n, changes in n affect the standard deviation. Patterns

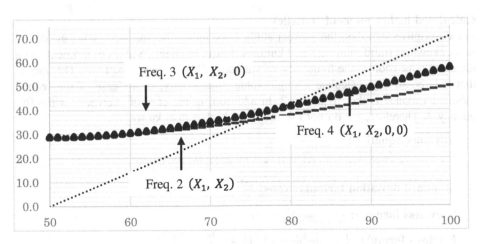

Fig. 1. Effects of values of 0 on standard deviation

including 0 in the frequency furthermore have larger standard deviations than those that do not. Such characteristics make it easy to determine the maximum value (X_1).

Considering these characteristics, this study considers all combination patterns that include 0.

3.3   Estimation of Combination Patterns Based on Descriptive Statistics

Calculated descriptive statistics

Descriptive statistics for calculated combination patterns by frequency are the frequency, mean, and higher-moment statistics (standard deviation, skewness, and kurtosis). From these statistics, we estimate the original data.

Table 4. Calculated and non-calculated values

Item	Target	Calculated	Not calculated	Supplementary information
Content	Maximum value	Frequency, mean, standard deviation, skewness, kurtosis	Max. & mini. values, range, percentiles	Intruder (second-largest value), median value

SDC presumes the existence of an intruder (with knowledge of the second-largest value) for estimating maximum value, and investigates the safety when the second-largest value is exposed. Statistics that are not calculated include maximum and minimum values, ranges, and percentiles (Table 4). In particular, because maximum values are the target of sensitivity rules, they do not allow for estimation of the combination patterns of the original data.

Calculated higher-moment statistics

For descriptive statistics, the present study uses moments: variance (standard deviation), skewness (third moment), and kurtosis (fourth moment). Values of moments with the same dimension as the frequency vary among all combination patterns. This means that it is possible to calculate a single value. Specifically, in the case of the second moment variance is the frequency 2. In addition, in the case of skewness is the frequency 3. Therefore, we do not use moments higher than kurtosis.

Variance equation: $\frac{\sum (x_i - \bar{x})^2}{(n-1)}$

For convenience, we convert variance to standard deviation.

Standard deviation formula: $\sqrt{\frac{\sum (x_i - \bar{x})^2}{(n-1)}}$

Skewness formula: $\frac{n}{(n-1)(n-2)} \sum \left(\frac{x_i - \bar{x}}{s} \right)^3$

Kurtosis formula: $\left\{ \frac{n(n+1)}{(n-1)(n-2)(n-3)} \sum \left(\frac{x_i - \bar{x}}{s} \right)^4 \right\} - \frac{3(n-1)^2}{(n-2)(n-3)}$

These calculations were performed using Microsoft Excel (version 2013).

Table 5. Maximum number of combinations for grouping frequencies $n = 10, 20, 50$ by statistic

SD	Skew.	Kurt.	Median	$n = 10$	$n = 20$	$n = 50$
✓				11,906	223,627	422,158
✓	✓			68	550	928
✓	✓	✓		6	15	<u>19</u>
✓	✓	✓	✓	5	9	<u>19</u>

Cases by frequency

We extract similar patterns from the combination pattern for the frequency in question, based on descriptive statistics of the original data.

When the total is set to 100, and all values for frequency n are integers, the standard deviation (variance), skewness, and kurtosis can be derived as a fully matched pattern. Multiple patterns can be derived, however, depending on the derived statistics.

Table 5 compares these values for various frequencies ($n = 10, 20$, and 50).

Table 5 shows that as the frequency increases from $n = 20$ to 50, the effect of the median decreases.

In the "19" pattern for standard deviation (variance), skewness, and kurtosis for frequency 50, these three statistics are all the same. These are 38 (19 patterns × 2 groups) of the 189,477,547 patterns for frequency $n = 50$.

The first group has ten patterns for an X_1 value of 11, and nine patterns for 12, while the second group has three patterns for 11, ten patterns for 12, and six patterns for 13. If we can extract from this a pattern with standard deviation (variance), skewness, and kurtosis similar to the original data, we can identify the maximum value (X_1).

Table 6. Range and mode of the maximum value (X_1) by frequency

Frequency	No. of combinations	X_1 (Largest variable in combination)						
		Max.	Mini.	Mode	# of combination of modes	Ratio (%)	Max. of SD	Mini. of SD
50	189,477,547	100	2	18	11,045,779	5.8	5.570	2.356
20	97,132,873	100	5	20	5,498,387	5.7	8.885	3.554
10	6,292,069	100	10	27	296,777	4.7	13.115	5.981

Summing the number of combinations of other multiple matching patterns gives 64,955,725 patterns, and there are 124,521,822 single combination patterns. We can thus identify slightly less than 70 % of combination patterns from the standard deviation (variance), skewness, and kurtosis.

Table 6 shows that the mode for the maximum value (X_1) is 18 for frequency $n = 50$. While there are many patterns, it is not difficult to determine the standard deviation (variance), skewness, and kurtosis.

To allow searching and storage of such a large number of patterns, in this study we saved the combination patterns by frequency in a database (Microsoft SQL Server 2014).

The next section estimates and evaluates the standard deviation (variance), skewness, and kurtosis for empirical analysis.

4 Empirical Analysis

Using a database in which all combination patterns by frequency are stored, we extract the optimal pattern following the values for standard deviation (variance), skewness, and kurtosis. All n values for frequency n are integers, so not all of these values will be identical. We therefore extract multiple patterns that are approximate to the original data. An intruder wishes to know the maximum value, so there is not necessarily a need to estimate all values in the original data. We therefore estimate the pattern in the by-frequency database that has the most similar pattern, and determine the maximum value from this estimation.

We define cases where the maximum value can be determined as an "unsafe" determination. We use a frequency of 20 and 50 for the empirical analysis. The data used are pseudo-individual data from the 2004 Japanese National Survey of Family Income and Expenditure.

Table 7. Statistics after conversion of the original data to a total value of 100

	Statistics	Living expenditure ($n = 20$)	Living expenditure ($n = 50$)
Original	Mean	195,624.80	312,999.5
	SD	59,892.60	112,814.1
	Skew	0.346	1.134
	Kurt	−1.004	1.862
Transformation	Mean	5.000	2.000
	SD	1.531	312,999.5
	Skew	0.346	112,814.1
	Kurt	−1.004	1.134

Data source: Pseudo-individual data from the 2004 National Survey of Family Income and Expenditure

In the analysis based on this microdata, we create a summary of the high-dimensional cross. Those results are transformed so that their total value is 100, and recalculate standard deviations based on this value of 100. Table 7 shows the statistics after conversion of the original data to a total value of 100. Note that skewness and kurtosis values are not changed here.

Database search conditions
Searches are performed by specifying a range containing the standard deviation, skewness, and kurtosis in the original data. A similar pattern is extracted according to

Table 8. Combinations of the maximum value (X_1) and X_2 for "living expenditure" ($n = 20$)

	Combination 1		Combination 2		SD Range	Skewness Range	Kurtosis Range
X_1 and X_2	8	8	8	7			
Estimation	14		8		1.522	-0.0369 – 1.195	0.398 – 1.653
Original Data	8.156		7.116		1.531	0.346	-1.004

differences in each pattern and the original data. The priority is standard deviation > skewness > kurtosis.

Verification of "living expenditure" ($n = 20$)

From the database of combination patterns for frequency 20, we extracted 22 similar patterns for standard deviation, skewness, and kurtosis. Table 8 shows combinations 1 and 2 of the maximum value (X_1) and X_2. Note that combination 2 has the most similar patterns of X_1 and X_2 for standard deviation, skewness, and kurtosis. An intruder with knowledge of the value for X_2 could extract combination 2, which has the same value of 8 for X_1.

Table 9. Combinations of the maximum value (X_1) and X_2 for "living expenditure" ($n = 50$)

	Combination 1		Combination 2		Combination 3		SD Range	Skewness Range	Kurtosis Range
X_1 and X_2	4	4	5	3	5	4			
Estimation	14		2		1		0.639 – 0.833	0.848 – 1.331	1.134 – 2.900
Original Data	4.412		3.749				0.721	1.134	1.862

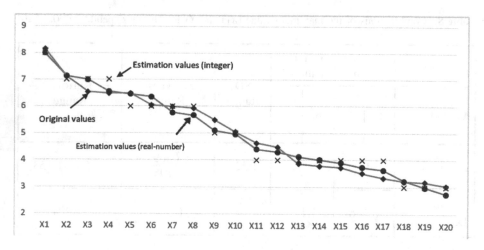

Fig. 2. Frequency 20 of original and estimation (integer, real-number) values

Verification of "living expenditure" ($n = 50$)
From the database of combination patterns for frequency 50, we extracted 17 similar patterns for standard deviation, skewness, and kurtosis. Table 9 shows combinations 1, 2 and 3 of the maximum value (X_1) and X_2. Combination 1 has the most similar patterns X_1 and X_2 for standard deviation, skewness, and kurtosis. An intruder with knowledge of the value for X_2 could extract combination 1, which has the same value of 4 for X_1.

Estimation of real-numbered values ($n = 20$)
After similar pattern extraction, we used Microsoft Excel's "solver" function (version 2013) to estimate patterns that include real-numbered values (Fig. 2). Although details of the estimation are omitted in this paper, it is possible to estimate real-numbered values for similar patterns of the same values for each statistic (standard deviation, skewness, and kurtosis).

5 Conclusion

We created by-frequency combinations patterns with integer values totaling 100. If all values are integers in the source data, too, then a fully matching pattern can be extracted. The empirical analysis contained real-number values, however, so a full match was impossible. We therefore extracted multiple similar patterns through searches over ranges that include the standard deviation, skewness, and kurtosis of the original data. We then used differences with the original data to derive the most similar pattern. While the results did not fully match the original data, they did allow determination of the maximum value (X_1), so care must be taken when revealing standard deviation, skewness, and kurtosis.

When determining whether to include values of 0 in frequency n, the case where values of 0 are included has a wider range of standard deviations. From this, estimation of frequency n including values of 0 from the standard deviation of frequency n is possible.

This paper considered a frequency of 20 and 50, but the method (with total values of 100) can be used to estimate maximum values and a maximum frequency of $n = 100$.

6 Future Topics

Here, we converted all values to total 100, including in the empirical analysis, and recalculated standard deviations based on this value of 100. Notably, we did not convert skewness or kurtosis. From these statistics, we derived the most similar combination pattern of integers, and saw that such a derivation was valid for a frequency of 50.

In future research, it is necessary to analyze which of the combination patterns up to 100 are "unsafe." It is also necessary to investigate frequencies exceeding 100. We will perform these investigations in the process of creating synthetic microdata that is suited to empirical analysis.

References

Brandt, M., Franconi, L., Guerke, C., Hundepool, A., Lucarelli, M., Mol, J., Ritchie, F., Seri, G., Welpton, R.: Guidelines for the checking of output based on microdata research. Final Report of ESSnet Sub-Group on Output SDC, Eurostat (2010)

Bring, J., Wang, Q.: Comparison of different sensitivity rules for tabular data and presenting a new rule – the *Interval Rule*. In: Domingo-Ferrer, J. (ed.) PSD 2014. LNCS, vol. 8744, pp. 36–47. Springer, Heidelberg (2014)

De Kort, S., Wathan, J.: Guide to Imputation and Perturbation in the Samples of Anonymised Records (2009) (unpublished)

Desai, T., Ritchie, F., Welpton, R.: Five Safes: Designing Data Access for Research. Economics Working Paper Series 1601, University of the West of England (2016)

Domingo-Ferrer, J., Torra, V.: Disclosure control methods and information loss for microdata. In: Doyle, P., et al. (eds.) Confidentiality, Disclosure and Data Access: Theory and Practical Applications for Statistical Agencies, pp. 91–110. Elsevier Science, Amsterdam (2001)

Duncan, G.T., Elliot, M., Salazar-González, J.: Statistical Confidentiality. Springer, New York (2011)

Giessing, S.: Survey on methods for tabular data protection in ARGUS. In: Domingo-Ferrer, J., Torra, V. (eds.) PSD 2004. LNCS, vol. 3050, pp. 1–13. Springer, Heidelberg (2004)

Hundepool, A., Domingo-Ferrer, J., Franconi, L., Giessing, S., Nordholt, E.S., Spicer, K., de Wolf, P.P.: Statistical Disclosure Control. Wiley, United Kingdom (2012)

Ito, S., Hoshino, N.: Assessing the effectiveness of disclosure limitation methods for census microdata in Japan. Paper presented at Joint UNECE/Eurostat Work Session on Statistical Data Confidentiality, Ottawa, Canada, pp. 1–10 (2013)

Ito, S., Hoshino, N.: Data swapping as a more efficient tool to create anonymized census microdata in Japan. Paper presented at Privacy in Statistical Databases 2014, Ibiza, Spain, pp. 1–14 (2014)

Loeve, A.: Note on Sensitivity Measures and Protection Levels. Technical report, Research Paper, Statistics Netherlands (2001)

Ritchie, F.: Secure access to confidential microdata: four years of the virtual microdata laboratory. Econ. Labour Mark. Rev. **2**(5), 29–34 (2008)

Ritchie, F., Welpton, R.: Operationalizing principles-based output statistical disclosure control. mimeo (2015)

Takemura, A.: Local recoding and record swapping by maximum weight matching for disclosure control of microdata sets. J. Official Stat. **18**(2), 275–289 (2002)

Willenborg, L., de Waal, T.: Elements of Statistical Disclosure Control. Springer, New York (2001)

Zayatz, L.: Disclosure avoidance practices and research at the U.S. Census Bureau: an update. J. Official Stat. **23**(2), 253–265 (2007)

A Second Order Cone Formulation
of Continuous CTA Model

Goran Lesaja[1]([⊠]), Jordi Castro[2], and Anna Oganian[1,3]

[1] Department of Mathematical Sciences, Georgia Southern University,
P.O. Box 8093, Statesboro, GA 30460-8093, USA
goran@georgiasouthern.edu
[2] Department of Statistics and Operations Research, Universitat Politècnica
de Catalunya, Jordi Girona 1–3, 08034 Barcelona, Catalonia, Spain
jordi.castro@upc.edu
[3] National Center for Health Statistics,
3311 Toledo Road, Hyatsville, MD 20782, USA
aoganyan@cdc.gov

Abstract. In this paper we consider a minimum distance Controlled
Tabular Adjustment (CTA) model for statistical disclosure limitation
(control) of tabular data. The goal of the CTA model is to find the closest safe table to some original tabular data set that contains sensitive
information. The measure of closeness is usually measured using ℓ_1 or
ℓ_2 norm; with each measure having its advantages and disadvantages.
Recently, in [4] a regularization of the ℓ_1-CTA using Pseudo-Huber function was introduced in an attempt to combine positive characteristics of
both ℓ_1-CTA and ℓ_2-CTA. All three models can be solved using appropriate versions of Interior-Point Methods (IPM). It is known that IPM
in general works better on well structured problems such as conic optimization problems, thus, reformulation of these CTA models as conic
optimization problem may be advantageous. We present reformulation
of Pseudo-Huber-CTA, and ℓ_1-CTA as Second-Order Cone (SOC) optimization problems and test the validity of the approach on the small
example of two-dimensional tabular data set.

Keywords: Statistical disclosure limitation (control) · Controlled tabular adjustment models · Pseudo-Huber Function · Convex optimization ·
Second-order cone optimization · Interior-Point Methods

1 Introduction

The *statistical disclosure limitation (control)* is the term that describes the theory and methods of protecting sensitive information when releasing statistical
microdata or tabular data. An up-to-date overview of theory and methods of
this field can be found in the monograph [19] and, for tabular data only, in the
survey [8]. An excellent reference is also [27].

© Springer International Publishing Switzerland 2016
J. Domingo-Ferrer and M. Pejić-Bach (Eds.): PSD 2016, LNCS 9867, pp. 41–53, 2016.
DOI: 10.1007/978-3-319-45381-1_4

Minimum-distance controlled tabular adjustment (CTA) methodology was first introduced in [7,15]. As indicated in [4] CTA can be formulated as the following problem: Given a table with sensitive cells, compute the closest safe table in which sensitive cells are modified to avoid re-computation, and the remaining cells are minimally adjusted to satisfy the table equations. The closeness of the original and modified table is measured by the weighted distance between the tables with respect to a certain norm. Most commonly used norms are ℓ_1 and ℓ_2 norms. Thus, the problem can be formulated as a minimization problem with the objective function being a particular weighted distance function and constraints being table equations and lower and upper bounds on the cell values.

In general, CTA is Mixed Integer Optimization Problem (MIOP) which is a difficult problem to solve especially for the large dimension problems. A priori fixing the values of binary variables reduces the problem to the continuous optimization problem which is easier to solve, however, the quality of the solution may be reduced. In addition, the values of the binary variables have to be assigned carefully otherwise the problem may become infeasible [12,13].

The objective function in continuous CTA is based on either the ℓ_1-norm or ℓ_2-norm. The formulation of ℓ_2-CTA leads to the Quadratic Programing (QP) problem, while ℓ_1-CTA can be formulated as the Linear Programming (LP) problem. However, the resulting LP has the number of variables that is twice the number of cells of the table as opposed to ℓ_2-CTA where the resulting QP problem has a number of variables equal to the number of cells. In general, the QP of ℓ_2-CTA is usually more efficiently solved than the LP of ℓ_1-CTA [4,7].

In [4] the Pseudo-Huber regularization of the ℓ_1-CTA is proposed. The Pseudo-Huber approximation of the ℓ_1-norm objective function leads to the convex optimization problem. However, the advantage is that the number of variables in Pseudo-Huber formulation of the ℓ_1-CTA remains the same as the number of cells. In [4] it is shown that Pseudo-Huber-CTA can be more efficiently solved than LP ℓ_1-CTA for certain types of tables and using an appropriate method that takes into account the structure of the problem.

All these models are solved using appropriate versions of the Interior-Point Method (IPM). These methods have been developed in recent years to efficiently solve different types of, often large, nonlinear (convex) optimization problems. It has been shown both theoretically and numerically that IPMs perform better on problems that have a certain structure, such as Conic Optimization (CO) problems, which are LP problems where variables are elements of cones. Most common cones are non-negative orthant, second order (quadratic) cone and semi-definite cone [2,3,25].

Hence, motivated by the above comment, in this paper we develop a new Second Order (Quadratic) Cone (SOC) formulation of the ℓ_1 and Pseudo-Huber-CTA. It is shown on the small example of a two-dimensional table that SOC CTA models are more efficiently solved than the original models. It is expected that the same will be the case for larger and more complex tables. Extensive numerical testing on various types of tables is beyond the scope of this paper; however, it is needed and it is forthcoming as a part of future research.

The paper is organized as follows. In Sect. 2 the general MIOP and then continuous CTA are formulated. Then the ℓ_1 and ℓ_2 continuous CTA are derived. The Pseudo-Huber-CTA formulation is considered in Sect. 3. The new SOC formulations of both Pseudo-Huber and ℓ_1 CTA are developed in Sect. 4. In Sect. 5 the SOC CTA models are applied to the small example of two-dimensional table and these instances are solved using MOSEK SOC solver. The concluding remarks are given in Sect. 6.

2   Formulation of the General CTA Model

The following CTA formulation is given in [4]: Given the following set of parameters:

(i) A set of cells $a_i, i \in \mathcal{N} = \{1, \ldots, n\}$. The vector $a = (a_1, \ldots, a_n)^T$ satisfies certain linear system $Aa = b$ where $A \in \mathbb{R}^{m \times n}$ is an $m \times n$ matrix and and $b \in \mathbb{R}^m$ is m-vector.

(ii) A lower, and upper bound for each cell, $l_{a_i} \leq a_i \leq u_{a_i}$ for $i \in \mathcal{N}$, which are considered known by any attacker.

(iii) A set of indices of sensitive cells, $\mathcal{S} = \{i_1, i_2, \ldots, i_s\} \subseteq \mathcal{N}$.

(iv) A lower and upper protection level for each sensitive cell $i \in \mathcal{S}$ respectively, lpl_i and upl_i, such that the released values must be outside of the interval $(a_i - lpl_i, \ a_i + upl_i)$.

(v) A set of weights, $w_i, i \in \mathcal{N}$ used in measuring the deviation of the released data values from the original data values.

A CTA problem is a problem of finding values $z_i, \ i \in \mathcal{N}$, to be released, such that $z_i, \ i \in \mathcal{S}$ are safe values and the weighted distance between released values z_i and original values a_i, denoted as $\|z - a\|_{l(w)}$, is minimized, which leads to solving the following optimization problem

$$\begin{aligned} \min_z \ & \|z - a\|_{l(w)} \\ s.t. \ & Az = b, \\ & l_{a_i} \leq z_i \leq u_{a_i}, \ i \in \mathcal{N}, \\ & z_i, \ i \in \mathcal{S} \text{ are safe values.} \end{aligned} \tag{1}$$

As indicated in the assumption (iv) above, safe values are the values that satisfy

$$z_i \leq a_i - lpl_i \text{ or } z_i \geq a_i + upl_i, \ i \in \mathcal{S}. \tag{2}$$

By introducing a vector of binary variables $y \in \{0, 1\}^s$ the constraint (2) can be written as

$$\begin{aligned} z_i &\geq -M(1 - y_i) + (a_i + upl_i) y_i, \ i \in \mathcal{S}, \\ z_i &\leq My_i + (a_i - lpl_i)(1 - y_i), \quad i \in \mathcal{S}, \end{aligned} \tag{3}$$

where $M \gg 0$ is a large positive number. Constraints (3) enforce the upper safe value if $y_i = 1$ or the lower safe value if $y_i = 0$.

Replacing the last constraint in the CTA model (1) with (3) leads to a mixed integer convex optimization problem (MIOP) which is in general a difficult problem to solve; however, it provides solutions with high data utility [11]. The alternative approach is to fix binary variables up front which leads to a CTA that is acontinuous convex optimization problem. The continuous CTA may be easier to solve; however, the obtained solution may have a lower data utility. Furthermore, a wrong assignment of binary variables may result in the problem being infeasible. Strategies on how to avoid this difficulty are discussed in [12,13].

In this paper we consider a continuous CTA where binary variables are fixed and vector z is replaced by the vector of *cell deviations*

$$x = z - a. \tag{4}$$

The CTA (1) with constraints (3) reduces to the following convex optimization problem:

$$\min_{x} \|x\|_{l(w)}$$
$$s.t.\ Ax = 0, \tag{5}$$
$$l \leq x \leq u,$$

where upper and lower bounds for x_i, $i \in \mathcal{N}$ are defined as follows:

$$l_i = \begin{cases} upl_i & \text{if } i \in \mathcal{S} \text{ and } y_i = 1 \\ l_{a_i} - a_i & \text{if } (i \in \mathcal{N} \setminus \mathcal{S}) \text{ or } (i \in \mathcal{S} \text{ and } y_i = 0) \end{cases} \tag{6}$$

$$u_i = \begin{cases} -lpl_i & \text{if } i \in \mathcal{S} \text{ and } y_i = 0 \\ u_{a_i} - a_i & \text{if } (i \in \mathcal{N} \setminus \mathcal{S}) \text{ or } (i \in \mathcal{S} \text{ and } y_i = 1). \end{cases} \tag{7}$$

The two most commonly used norms in problem (5) are the ℓ_1 and ℓ_2 norms. For the ℓ_2-norm the problem, (5) reduces to the following ℓ_2-CTA model which is a QP problem:

$$\min_{x} \sum_{i=1}^{n} w_i x_i^2$$
$$s.t.\ Ax = 0, \tag{8}$$
$$l \leq x \leq u.$$

For the ℓ_1-norm the problem, (5) reduces to the following ℓ_1-CTA model:

$$\min_{x} \sum_{i=1}^{n} w_i |x_i|$$
$$s.t.\ Ax = 0, \tag{9}$$
$$l \leq x \leq u.$$

The above ℓ_1-CTA model (9) is a convex optimization problem; however, the objective function is not differentiable at $x = 0$. Since most of the algorithms, including IPMs, require differentiability of the objective function, problem (9) needs to be reformulated. The reformulations that have been considered in [4] are reviewed in the next section.

3   LP and Pseudo-Huber Formulation of ℓ_1-CTA

The ℓ_2-CTA model (8) is a standard QP problem that can be efficiently solved using IPM or other methods. However, as noted at the end of the previous section, the ℓ_1-CTA model (9) needs reformulation in order to be efficiently solved by IPM or some other method. The standard reformulation is the transformation of model (9) to the following LP model:

$$
\begin{aligned}
\min_{x^-,x^+} \ & \sum_{i=1}^{n} w_i \left(x_i^+ + x_i^- \right) \\
s.t. \ & A \left(x_i^+ - x_i^- \right) = 0, \\
& l^+ \le x^+ \le u^+, \\
& l^- \le x^- \le u^-,
\end{aligned}
\tag{10}
$$

where

$$
x^+ = \begin{cases} x & \text{if } x \ge 0 \\ 0 & \text{if } x < 0, \end{cases} \qquad x^- = \begin{cases} 0 & \text{if } x > 0 \\ -x & \text{if } x \le 0, \end{cases}
\tag{11}
$$

and lower and upper bounds for x_i^- and x_i^+, $i \in \mathcal{N}$ are as follows:

$$
l_i^+ = \begin{cases} upl_i & \text{if } i \in \mathcal{S} \text{ and } y_i = 1 \\ 0 & \text{if } (i \in \mathcal{N} \setminus \mathcal{S}) \text{ or } (i \in \mathcal{S} \text{ and } y_i = 0) \end{cases}
$$

$$
u_i^+ = \begin{cases} 0 & \text{if } i \in \mathcal{S} \text{ and } y_i = 0 \\ u_{a_i} - a_i & \text{if } (i \in \mathcal{N} \setminus \mathcal{S}) \text{ or } (i \in \mathcal{S} \text{ and } y_i = 1) \end{cases}
$$

$$
l_i^- = \begin{cases} lpl_i & \text{if } i \in \mathcal{S} \text{ and } y_i = 0 \\ 0 & \text{if } (i \in \mathcal{N} \setminus \mathcal{S}) \text{ or } (i \in \mathcal{S} \text{ and } y_i = 1) \end{cases}
\tag{12}
$$

$$
u_i^- = \begin{cases} 0 & \text{if } i \in \mathcal{S} \text{ and } y_i = 1 \\ a_i - l_{a_i} & \text{if } (i \in \mathcal{N} \setminus \mathcal{S}) \text{ or } (i \in \mathcal{S} \text{ and } y_i = 0). \end{cases}
$$

Problem ℓ_1-CTA (10) is an LP problem; however, it has twice the number of variables as the QP problem (8) and twice the number of box constraints. As indicated in [4], the splitting of the variables $x = x^+ - x^-$ and the increased dimension of the model may cause problems. In order to overcome these difficulties in [4] it was suggested to use a regularization of problem (9) by approximating absolute value with the Pseudo-Huber function that has the same number of variables as in the QP formulation (8).

The original Huber function $\varphi_\delta : \mathbb{R} \longrightarrow \mathbb{R}_+$ is defined as

$$
\varphi_\delta(x_i) = \begin{cases} \frac{x_i^2}{2\delta} & |x_i| \le \delta \\ |x_i| - \frac{\delta}{2} & |x_i| \ge \delta. \end{cases}
\tag{13}
$$

It approximates $|x_i|$ for small values of $\delta > 0$; the smaller the δ, the better the approximation. The Huber function is continuously differentiable; however, the second derivative is not continuous at $|x_i| = \delta$ which may cause problems when this function is used in second order optimization algorithms, such as IPMs. Hence, it is better to consider the Pseudo-Huber function $\phi_\delta : \mathbb{R} \longrightarrow \mathbb{R}_+$

$$\phi_\delta(x_i) = \sqrt{\delta^2 + x_i^2} - \delta \tag{14}$$

whose first and second derivatives are bounded and Lipschitz continuous [17]. Again, the smaller the δ the better the approximation.

Now, the ℓ_1-CTA problem (9) can be approximated by the following convex optimization problem

$$\begin{aligned}
\min_x \ & \sum_{i=1}^n w_i \phi_\delta(x_i) \\
s.t. \ & Ax = 0, \\
& l \le x \le u.
\end{aligned} \tag{15}$$

The advantage of the Pseudo-Huber-CTA model (15) is that it has the same number of variables as ℓ_2-CTA and the same feasible region, the only difference is that the quadratic objective function is replaced by a strictly convex function.

Optimization problems (8), (10) and (15) can be solved with appropriate versions of the Interior-Point Methods (IPM). Since IPMs are the methods of choice to solve different CTA models, in the rest of the section we describe the main ideas of IPMs, only on a conceptual level, and then we discuss their application on given CTA models.

IPMs have in many ways revolutionized the optimization theory and practice in the past three decades since the appearance of the Karmarkar's breakthrough paper [20]. Since then, the field of IPMs has been a very active area of research with literary thousands of papers published as well as numerous excellent monographs and textbooks. The general theory of IPMs for convex optimization problems can be found in the seminal monograph of Nesterov and Nemirovskii [26]. In addition to this monograph, and without any attempt to be complete, we mention a few other relevant references [22, 28, 29]. The reason for such an interest is that IPMs have proven to be very efficient in solving large linear and non-linear (convex) optimization problems which were previously hard to solve. Now-days almost every relevant optimization software, whether commercial or open source, contains an IPM solver which is capable of solving at least LP problems and in many cases QP problems, and, less frequently, conic optimization problems. In the case of LP there are plenty of numerical studies showing that IPMs are at least as efficient, if not more, as the classical Simplex Method (SM) on large scale LP problems.

The basic idea of path-following IPMs, that are most commonly used and studied, is centered around approximately following the parametric trajectory that is called *central path* which leads to the solution of the problem when a parameter is approaching zero. The points on the central path are called *μ-centers* and are obtained as solutions of the Karush-Kuhn-Tucker (KKT) optimality conditions of the problem where a (the) complementarity equation(s) is (are) perturbed by a positive parameter $\mu > 0$. In particular, the perturbed KKT system for Pseudo-Huber-CTA is explicitly listed in [4].

The solution of the problem, which is obtained when $\mu = 0$, is found by tracing the central path while gradually reducing μ to zero. However, tracing the central path exactly would be prohibitively inefficient. The main achievement of IPMs have been to show that it is sufficient to trace the central path approximately; as long as the iterates are in the certain neighborhood of the central path, it is still possible to prove global convergence and, moreover, show that the ϵ-approximate solution of the problem, according to the appropriate proximity measure, can be obtained in polynomial number of iterations with the best theoretical upper bound being $O\left(\sqrt{n}\log\frac{n}{\epsilon}\right)$, where n represents the number of variables of the problem at hand.

However, practical behavior of IPM heavily depends on many factors, such as the structure of the problem, the starting point, the accuracy needed, etc. As reported in [4], Pseudo-Huber-CTA (15) can be difficult to solve with a general convex optimization solver even for small instances if the solver is not 'appropriately tuned'. However, for problems that exhibit a special structure such as 3-D tables whose constraints have a block-angular structure, the specialized block-angular IPM of J. Castro [5,9,10] solves Pseudo-Huber-CTA more efficiently than ℓ_1-CTA while ℓ_2-CTA has by far the best CPU time. Hence, Pseudo-Huber-CTA is a viable option for solving ℓ_1-CTA; however, the IPM have to be implemented with care and, in addition, the specialized IPM may not work efficiently for other types of tables. As indicated in [4], modifications and tuning of the Block-angular IPM so it can handle large and complex tables of different types is a direction for future research.

Another direction in searching how to efficiently solve Pseudo-Huber-CTA and ℓ_1-CTA is to investigate whether these models can be transformed into the conic optimization (CO) problems. The motivation for such investigation comes from the fact that it has been established both theoretically and numerically that IPMs perform better on the well structured problems such as CO problems than on general convex optimization problems [2,3,25]. CO problems are LP problems over cones, that is, variables belong to certain types of cones. Most common cones are either non-negative orthant, second-order (quadratic) cone or semidefinite cone definitions; of which are listed in the next section. Thus, formulating Pseudo-Huber and ℓ_1-CTA as CO problems would be advantageous. In the next section we develop SOC formulation of both Pseudo-Huber and ℓ_1 CTA.

4   SOC Formulation of Pseudo-Huber and ℓ_1 CTA

In this section we investigate how Pseudo-Huber and ℓ_1 CTA can be formulated as SOC models.

The CO problems can be formulated as

$$\min_{x} c^T x$$
$$s.t.\ Ax = b, \tag{16}$$
$$x \in \mathcal{K},$$

where $\mathcal{K}$ is a cone of the following three types:

1. The linear cone or non-negative orthant:

$$\mathcal{K} = \mathbb{R}_+^n := \{x \in \mathbb{R}^n : x_i \geq 0, \ i = 1, \ldots, n\}.$$

2. The positive semidefinite cone:

$$\mathcal{K} = \mathbf{S}_+^n := \{X \in \mathbf{S}^n : X \succeq 0\},$$

where $\succeq$ means that X is positive semidefinite matrix and $\mathbf{S}^n$ is a set of symmetric n-dimensional matrices.

3. The quadratic or second-order cone:

$$\mathcal{K} = \mathcal{L}^n = \{x \in \mathbb{R}^n : x_i \geq \sqrt{x_1^2 + \cdots + x_{i-1}^2 + x_{i+1}^2 + \cdots + x_n^2}\}.$$

More generally, $\mathcal{K}$ can be a Cartesian product of the above mentioned cones. It is also worth mentioning that the cones defined above are examples of symmetric cones, thus problem (16) can be considered in a more general framework of Symmetric Optimization (SO) problems, see [16, 18, 24] and references therein.

In what follows, we present a reformulation of Pseudo-Huber-CTA problem (15) as a SOC problem. Consider Pseudo-Huber Function (14)

$$\phi_\delta(x_i) = \sqrt{\delta^2 + x_i^2} - \delta.$$

Let's define

$$t_i := \sqrt{\delta^2 + x_i^2} \quad \text{and} \quad y_i := \delta, \quad i = 1, \ldots, n. \tag{17}$$

Hence, we have

$$t_i = \sqrt{x_i^2 + y_i^2}$$

which is the boundary of the second-order (quadratic) cone

$$\mathcal{K}_i = \left\{(x_i, y_i, t_i) \in \mathbb{R}^3 \ : \ t_i \geq \sqrt{x_i^2 + y_i^2}\right\}.$$

Now, the reformulation of the Pseudo-Huber-CTA (15) as a SOC problem follows

$$
\begin{aligned}
\min_x \ & \sum_{i=1}^n w_i (t_i - y_i) \\
\text{s.t.} \ & Ax = 0, \\
& y_i = \delta; \quad i = 1, \ldots, n, \\
& (x_i, y_i, t_i) \in \mathcal{K}_i; \quad i = 1, \ldots, n, \\
& l \leq x \leq u.
\end{aligned}
\tag{18}
$$

This model is valid even for $\delta = 0$. In that case we obtain a SOC formulation of the $l1$-CTA (9)

$$
\begin{aligned}
\min_x \ & \sum_{i=1}^n w_i t_i \\
\text{s.t.} \ & Ax = 0, \\
& (x_i, t_i) \in \mathcal{K}_i; \quad i = 1, \ldots, n, \\
& l \leq x \leq u.
\end{aligned}
\tag{19}
$$

This model could have been obtained directly from $l1$-CTA (9) because the absolute value has an obvious second-order cone representation since the epigraph of the absolute value function is exactly second-order cone, that is,

$$t_i = |x_i| \quad \longrightarrow \quad \mathcal{K}_i = \left\{ (x_i, t_i) \in \mathbb{R}^2 \; : \; t_i \geq \sqrt{x_i^2} \right\}.$$

It is well known that the solutions of SOC problems (18) and (19) achieve solutions at the boundary of the cones, hence, Eq. (17) will hold at the solution [2,3]. Thus, it is not necessary to enforce these equations in SOC models; in fact, their inclusion would lead to noncovex problems that would be difficult to solve.

An IPM for SOC can now be used to find an ϵ-approximate solutions to SOC Pseudo-Huber and ℓ_1 CTA models. We have used MOSEK SOC solver [1] that is considered one of the best, if not the best, SOC solver available on the market today.

5   Numerical Results for the Small Example

In this section an example of the small two-dimensional table stated in Fig. 3 in [4] is considered. The table is listed in Fig. 1 below as the table (a).

The continuous CTA model based on the table (a) is formulated in the following way:

- The linear constraints are obtained from the requirement that the sum of the elements in each row (or column) remains constant and is equal to the corresponding component in the last column (or row) of table (a).
- The sensitive cells are cells a_1 and a_{12}. For both of them the upper safe values are enforced, which are listed in the parentheses in the lower right corners of the cells, $upl_1 = 3$ and $upl_{12} = 5$ respectively. Hence, in the transformed tables the upper safe value of the cell a_1 should be 13 or above and for a_{12} the upper safe value should be 18 or above.
- For the nonsensitive cells the lower and upper bounds are set to be zero and positive infinity respectively, that is, $l_{a_i} = 0$ and $u_{a_i} = \inf$ for $i = 2, \ldots, 11$.
- The weights in the objective function are set to have the value one, that is, $w_i = 1$ for $i = 1, \ldots, 12$.

From this basic CTA model different CTA models discussed in the paper were formulated and then these models were solved using appropriate IPM solvers. The results are listed in Fig. 1.

In [4] it was observed that ℓ_2-CTA had the fastest execution. Hence, we replicated the solution of the ℓ_2-CTA instance of the example and compared its performance with SOC models instances. The calculations were carried out on a Lenovo ThinkPad W530 computer with Intel(R) CORE i7-3740QM 2.70 GHz processor. The results are given in Table 1.

From Table 1 we can observe that SOC versions are comparable to the ℓ_2 version both in number of iterations and CPU time; SOC ℓ_1 was slightly faster

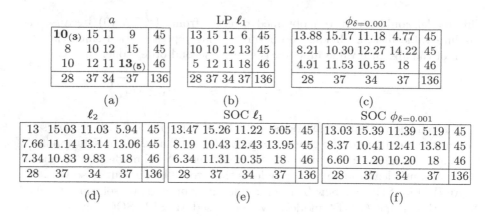

Fig. 1. Results of the small example (rounded to two decimal places).

Table 1. Results for ℓ_2 and SOC CTA

CTA model	Obj. funct.	It. no.	CPU
ℓ_2	20.69	6	0.08
SOC-ℓ_1	20	7	0.07
SOC Pseudo-Huber	20	9	0.09

than ℓ_2 while SOC Pseudo-Huber was slightly slower, which is the expected result. Hence, the SOC models are more effective than the LP ℓ_1 and Pseudo-Huber-CTA models for this example.

Furthermore, for LP ℓ_1, Pseudo-Huber $\phi_{0.001}$, SOC ℓ_1, and SOC Pseudo-Huber $\phi_{0.001}$ CTA instances the optimal values of their respective objective functions are the same, namely, the value is 20, while for ℓ_2-CTA instance it is 20.69. Thus, the objective values for SOC Pseudo Huber and ℓ_1-CTA instances are the same as for the original non-SOC instances, namely 20, which was expected.

These results are in line with plenty of other evidence that it is advantageous to solve the SOC formulation of the problem by IPM, rather than using IPM to the original formulation of the problem (see for example [2,3,23,25]). We are confident that the advantages of the SOC models will be even more visible when applied to larger tabular data sets. Moreover, the SOC IPM is robust and flexible enough to handle different types of tables.

6 Concluding Remarks

The main goal of the paper is mainly theoretical, that is, to present a Second Order Cone (SOC) formulation of the Pseudo-Huber and the ℓ_1 CTA models, (18) and (19) respectively as an alternative to the original Pseudo-Huber and LP ℓ_1 CTA models, (15) and (10) respectively. The application of the SOC models to the small example in Sect. 5 shows promise to be an effective alternative to the application of the original models to the small example. More numerical testing is needed and is forthcoming as a future research topic where SOC models would be implemented and tested on the different types of tables of large dimensions mentioned in Conclusion of [4].

From Fig. 1, it can be observed that the resulting tables for all the models except LP ℓ_1 change most of the cells of the original table (a) that are not fixed. The reason lays in the nature of IPMs. In these methods, the iterates approximately follow the central path that converges to the analytic center of the optimal set which implies that most of the cells will be changed, while the IPM with crossover or alternatively the Simplex Method, for LP ℓ_1 finds the basic solution which implies fewer cells will be changed. Hence, if there is a requirement to minimize the number of non-sensitive cells that are changed, then the LP ℓ_1 models solved with SM or IPM with crossover is the right approach. However, if the number of nonsensitive cells changed is not an issue such as for certain types of magnitude tables, then the suggested approach is to use either the SOC ℓ_1 model or the ℓ_2 model because they are faster. Unless prior regularization of the ℓ_1 model is necessary, which then leads to the Pseudo-Huber model and related SOC Pseudo-Huber model, it is more efficient to use the SOC ℓ_1 model directly.

As noted in [4], it has been empirically shown that CTA in general exhibits a low disclosure risk [6] and, at the same time, high data utility [13,14] (see also [21]). However, the study of the disclosure risk and data utility of tables protected by the Pseudo-Huber-CTA model and the SOC CTA models is lacking and is certainly an interesting future research topic.

Acknowledgments. The first author would like to thank Erling Andersen and Florian Jarre for the constructive discussion regarding the SOC model and Iryna Petrenko for her help in performing the calculations described in Sect. 5.

The authors would like to express their appreciation to Donald Malec for his careful reading of the paper and many useful suggestions.

Any opinions, findings, and conclusions or recommendations expressed in this publication are those of the authors only and do not necessarily reflect the views of the Centers for Disease Control and Prevention.

References

1. Andersen, E.D.: MOSEK solver (2016). https://mosek.com/resources/doc
2. Alizadeh, F., Goldfarb, D.: Second-order cone programming. Math. Program. **95**(1), 3–51 (2003)
3. Andersen, E.D., Roos, C., Terlaky, T.: On implementing a primal-dual interior-point method for conic quadratic optimization. Math. Program. **95**(2), 249–277 (2003)
4. Castro, J.: A CTA model based on the huber function. In: Domingo-Ferrer, J. (ed.) PSD 2014. LNCS, vol. 8744, pp. 79–88. Springer, Heidelberg (2014)
5. Castro, J.: An interior-point approach for primal block-angular problems. Comput. Optim. Appl. **36**, 195–219 (2007)
6. Castro, J.: On assessing the disclosure risk of controlled adjustment methods for statistical tabular data. Int. J. Uncertainty Fuzziness Knowl. Based Syst. **20**, 921–941 (2012)

7. Castro, J.: Minimum-distance controlled perturbation methods for large-scale tabular data protection. Eur. J. Oper. Res. **171**, 39–52 (2006)
8. Castro, J.: Recent advances in optimization techniques for statistical tabular data protection. Eur. J. Oper. Res. **216**, 257–269 (2012)
9. Castro, J., Cuesta, J.: Quadratic regularization in an interior-point method for primal block-angular problems. Math. Program. **130**, 415–445 (2011)
10. Castro, J., Cuesta, J.: Solving ℓ_1-CTA in 3D tables by an interior-point method for primal block-angular problems. TOP **21**, 25–47 (2013)
11. Castro, J., González, J.A.: Assessing the information loss of controlled adjustment methods in two-way tables. In: Domingo-Ferrer, J. (ed.) PSD 2014. LNCS, vol. 8744, pp. 11–23. Springer, Heidelberg (2014)
12. Castro, J., Gonzalez, J.A.: A fast CTA method without complicating binary decisions. In: Documents of the Joint UNECE/Eurostat Work Session on Statistical Data Confidentiality, Statistics Canada, Ottawa, pp. 1–7 (2013)
13. Castro, J., Gonzalez, J.A.: A multiobjective LP approach for controlled tabular adjustment in statistical disclosure control. Working paper, Department of Statistics and Operations Research, Universitat Politecnica de Catalunya (2014)
14. Castro, J., Giessing, S.: Testing variants of minimum distance controlled tabular adjustment. In: Monographs of Official Statistics, Eurostat-Office for Official Publications of the European Communities, Luxembourg, pp. 333–343 (2006)
15. Dandekar, R.A., Cox, L.H.: Synthetic tabular data: an alternative to complementary cell suppression. Manuscript, Energy Information Administration, U.S. (2002)
16. Faraut, J., Koranyi, A.: Analysis on Symmetric Cones. Oxford University Press, New York (1994)
17. Fountoulakis, K., Gondzio, J.: A second-order method for strongly convex L1-regularization problems. Technical report ERGO-14-005, School of Mathematics, The University of Edinburgh (2014)
18. Gu, G.: Interior-point methods for symmetric optimization. Ph.d. thesis, TU Delft (2009)
19. Hundepool, A., Domingo-Ferrer, J., Franconi, L., Giessing, S., Schulte Nordholt, E., Spicer, K., De Wolf, P.-P.: Statistical Disclosure Control. Wiley, Chichester (2012)
20. Karmarkar, N.: A polynomial-time algorithm for linear programming. Combinatorica **4**, 373–395 (1984)
21. Karr, A.F., Kohnen, C.N., Oganian, A., Reiter, J.P., Sanil, A.P.: A framework for evaluating the utility of data altered to protect confidentiality. Am. Stat. **60**(3), 224–232 (2006)
22. Lesaja, G.: Introducing interior-point methods for introductory operations research courses and/or linear programming courses. Open Oper. Res. J. **3**, 1–12 (2009)
23. Lesaja, G., Slaughter, V.: Interior-point algorithms for a class of convex optimization problems. Yugoslav J. Oper. Res. **19**(3), 239–248 (2009)
24. Lesaja, G., Roos, C.: Kernel-based interior-point methods for monotone linear complementarity problems over symmetric cones. J. Optim. Theor. Appl. **150**(3), 444–474 (2011)
25. Ben-Tal, A., Nemirovski, A.: Lectures in Modern Convex Optimization: Analysis, Algorithms and Engineering Applications. MPS/SIAM Series in Optimization. SIAM, Philadelphia (2001)
26. Nesterov, Y., Nemirovski, A.: Interior-Point Polynomial Algorithms in Convex Programming. SIAM Studies in Applied Mathematics, vol. 13. SIAM, Philadelphia (1994)

27. Oganian, A.: Security and information loss in statistical database protection. Ph.d. thesis, Universitat Politecnica de Catalunya (2003)
28. Roos, C., Terlaky, T., Vial, J.P.: Theory and Algorithms for Linear Optimization. An Interior-Point Approach. Springer, Heidelberg (2005)
29. Wright, S.J.: Primal-Dual Interior-Point Methods. SIAM, Philadelphia (1996)

Microdata and Big Data Masking

Anonymization in the Time of Big Data

Josep Domingo-Ferrer[✉] and Jordi Soria-Comas[✉]

UNESCO Chair in Data Privacy, Department of Computer Engineering
and Mathematics, Universitat Rovira i Virgili, Av. Països Catalans 26,
43007 Tarragona, Catalonia, Spain
{josep.domingo,jordi.soria}@urv.cat

Abstract. In this work we explore how viable is anonymization to prevent disclosure in structured big data. For the sake of concreteness, we focus on k-anonymity, which is the best-known privacy model based on anonymization. We identify two main challenges to use k-anonymity in big data. First, confidential attributes can also be quasi-identifier attributes, which increases the number of quasi-identifier attributes and may lead to a large information loss to attain k-anonymity. Second, in big data there is an unlimited number of data controllers, who may publish independent k-anonymous releases on overlapping populations of subjects; the k-anonymity guarantee does not longer hold if an observer pools such independent releases. We propose solutions to deal with the above two challenges. Our conclusion is that, with the proposed adjustments, k-anonymity is still useful in a context of big data.

Keywords: Data anonymization · Big data · k-anonymity · Curse of dimensionality · Multiple releases

1 Introduction

When releasing microdata files for secondary use, the privacy of the subjects to whom the data refer must be taken into consideration. Statistical disclosure control (SDC) addresses this by altering the original data with the aim of protecting the privacy of the subjects while preserving sufficient utility for statistical analyses. In other words, a data user should not be able to accurately infer anything about a specific subject, but she should be able to accurately infer properties of subgroups of subjects. The SDC literature is quite rich; see [6,10,20] for comprehensive surveys.

Among the SDC methods used, anonymization is of particular relevance. The goal of anonymization of a microdata set is to blur the relation between each record and the corresponding subject. The usual anonymization approach is to classify the attributes into two categories: quasi-identifier attributes (QI) and confidential attributes. Confidential attributes are those that contain the sensitive information whose disclosure we want to prevent. An attribute is classified as a quasi-identifier attribute if it can help re-identify the subject corresponding to a record (link the record to a specific subject). The usual quasi-identifier attributes

© Springer International Publishing Switzerland 2016
J. Domingo-Ferrer and M. Pejić-Bach (Eds.): PSD 2016, LNCS 9867, pp. 57–68, 2016.
DOI: 10.1007/978-3-319-45381-1_5

are socio-demographic attributes such as age, sex, gender, ZIP code, etc. More strictly, any attribute that is known to be externally available in association with a specific subject should be classified as a quasi-identifier attribute. The idea underlying anonymization is to alter the values of quasi-identifier attributes in the released data set in order to prevent linkage of its records to subjects in external databases containing identifiers and some of the quasi-identifier attributes; the ultimate goal is to prevent observers from learning the confidential attribute values of identified subjects.

With the development of information and communication technologies, data collection capabilities have witnessed a dramatic growth. The large amount of data available, together with the use of innovative data analysis techniques (whose goal is to generate knowledge rather than test hypothesis), have led to the development of big data. Previously, data collection (for example, by national statistical institutes) was a planned process that culminated in a structured data set. In a time when data were costly to acquire, the planning of data collection aimed at minimizing the resources required to collect the data, while ensuring the latter would be valid to make inferences on the population. In essence, we had scarce but accurate data. With big data, the situation is the opposite. Data are not scarce anymore; rather, they effortlessly and routinely flow in from different (and apparently unrelated) sources. There is no need for planning expensive data gathering processes: by living their normal lives, subjects generate lots of data (through queries submitted to a web search engine, emails, social media activities, web browsing, spatiotemporal traces of mobile devices, etc.). Big data may not be as accurate as traditional data, and they may be more complex to analyze (they may require automated analysis of textual information, merging of data coming from different sources, etc.). However, the sheer volume of information available makes up for the lost accuracy, and the very diverse nature of the data can make them suitable for very diverse analyses.

In a big data context, important assumptions previously made in data anonymization procedures seem unrealistic. The determination of the quasi-identifier attributes was already controversial in the past because an observer with privileged knowledge of a target subject could use attributes not publicly known to most people as quasi-identifier attributes to re-identify the target's record. Despite this possibility, the fact that there were relatively few data controllers, that these were trusted, and that most of them collected the data under a strong pledge of privacy made it reasonable to classify attributes into two disjoint sets, the quasi-identifier attributes and the confidential attributes. With the development of big data, the number of data controllers has grown dramatically and it is unrealistic to view all of them as trusted. In fact, an untrusted controller can unduly leak confidential attributes, so that they can be used as quasi-identifier attributes by whoever acquires them. This new scenario must be taken into account when using SDC to protect a data set before release; otherwise the privacy protection may be insufficient. In particular, in the case of anonymization, we must be aware that any attribute is a potential quasi-identifier attribute.

Even though big data may seem to undermine the foundations of data anonymization by turning each attribute into a potential quasi-identifier attribute, anonymization remains an important desideratum in day-to-day practice. It is not only that its intuitiveness makes it a very appealing SDC method but also that many data protection regulations [1–3] are built around anonymization. Since this situation is unlikely to change, we should strive to carry out anonymization in a safe manner.

Contribution and plan of this paper

In this work, we analyze the impact of big data on current anonymization practices, most notably on the best-known anonymization-based privacy model, namely k-anonymity. We propose alternatives to maintain the disclosure control guarantees of k-anonymity in the big data world. Our aim is not so much to propose novel anonymization methods that are resilient to observers in a big data environment, but to find ways to adapt the current k-anonymity algorithms so that they can better deal with the new state of things. In other words, we do not seek to propose k-anonymity methods capable of dealing with all big data formats (e.g. textual data, video, audio, etc.) but rather enhance existing methods for structured data so that they remain effective in a big data environment.

The remainder of this paper is structured as follows. Section 2 gives background on anonymization concepts, data protection regulations, k-anonymity and some shortcomings of k-anonymity. Section 3 analyzes the two main challenges encountered by k-anonymity in a big data setting: first, confidential attributes are likely to be also QI attributes, which may lead to a large number of quasi-identifiers and hence to unacceptable utility loss; second, different data controllers may disseminate multiple independent k-anonymous releases on overlapping populations of subjects, which compromises the k-anonymity guarantee. This section does not only identify the two aforementioned challenges, but it also proposes possible solutions. Finally, conclusions and future research issues are gathered in Sect. 4.

2 Background

2.1 Anonymization

Anonymization is the process of masking the correspondence between records in a data set and the subjects to whom they correspond. It entails two main steps:

- *De-identification.* This consists of suppressing the identifier attributes (those that unambiguously identify a subject). Example identifier attributes include: name, social security number, email address, etc.
- *Quasi-identifier masking.* Quasi-identifier attributes can also be used to re-identify the subject to whom a record corresponds. However, unlike identifiers, quasi-identifier attributes should not be removed in general, because of the analytical value they convey. In fact, most attributes have the potential of being quasi-identifier attributes, depending on the observer's knowledge. Hence, masking them is much better than removing them.

2.2   Data Protection Regulations and Anonymization

The general view of data protection regulations is that anonymized data are not personal (re-identifiable) anymore. Two remarkable examples of such regulations are:

- U.S. Health Insurance Portability and Accountability Act (HIPAA) [1]. It was enacted in 1996 by the U.S. Congress in an attempt to improve healthcare and health insurance. According to Title II of the HIPAA, health information is not re-identifiable if either (i) a person with appropriate knowledge justifies that the risk is very small, or (ii) several attributes are removed or generalized to a specified level (e.g. names, telephone numbers, and social security numbers are removed; geographic units and dates are generalized).
- E.U. General Data Protection Regulation (GDPR) [3]. This is the European Union's forthcoming data protection regulation. In comparison with the current Directive 95/46/EC [2], that it will replace, the GDPR is more ambitious in terms of uniformity across the E.U., enforcement mechanisms and fines. The GDPR targets any personal (re-identifiable) data whatever the nature of the data is. Thus, non-re-identifiable (e.g. anonymized) data are not under the umbrella of the GDPR. However, the GDPR does not specify when data are (or are not) re-identifiable.

2.3 k-Anonymity

k-Anonymity [15, 16] is a well-known data privacy model that is based on anonymization. k-Anonymity seeks to guarantee at least a certain level of anonymity for the records in a data set. To that end, k-anonymity assumes that the attributes an observer can use to re-identify a record can be determined beforehand by the data controller, and makes each record indistinguishable with regard to these attributes within a group of k records.

In k-anonymity, the term *quasi-identifier* refers to a *set of attributes* whose combined values can lead to record re-identification. Thus, it focuses on controlling the release of such attributes. This is similar to the above-mentioned notion of quasi-identifier attribute but it is not equivalent. The attributes that form a quasi-identifier are quasi-identifier attributes, but a quasi-identifier attribute alone is not necessarily a quasi-identifier.

Definition 1 (Quasi-identifier). *Let X be a data set with attributes $X_1, \ldots, X_m$. A quasi-identifier of X is a subset of attributes $\{X_{i_1}, \ldots, X_{i_j}\} \subset \{X_1, \ldots, X_m\}$ whose combined values can lead to re-identification. Denote by QI_X the set of quasi-identifiers of X.*

Whether a set of attributes should be considered a quasi-identifier depends on the external information available to observers. The set of quasi-identifiers usually includes all the combinations of attributes that the data releaser reasonably assumes to be publicly available in external non-de-identified data sets (electoral rolls, phone books, etc.).

To prevent re-identification of records based on a quasi-identifier, k-anonymity requires each combination of values of the quasi-identifier attributes to be shared by k or more records.

Definition 2 (k-**Anonymity**). *Let X be a data set and let QI be the maximal quasi-identifier in QI_X, that is, the one containing all quasi-identifier attributes. The data set X is said to be k-anonymous if each combination of values of the attributes in QI that appears in X is shared by k or more records.*

In a k-anonymous data set, each group of k or more records sharing the values of all quasi-identifier attributes is called a k-*anonymous class*.

Several algorithms to generate k-anonymous data sets have been proposed. The most common ones are based on generalization and suppression [11,16] and on microaggregation [7].

2.4   Shortcomings of k-anonymity

The kind of disclosure risk limitation offered by k-anonymity is very intuitive: if we prevent records from being linked to the corresponding subjects, releasing the data set will not result in any disclosure. Indeed, this view of disclosure risk limitation is quite effective when the observer cannot decide by looking at the k-anonymous classes if a given subject is in the data set or not. However, the usual observer considered in k-anonymity knows that his target subject is in the k-anonymous data set. Under this observer model, the target subject can be linked to a k-anonymous class.

When all records in a k-anonymous class have similar values for the confidential attributes, an observer linking a subject to that class can infer confidential information on that subject. In other words, while k-anonymity provides protection against identity disclosure (at most a subject can be linked to her record's class), it cannot prevent attribute disclosure when the values of the confidential attributes are similar within a k-anonymous class. Two simple attacks have been proposed in the literature that exploit the lack of variability in the confidential attributes: the homogeneity attack and the background knowledge attack [13]. In the former, all the records in a k-anonymous class share the same value for the confidential attributes; thus, k-anonymity fails to provide any protection. In the background knowledge attack, the variability of the confidential attributes within the class is small and the observer possesses some background information that allows him to further restrict the feasible values of the confidential attributes for the target subject. To achieve protection against attribute disclosure, several refinements of the basic notion of k-anonymity have been proposed (e.g. l-diversity [13], t-closeness [12], β-likeness [5], etc.). All these require at least a certain level of variability in the confidential attributes within each k-anonymous class. Although attribute disclosure is an important issue in k-anonymity, it is not specifically related to the big data scenario and it has already been extensively treated in the literature (for example, in [8], in addition to the previous references).

How to safely release multiple k-anonymous data sets sharing some subjects and some confidential attributes is an open issue [14,19]. An example would be when two hospitals with possibly some common patients release respective k-anonymous patient discharge data sets including confidential attributes such as diagnosis and/or measurements on the patients. In this situation, the k-anonymity disclosure limitation guarantee is preserved over the two pooled releases only if any two k-anonymous classes coming from different k-anonymous releases have either all subjects in common or no subject in common. If two k-anonymous classes have k' subjects in common, with $0 < k' < k$, then only k'-anonymity is provided for those common subjects and only $(k - k')$-anonymity for the non-common subjects (note that comparing the confidential attribute values is likely to help determine the common subjects).

Hence, preserving k-anonymity in case of multiple releases requires a coordination that is easiest if all releases are performed by the same data controller and they all relate to the same population of subjects. When the data controller is the same but the populations of subjects in the releases are neither disjoint nor identical, the problem is more complex but still manageable. In particular, [14] proposes an approach to release k-anonymous incremental updates, which may modify the population by adding or removing some subjects, based on coarsening subsequent data releases given the previous ones to ensure that k-anonymity is still satisfied. Since the focus of our work here is big data, we are more interested in k-anonymous data releases independently performed by different data controllers.

3 k-Anonymity in a Big Data Context

We leave aside the difficulties of k-anonymity with low-variability confidential attributes and with multiple releases reviewed in the previous section, and we move to challenges inherent to big data. In a big data setting, there is a very large number of data controllers, so the most likely option in case of multiple releases on non-disjoint populations is that they are performed by different controllers. Let us explore the ramifications of this situation.

3.1   Confidential Attributes as QI Attributes

Like most privacy models, k-anonymity makes some assumptions on the side knowledge available to the observer. The disclosure risk limitation procedure is then tailored to reach the privacy guarantee stated by the model against such an observer. If the assumptions were too optimistic (that is, the actual observer has more side knowledge than assumed), the intended privacy guarantee may not be attained.

A key assumption in k-anonymity is that there is a clear distinction between quasi-identifiers and confidential attributes. In a big data setting, we argue that such a distinction is no longer tenable. There is a large number of data controllers, a lot of which collect data unawares of the subjects (by inspecting emails, tracking web navigation, recording web searches, recording geospatial information available from mobile devices, etc.). Hence, it is unrealistic to think that all controllers can be trusted. In turn, untrusted controllers may share, leak or sell any confidential attribute, which can then be used by the party acquiring it as a quasi-identifier attribute to improve re-identification attacks. Even if an untrusted controller does not share confidential attributes, he can use them himself to mount re-identification attacks against k-anonymous data sets released by other controllers.

Another usual assumption in k-anonymity is that there is a single confidential attribute. This simplification is innocuous when the confidential data cannot be used to improve record re-identification. However, when they *can* be used, as we argue is the case for big data, this simplification may mask the problem: if there is a single confidential attribute, an observer using it to improve record re-identification does not learn much, because he already knew the confidential data in the first place. To avoid this kind of (wrong) justification for excluding confidential attributes from quasi-identifiers, we consider a data set with multiple confidential attributes. Hence, our data set definition is as follows.

Definition 3. *The original data set is a table* $T(A_1, \ldots, A_r, C_1, \ldots, C_s)$ *where* $A_1, \ldots, A_r$ *are quasi-identifier attributes and* $C_1, \ldots, C_s$ *are confidential and quasi-identifier attributes at the same time.*

The usual approach in k-anonymity is to mask the QI attributes in such a way that each subject can be linked to k records with the same probability. For instance, when using generalization, we look for a minimal generalization of the QI that satisfies the k-anonymity requirement. Because the confidential attributes are also QI, all the attributes must be generalized. This naive approach has an important drawback: the curse of dimensionality, which states that, as the number of QI attributes grows, all the discriminatory information in the data is lost in order to achieve k-anonymity [4]. Thus, adding all confidential attributes to the set of QI attributes can be devastating for the data utility. To avoid increasing the number of QI attributes so much, we propose to release a separate k-anonymous data set for each confidential attribute. That is, rather than generating a k-anonymous $T_k(A_1, \ldots, A_r, C_1, \ldots, C_s)$, we generate separate k-anonymous versions of the s following data sets:

$$T_k^1(A_1, \ldots, A_r, C_1), \ldots, T_k^s(A_1, \ldots, A_r, C_s). \tag{1}$$

It might seem that we can release the data sets given by Expression (1) without generalizing the confidential attribute in each data set. After all, if the observer uses the confidential attribute to re-identify a subject, then it means that he already knew the value of the confidential attribute and, therefore, he does not learn anything. The previous reasoning is a (also flawed) variant of the above-mentioned justification to exclude confidential attributes from the set of QI attributes. It overlooks that the observer can leverage the knowledge of the confidential attribute value of a subject to improve the re-identification of the other subjects in the same k-anonymous class: this amounts to downgrading k-anonymity to $(k-1)$-anonymity. Therefore, when generating the k-anonymous versions of $T_k^1, \ldots, T_k^s$, the confidential attribute must also be generalized.

On the other hand, in Sect. 2.4 we said that, when releasing multiple k-anonymous data sets corresponding to original data sets sharing some subjects *and some confidential attributes*, two k-anonymous classes coming from two different releases should either have all subjects in common or no subject in common. The reason was that, if two classes had $0 < k' < k$ subjects in common, only k'-anonymity was guaranteed for the common subjects and only $(k-k')$-anonymity was guaranteed for the non-common subjects. When stating so, there was an implicit assumption that the records corresponding to the k' common subjects could be determined within each class, by linking them via the confidential attributes. However, in the k-anonymous releases of the datasets in Expression (1), there are *no common* confidential attributes among releases. If, beyond being different, the confidential attributes C_i and C_j in two releases are uncorrelated, then common subjects cannot be determined. Therefore, in case of uncorrelated confidential attributes, the data sets in Expression (1) can be k-anonymized independently, using standard k-anonymization algorithms in the literature.

The situation changes if there is a strong correlation among the confidential attributes in the different releases. In this case, the partition of subjects into k-anonymous classes in T_k^i and T_k^j must be the same, for all $i \neq j$. This constraint limits our ability to use existing algorithms out of the box: the quasi-identifier is not the same in both tables, because the confidential attributes are also QI attributes and they are different. Thus, using an existing k-anonymization algorithm independently on both tables is likely to yield different k-anonymous classes. To overcome this issue, we propose to perform the generalization in two steps: the first step generates k-anonymous classes taking into account only attributes $A_1, \ldots, A_r$; the second step generalizes the confidential attribute for it to have a single value within each class generated in the previous step. Algorithm 1 formalizes this two-step k-anonymization.

To guard against attribute disclosure, it is relatively simple to adjust Algorithm 1 to ensure a certain level of variability of the confidential attributes $C_1, \ldots, C_s$ within each k-anonymous class. This minimum level of variability must be taken into account when generating the k-anonymous classes G_k.

Even though Algorithm 1 is not needed to enforce k-anonymity when the confidential attributes are uncorrelated, there is an advantage in using it always: we can merge the k-anonymous data sets $T_k^1, \ldots, T_k^s$ into a single k-anonymous data set that contains all the confidential attributes.

3.2   Independent k-anonymous Data Releases on Non-disjoint Populations

k-Anonymity was designed to limit the disclosure risk of a single data release. In general, if several k-anonymous releases are pooled by the observer, there is no guarantee that k-anonymity holds anymore and attribute disclosure can

Algorithm 1. Masking of an original data set into several k-anonymous data sets having the same subject partition and the same non-confidential quasi-identifier attributes $A_1, \ldots, A_r$ and having each a single non-common confidential quasi-identifier attribute.

---

let $T(A_1, \ldots, A_r, C_1, \ldots, C_s)$ be a table where $C_1, \ldots, C_s$ are confidential attributes.

let G_k be the set of k-anonymous classes generated for the attributes $A_1, \ldots, A_r$
for $i = 1$ **to** s
$\quad T_k^i(A_1, \ldots, A_r, C_i) := G_k +$ generalization of C_i values within each k-anonymous class
end for

return $T_k^1, \ldots, T_k^s$

---

occur. A paradigmatic example consists of two k-anonymous data releases in which the confidential attributes are not considered to be QI attributes (and hence stay unmodified) and there are two k-anonymous classes that differ in one subject. By comparing the values of the confidential attributes in these k-anonymous classes, an observer can learn the value of the confidential attributes of that subject (attribute disclosure). Furthermore, by comparing the way the QI attribute values for that subject have been modified in the two k-anonymous releases, the observer may gain some information on the original values of those QI attributes, which may facilitate re-identifying that subject with probability greater than $1/k$ (re-identification disclosure).

As said in Sect. 2.4, if all the data releases are performed by the same data controller, the above attack can be avoided by either making sure that the same k-anonymous class of records are used when the population does not change between data releases or by adjusting the classes of records to satisfy k-anonymity when the population changes. The thorniest scenario is when independent k-anonymous data releases are performed by different data controllers, which is precisely the most common case in big data settings. We tackle this scenario next.

The example given at the beginning of this section shows that controlling the disclosure risk across multiple and independent k-anonymous data releases is not feasible in general. Even if we follow the approach suggested in Sect. 3.1 for the case of confidential attributes that are also QI attributes (thus making sure that every subject can be linked to k records), *the change in the generalization of the confidential attribute between k-anonymous classes of records that differ in one subject may reveal the actual confidential attribute value of that subject.* In particular, for the case of a continuous confidential attribute that has been generalized as an interval encompassing all values of the confidential attribute in the k-anonymous class, the confidential attribute may be disclosed for the subjects that have the largest or the smallest confidential attribute value in the class.

To minimize the risk of disclosing confidential information, we propose to replace the generalization of the confidential attribute by an alternative aggregation operation that is less sensitive to changes in the components of the k-anonymous class. We list below two possible alternatives:

- Mean of the values of the confidential attribute. By using the mean, we make sure that each subject contributes to the aggregation by a proportion of $1/k$. Despite this fact, the mean can still be sensitive to the presence of subjects with very extreme values.
- Median of the values of the confidential attribute. In comparison to the mean, the median is less sensitive to the extreme values. However, because the median reflects the value of a specific subject, a change in the median between k-anonymous classes of records that differ in one subject may reveal the exact value of the confidential attribute for that subject.

Apart from the above-mentioned respective shortcomings of the mean and the median, the fact that both are deterministically computed from the values in the k-anonymous classes introduces another problem, described next. In a big data scenario and given a target subject, it is not unrealistic to assume that the values of the confidential attribute are available to the observer for all the subjects other than the target subject in latter's k-anonymous class. In this case, when using the mean, the observer can determine the target subject's confidential attribute value; when using the median, the observer can determine it in some cases.

To prevent aggregation-related disclosures, we should insert some uncertainty in the aggregation. Differential privacy [9] can be used to introduce such an uncertainty in the case of the mean. Differential privacy aims at making the result of a computation similar between data sets that differ in one subject (neighbor data sets). By adding noise to the result of the computation, differential privacy makes sure that the probability of getting a specific result is similar between neighbor data sets. Usually, one adds Laplace-distributed noise with 0 mean and scale parameter that depends on the maximum change in the computation between neighbor data sets (a.k.a. the global sensitivity). For the case of adding differentially private noise to the mean of k-anonymous classes, the global sensitivity is the size of the domain of the confidential attribute divided by k [17, 18].

4   Conclusions and Future Work

This work has tried to answer the question of whether anonymization, and more precisely k-anonymity, is still a valid approach to prevent disclosure in the current big data situation. We have analyzed from a qualitative perspective how k-anonymity can be adapted for statistical disclosure control of big data. In particular, we have addressed the following two challenges and we have proposed solutions for them:

- In big data, any attribute, including the confidential attributes, can be a QI attribute. However, reaching k-anonymity with so many QI attributes in a straightforward way is subject to the so-called curse of dimensionality: a lot of information is lost. We have given a k-anonymization algorithm that circumvents this dimensionality problem.
- In big data there is a great number of data controllers and these are likely to generate multiple independent k-anonymous releases on overlapping populations of subjects. Keeping the disclosure risk at the level that is guaranteed for a single k-anonymous data release is not feasible: for example, an observer can compare k-anonymous data sets that differ in one subject and discover the confidential attribute values of that subject. To minimize the risk of disclosure, we have proposed to replace the usual generalization performed within each k-anonymous class by an alternative aggregation operation that is less sensitive to changes in the subjects of the class. In particular, we have explored using the mean and the median. The fact that these are computed deterministically from the confidential values of the subjects in the k-anonymous class is also a potential source of disclosure. For the case of the mean, we have proposed to neutralize such disclosure using a ε-differentially private mean.

Our qualitative analysis seems to indicate that anonymization and k-anonymity are still useful in our big data world. As future work, we expect to come up with a more quantitative analysis, where we will evaluate the utility that can be achieved for a given anonymity guarantee. We will also explore how to randomize other aggregation operators beyond the mean for the case of multiple independent k-anonymous releases.

Acknowledgments and Disclaimer. Partial support to this work has been received from the European Commission (projects H2020-644024 "CLARUS" and H2020-700540 "CANVAS"), from the Government of Catalonia (ICREA Acadèmia Prize to J. Domingo-Ferrer and grant 2014 SGR 537), and from the Spanish Government (project TIN2014-57364-C2-1-R "SmartGlacis" and TIN 2015-70054-REDC). The authors are with the UNESCO Chair in Data Privacy, but the views in this paper are the authors' own and are not necessarily shared by UNESCO.

References

1. U.S. Health Insurance Portability and Accountability Act (HIPAA, Pub. L 104–191, 110 Stat. 1936), 21 August 1996
2. Directive 95/46/EC of the European Parliament and of the Council of 24 October 1995 on the protection of individuals with regard to the processing of personal data and on the free movement of such data. Official J. Eur. Communities, 31–50, 23 November 1995
3. European Parliament legislative resolution of 14 April 2016 on the Council position at first reading with a view to the adoption of a regulation of the European Parliament and of the Council on the protection of natural persons with regard to the processing of personal data and on the free movement of such data, and repealing Directive 95/46/EC (General Data Protection Regulation), April 2016

4. Aggarwal, C.C.: On k-anonymity and the curse of dimensionality. In: Proceedings of the 31st International Conference on Very Large Data Bases, VLDB 2005, pp. 901–909 (2005)
5. Cao, J., Karras, P.: Publishing microdata with a robust privacy guarantee. Proc. VLDB Endowment **5**(11), 1388–1399 (2012)
6. Domingo-Ferrer, J., Sanchez, D., Soria-Comas, J.: Database Anonymization: Privacy Models, Data Utility, and Microaggregation-Based Inter-model Connections. Morgan & Claypool, San Rafael (2016)
7. Domingo-Ferrer, J., Torra, V.: Ordinal, continuous and heterogeneous k-anonymity through microaggregation. Data Minining Knowl. Disc. **11**(2), 195–212 (2005)
8. Domingo-Ferrer, J., Torra, V.: A critique of k-anonymity and some of its enhancements. In: Proceedings of the 3rd International Conference on Availability, Reliability and Security-ARES 2008, pp. 990–993. IEEE (2008)
9. Dwork, C.: Differential privacy. In: Bugliesi, M., Preneel, B., Sassone, V., Wegener, I. (eds.) ICALP 2006. LNCS, vol. 4052, pp. 1–12. Springer, Heidelberg (2006)
10. Hundepool, A., Domingo-Ferrer, J., Franconi, L., Giessing, S., Nordholt, E.S., Spicer, K., de Wolf, P.P.: Statistical Disclosure Control. Wiley, New York (2012)
11. LeFevre, K., DeWitt, D.J., Ramakrishnan, R.: Mondrian multidimensional k-anonymity. In: Proceedings of the 22nd International Conference on Data Engineering, ICDE 2006, pp. 25–37. IEEE (2006)
12. Li, N., Li, T., Venkatasubramanian, S.: t-closeness: privacy beyond k-anonymity and l-diversity. In: Proceedings of the 23rd International Conference on Data Engineering, ICDE 2007, pp. 106–115. IEEE (2007)
13. Machanavajjhala, A., Kifer, D., Gehrke, J., Venkitasubramaniam, M.: l-diversity: privacy beyond k-anonymity. ACM Trans. Knowl. Disc. Data **1**(1), 3 (2007)
14. Pei, J., Xu, J., Wang, Z., Wang, W., Wang, K.: Maintaining k-anonymity against incremental updates. In: 19th International Conference on Scientific and Statistical Database Management, SSBDM 2007. IEEE (2007)
15. Samarati, P.: Protecting respondents' identities in microdata release. IEEE Trans. Knowl. Data Eng. **13**(6), 1010–1027 (2001)
16. Samarati, P., Sweeney, L.: Protecting Privacy when Disclosing Information: k-Anonymity and its Enforcement through Generalization and Suppression. Technical report, SRI International (1998)
17. Sánchez, D., Domingo-Ferrer, J., Martínez, S.: Improving the utility of differential privacy via univariate microaggregation. In: Domingo-Ferrer, J. (ed.) PSD 2014. LNCS, vol. 8744, pp. 130–142. Springer, Heidelberg (2014)
18. Sánchez, D., Domingo-Ferrer, J., Martínez, S., Soria-Comas, J.: Utility-preserving differentially private data releases via individual ranking microaggregation. Inf. Fusion **30**, 1–14 (2016)
19. Stokes, K., Torra, V.: Multiple releases of k-anonymous data sets and k-anonymous relational databases. Int. J. Uncertainty Fuzziness Knowl.-Based Syst. **20**(6), 839–853 (2012)
20. Willenborg, L., DeWaal, T.: Elements of Statistical Disclosure Control. Springer, New York (2001)

Propensity Score Based Conditional Group Swapping for Disclosure Limitation of Strata-Defining Variables

Anna Oganian[1,2](✉) and Goran Lesaja[2]

[1] National Center for Health Statistics, 3311 Toledo Rd, Hyatsville, MD 20782, USA
aoganyan@cdc.gov
[2] Department of Mathematical Sciences, Georgia Southern University,
P.O. Box 8093, Statesboro, GA 30460-8093, USA
goran@georgiasouthern.edu

Abstract. In this paper we propose a method for statistical disclosure limitation of categorical variables that we call Conditional Group Swapping. This approach is suitable for design and strata-defining variables, the cross-classification of which leads to the formation of important groups or subpopulations. These groups are considered important because from the point of view of data analysis it is desirable to preserve analytical characteristics within them. In general data swapping can be quite distorting [13,16,20], especially for the relationships between the variables not only within the subpopulations but for the overall data. To reduce the damage incurred by swapping, we propose to choose the records for swapping using conditional probabilities which depend on the characteristics of the exchanged records. In particular, our approach exploits the results of propensity scores methodology for the computation of swapping probabilities. The experimental results presented in the paper show good utility properties of the method.

Keywords: Statistical disclosure limitation (SDL) · Group swapping · Strata · Subpopulations · Propensity scores

1 Introduction

Statistical agencies have an obligation by law to protect privacy and confidentiality of data subjects while preserving important analytical features in the data they provide. Privacy and confidentiality are not guaranteed by removal of direct identifiers, such as names, addresses and social security numbers, from the microdata file. Re-identification of individuals in the data is still possible by linking the file without direct identifiers to external databases. That is why in addition to the removal of direct identifiers, released microdata are typically modified, in order to make disclosure more difficult; that is, statistical disclosure limitation (SDL) methods are applied to the data prior to their release. The goal of such a modification is two-fold: to reduce the risk of re-identification and

© Springer International Publishing Switzerland 2016
J. Domingo-Ferrer and M. Pejić-Bach (Eds.): PSD 2016, LNCS 9867, pp. 69–80, 2016.
DOI: 10.1007/978-3-319-45381-1_6

at the same time to preserve important distributional properties of the original microdata file. Although it is not possible to know all the uses of the data beforehand, some of the relationships of interest to the user may be known. For example, some surveys oversample particular groups of individuals with the goal of obtaining better estimates for these groups. This requires special sample design and allocation of additional funds to obtain bigger samples for these groups. It would be particularly undesirable and counterproductive if SDL methods significantly change the estimates within these groups and/or considerably increase their standard errors. So every scenario of data release is different and disclosure limitation methods should be chosen accordingly. In this paper, we have focused on the situation of data release when the data protector has to modify categorical variables that define strata or subpopulations, but at the same time wants to minimize the distortion to the analytical structure within these strata.

To accomplish his/her task, the data protector can choose from among a wide variety of methods which can be divided in two groups: masking methods which release a modified version of the original microdata, and synthetic methods which generate synthetic records or values for specific variables from the distribution representing the original data.

A few examples of masking methods are: additive or multiplicative noise [1,14,15,20,21], in which noise is applied to numerical data values to reduce the likelihood of exact matching on key variables or to distort the values of sensitive variables; microaggregation, a technique similar to data binning (see [4,9,10,26]) and data swapping [2], in which data values are swapped for selected records. There are many variants of swapping, some examples are [2,8,17,22,24]. Data swapping is popular among government agencies since it preserves marginal distributions, and it is often implemented as a simple random swapping [8] in which a prespecified percentage of randomly selected records is swapped with some other randomly selected records for specific variables.

To measure the utility of masked data, the data protector can use either analysis-specific utility measures, tailored to specific analyses, or broad measures reflecting global differences between the distributions of original and the masked data [13,19,28]. One example of an analysis-specific measure tailored for regression analysis is an overlap in the confidence intervals for the regression coefficients estimated with the original and masked data [13]. An example of broad measure is the propensity score measure proposed in [28]. It compares favorably with others and it is suitable for data sets with mixed attributes [5,28]. Below we will review this measure in more detail because it is used as a part of our Conditional Group Swapping method described in Sect. 2.

Propensity score measure

First, let us recall the definition of a propensity score. The propensity score is the probability that an observation i is assigned to a particular group, call it a treatment group, given covariate values x_i. We denote $T = 1$ if a record is assigned to a treatment group and $T = 0$ otherwise. As [23] shows, T and

x are conditionally independent given the propensity score. Thus, when two groups have the same distributions of propensity scores, the groups should have similar distributions of x. This theory was used in [28] to measure data utility of disclosure protected data. In particular, [28] suggested the following approach. First, merge (by "stacking") the records from groups A and B that are being compared in their distributions. Then add an indicator variable T that equals one for all records from B and zero otherwise. Secondly, for each record i in the merged set, compute the propensity score, that is the probability of being in B given x_i - the values of the variables for this record. Propensity scores can be estimated via a logistic regression of the variable T on functions of all variables x in the data set. Thirdly, compare the distributions of the propensity scores in groups A and B. If the propensity scores are approximately the same for all the records in groups A and B, then the distributions of x in these groups are approximately the same This is an implication of the conditional independence of T and x_i given the propensity score (see [23, 28]). The propensity score distance measure proposed in [28] is

$$U_p = \frac{1}{N} \sum_{i=1}^{N} [\hat{p}_i - c]^2 \tag{1}$$

where N is the total number of records in the merged data set, $\hat{p}_i$ is the estimated propensity score for unit i, and c equals the proportion of B units in the merged data set.

1.1 Contribution and Plan of This Paper

In this paper, we have focused on a non-synthetic approach for disclosure limitation suitable for categorical strata-defining variables, the cross-classification of which leads to the formation of important groups for a data analyst. We present the Conditional Group Swapping method designed to minimize the distortion incurred by swapping, to the relationships between the variables, particularly those that involve categorical strata-defining variables. The idea of the method is described in Sect. 2. The results of the numerical experiments are reported in Sect. 3. Section 4 provides a concluding discussion and sketches lines for future work.

2 Propensity Score Based Conditional Group Swapping

In this section we describe the algorithm of our Conditional Group Swapping approach, hereafter, abbreviated as CGS. Below are the main steps of the method.

1. Compute pairwise distances between all the strata using the propensity score metric (1) described in Sect. 1. Note that the interpretation of the absolute value of this metric is not relevant here. The goal is to identify the pairs of the closest strata.

2. Compute swapping probabilities, that is the probabilities of moving records from one stratum to another, for the records in two closest strata. This will be done as follows. Suppose the distance between stratum A and stratum B is the smallest among all pairwise distances. Let n_s be the desired swapping rate, that is the number of records that will be moved from one stratum to another. To compute the swapping probabilities, first combine together all the records from A and B (by "stacking") and add an indicator variable T, $T = 1$ for all the records from stratum B and 0 for all the records from stratum A. Next, for every record i in the combined set compute the probability that this record is assigned to stratum B given the values of the variables for this record, x_i. In other words we compute the propensity scores, denoting them as $P_{AB}(i \rightarrow B|x_i)$.

3. Select n_s records from stratum A with the probabilities proportional to their propensity scores $P_{AB}(i \rightarrow B|x_i)$ and change their stratum indicator to B. For example, if in stratum A there are residential hospital records and in stratum B - multi-service hospitals, then for the selected n_s residential hospitals we will change their hospital type indicator to multi-service.

4. Select n_s records from stratum B with probabilities proportional to $1 - P_{AB}(i \rightarrow B|x_i)$ and "move" them to stratum A. The records that arrived from stratum A on the previous step will be excluded from the selection.

5. Repeat steps 3 and 4 for another pair of strata with the next closest distance.

6. Repeat step 5 until there are no strata that have not been swapped.

3   Numerical Experiments

The procedure described above was implemented and evaluated on several data sets. We experimented with genuine and simulated data. In this section we present only the results obtained on two genuine data sets. Simulated data results were very similar, so we omit them for brevity of the exposition. Below is the description of the two genuine data sets we used.

- The Titanic data is a public data set that was obtained from the Kaggle website [12]. This is a collection of records of 889 passengers of the Titanic, the British passenger liner that sank in the North Atlantic Ocean on April 15th 1912. The variables in this data set are: Survived - survival status (0=No; 1=Yes), Pclass - passenger class (1=1st; 2=2nd; 3=3rd), Sex - sex, Age - age in years, SibSp - number of siblings/spouses aboard, Parch - number of parents/children aboard, Fare - passenger fare, Embarked - port of embarkation (C= Cherbourg; Q=Queenstown; S= Southampton). The original file from Kaggle also contained names of the passengers, their ticket numbers and cabin number. These variables are irrelevant for our analysis, so they were excluded.
- The 1998 Survey of Mental Health Organizations (abbreviated as SMHO). This sample contains 874 hospitals. It is publicly available and can be obtained from the PracTools R package [27]. The 1998 SMHO was conducted by the US Substance Abuse and Mental Health Services Administration, which collected

data on mental health care organizations and general hospitals that provide mental health care services. The goal of the survey was to provide estimates for total expenditure, full-time equivalent staff, bed count, and total scaled by type of organization. For this data it is desirable to preserve as much as possible the estimates of these variables within the strata defined by the type of hospital. There are five types of categories for the variable hosp.type: (1) Psychiatric, (2) Residential/Veterans hospitals, (3) General, (4) Outpatient/Partial care and (5) Multi-service/Substance abuse. Other variables in the data are: Exptotal - total expenditures in 1998, Beds - total inpatient beds, Seencnt - unduplicated client/patient count seen during year, Eoycnt - end of year count of patients on the role, Findirct - money hospital receives from the statement health agency (1=yes; 2=no).

We applied the approach described in Sect. 2 to these data sets. For the Titanic data one of the relevant analyses is to check what sorts of people were likely to survive. In fact, since the sinking of the Titanic, there has been a widespread belief that the social norm of women and children first gives women a survival advantage over men in maritime disasters, and that captains and crew members give priority to passengers. However, [6] presented an interesting study of historical records, spanning over three centuries, that suggests that in maritime disasters women and children die at significantly higher rates than male passengers and crew members. Their findings suggest that the events on the Titanic, where 20 percent of men and 70 percent of women and children survived, were highly unusual, if not unique. Besides gender, the class of the Titanic passengers was also related to their survival status.

Based on these considerations, we divided the data in six strata according to the cross-classification of the variables Pclass and Sex: (1) first class male passengers, (2) first class females, (3) second class males, (4) second class females, (5) third class males, (6) third class females.

The first step of the CGS procedure identified the following strata as closest: first class males and first class female, second class males and second class females, and third class males and third class females. For the measure of distance between the distributions of different strata (specifically, between the multivariate distributions of Survived, Age, Fare, SibSp and Parch for each stratum), we used the following model to estimate propensity scores: the main effects for the variables Survived, Age, Fare, SibSp and Parch and the interactions between Survived and Fare, Survived and Age, Survived and SibSp, Survived and Parch. We didn't include all the main terms and interactions because otherwise the totality of the estimated parameters would not be supported by the sample size.

Because the goal of our experiments is to test the potential benefits of using conditional probabilities for swapping and more specifically to estimate the effect of such probabilities on the quality of different statistical estimates, we compared the outcome of Conditional Group Swapping to the outcome of a similar approach which is characterized by uniform swapping probabilities. For the later approach the values of the variables of the records do not influence the probabilities of these records being swapped. We call it Random Group Swapping, hereafter, abbreviated as RGS. In a sense, RGS reflects the idea of the traditional approach for swapping. To make a fair comparison and to estimate the effect of using conditional swapping probabilities, RGS and CGS were implemented in the same way (as described in Sect. 2), except for the way how the probabilities of swapping were computed: for CGS they were proportional to the propensity scores, as described in Sect. 2, but for RGS they were uniform as we mentioned above.

We experimented with two swapping rates: $n_s = 20$ and $n_s = 40$ records exchanged between the strata. This corresponds respectively to about 15 and 35 percent of records swapped for each stratum. For each swapping rate, we generated 100 realizations of swapped data using Random and Conditional Groups Swapping.

Next, we compared the results of several statistical analyses based on the original and swapped data. One of them was logistic regression fitted to the complete Titanic data with Survived as the predicted variable and Pclass, Sex and Age as predictors. Hereafter, we will use R notation for the models. For the aforementioned regression it will be: Survived ~ Pclass + Sex + Age. Denote this model Reg1. We used this set of predictors in Reg1 because they were identified as being statistically significant based on the original data.

We also fitted logistic regressions within each stratum: Survived ~ Age + Fare. Denote this model Reg2

Next, we compared confidence intervals of regression coefficients for these regressions based on the original and swapped data. There were five regression coefficients for Reg1, including intercept, coefficient for Age, coefficients for dummy variables Pclass=2, Pclass=3 and for Sex=male and three regression coefficients for Reg2 (intercept and coefficients for Age and Fare).

As a measure of comparison we used the relative confidence interval overlap similar to the one used in [13]. Let $(L_{\text{orig},k}, U_{\text{orig},k})$ and $(L_{\text{swap},k}, U_{\text{swap},k})$ be the lower and upper bounds for the original and masked confidence intervals for the coefficient k. Let $L_{\text{over},k} = \max(L_{\text{orig},k}, L_{\text{swap},k})$ and $U_{\text{over},k} = \min(U_{\text{orig},k}, U_{\text{swap},k})$. When the original and masked confidence intervals overlap, $L_{\text{over},k} < U_{\text{over},k}$ and $(L_{\text{over},k}, U_{\text{over},k})$ represent the lower and the upper bounds of the overlapping region. When these confidence intervals do not overlap, $L_{\text{over},k} > U_{\text{over},k}$ and $(L_{\text{over},k}, U_{\text{over},k})$ represent the upper and the lower bounds of the non-overlapping region between these intervals. The measure of relative confidence interval overlap for the coefficient k is defined as follows:

$$J_k = \frac{1}{2} \left[\frac{U_{\text{over},k} - L_{\text{over},k}}{U_{\text{orig},k} - L_{\text{orig},k}} + \frac{U_{\text{over},k} - L_{\text{over},k}}{U_{\text{swap},k} - L_{\text{swap},k}} \right] \qquad (2)$$

When confidence intervals overlap, $J_k \in (0, 1]$ and $J_k = 1$ when the intervals exactly coincide. In case one of the confidence intervals is "contained" in the other, the relative confidence interval measure will capture such a discrepancy, and $0 < J_k < 1$. When intervals don't overlap, $J_k \leq 0$. In this case, J_k measures non-overlapping area (between the intervals) relative to their lengths. We also report an average confidence interval overlap over all the coefficients defined as $J = (1/p) \sum_{i=1}^{p} J_k$.

Table 1 presents the results of the experiments. The first column of the table is the type of analysis, Reg1 or Reg2, for which the outcome is compared between the original and masked data. The second column is the swapping rate. Columns "Average" and "Range" display average confidence interval overlap J and the range of variation of individual confidence interval overlaps J_k over all 100 realizations and all the coefficients. Range of variation is reported for the central 90 % of the distribution of J_k. Column "# non-over" displays the fraction of times the intervals didn't overlap over all the realizations and coefficients. For example, 100/500 means that 100 out of 500 intervals didn't overlap (i.e. the number of times $J_k < 0$). For Reg1 the number of computed intervals is 500 = 100 realizations×5 coefficients; and for Reg2 it is 1800 = 3 coefficients×6 strata × 100 realizations

Table 1. The Titanic data results: original and masked confidence interval overlaps.

	Rate	CGS			RGS		
		Average	Range	# non-over	Average	Range	# non-over
Reg1	20	0.88	$[0.6, 0.99]$	0/500	0.52	$[-0.53, 0.95]$	100/500
	40	0.65	$[-0.15, 0.96]$	51/500	0.16	$[-1.86, 0.92]$	106/500
Reg2	20	0.85	$[0.60, 0.98]$	1/1800	0.76	$[0.42, 0.97]$	1/1800
	40	0.79	$[0.41, 0.97]$	0/1800	0.69	$[0.28, 0.95]$	3/1800

As can be seen from the table, the average confidence interval overlap J for Reg1 are relatively high for CGS (0.88 and 0.65 for $n_s = 20$ and 40 respectively). Moreover, the values of J are considerably higher for CGS than for RGS. The range of variation is also narrower for CGS. The lower bounds for the range of variation correspond to the worst cases of the confidence interval overlaps. These smallest overlaps are still quite larger for CGS than for RGS. The upper bounds of the range of variation are similar for both methods, although still slightly larger for the CGS.

Regarding individual coefficients overlap measures J_k, we observed that they were similar in values for different coefficients, except the coefficient for Sex. In particular, the average J_k values over 100 realizations were smaller for Sex than

Table 2. The Titanic data results: ratios of means and ratios of covariance matrices based on the original and masked data.

	Rate	CGS			RGS		
		Average	Range	# sign change	Average	Range	# sign change
Mean ratio	20	1.003	$[0.91, 1.10]$	N/A	1.002	$[0.81, 1.21]$	N/A
	40	1.004	$[0.87, 1.15]$	N/A	1.02	$[0.72, 1.37]$	N/A
Cov. ratio	20	1.02	$[0.61, 1.67]$	208/9600	1.03	$[0.51, 1.62]$	238/9600
	40	1.5	$[0.52, 1.68]$	238/9600	0.99	$[0.34, 1.68]$	364/9600

for other coefficients (it was equal to 0.5 for CGS). Confidence intervals for Sex overlapped for all 100 realizations of CGS for the swapping rate 20. However, confidence intervals for Sex never overlapped for RGS. There is an explanation to that. In particular, in both cases swapping was done between the strata which were identified as closest to each other. The closeness was estimated for the multivariate distribution of Survived, Age, Fare, SibSp and Parch. The closest strata happened to be the ones that have the same passenger class Pclass but different Sex, *e.g.*, 1st class male and 1st class females, and so on. So, it was Sex that was actually swapped for the selected records. The selection probabilities of RGS are independent of the values of the variables, so it is not surprising that the relationships between Sex and other variables, and in particular Sex and survival status, are particularly affected. On the other hand, when swapping probabilities are proportional to the propensity scores, as in the CGS method, the relationship between Sex and survival status is taken into account (through propensity scores), so the swapped and original data confidence intervals for Sex are much more similar.

In addition to confidence interval comparisons, we also computed the element-wise ratios of original and swapped data means and covariance matrices for numerical variables Age, Fare, SibSp and Parch within each stratum. The results of these comparisons are presented in Table 2. We can see that the ratios of original and masked means are very similar for CGS and RGS. The range of variation is, however, larger for Random Swapping, which is an indicator of larger disturbances introduced by Random Swapping. In column "# sign change" we display the fraction of times an element in the covariance matrix changed in sign. These changes occurred predominately for the variables with covariances close to zero. These sign changes happened more often for RGS than for CGS.

For our second data set, SMHO, we fitted a logistic regression of Findirct (hospital receives money from the statement health agency) on all other variables, denote it Reg3 and a regression of Exptotal (total expenditures in 1998) on all other variables, denote it Reg4. Both regressions were fitted to the complete data. Within strata, analyses included regressions: Findirct on all other variables (Reg5) and Exptotal on all other variables (Reg6). Hospital type was not included in the predictor set of Reg5 or Reg6 because it was the same value for all the records in a particular stratum. The results are presented in Table 3. Just as with the Titanic data, we also computed mean and covariance matrices ratios based on the original and swapped data within each stratum. These comparisons are presented in Table 4.

As can be seen from Tables 3 and 4, original and swapped confidence intervals overlap at a higher level for CGS than for RGS, and discrepancies in means and covariance matrices are smaller for CGS. Interestingly, when fitting logistic regression Reg5 within strata, we noticed that in 3 out of 100 realizations of swapped data using CGS the regression coefficient for variable Bed was not estimable within the Outpatient stratum. A closer examination of the swapping results of these three realizations showed that the values for the variable Bed in stratum Outpatient were predominately zeros in the original data. So, most of

Table 3. The SMHO data results: original and masked confidence interval overlaps.

	Rate	Conditional swap			Random swap		
		Average	Range	# non-over	Average	Range	# non-over
Reg3	20	0.91	[0.72, 0.99]	0/900	0.72	[0.2, 0.99]	2/900
	40	0.84	[0.84, 0.99]	0/900	0.64	[−0.29, 0.97]	100/900
Reg4	20	0.94	[0.81, 1]	0/900	0.84	[0.58, 0.97]	0/900
	40	0.92	[0.77, 0.99]	0/900	0.71	[0.26, 0.95]	11/900
Reg5	20	0.85	[0.56, 0.99]	5/2500	0.64	[−0.25, 0.97]	196/2500
	40	0.82	[0.45, 0.98]	14/2500	0.47	[−0.7, 0.95]	393/2500
Reg6	20	0.81	[0.45, 0.98]	0/2500	0.72	[0.26, 0.96]	25/2500
	40	0.73	[0.15, 0.97]	71/2500	0.61	[−0.06, 0.94]	158/2500

Table 4. The SMHO data results: ratios of means and covariance matrices based on the original and masked data.

	Rate	Conditional swap			Random swap		
		Average	Range	# sign change	Average	Range	# sign change
Mean ratio	20	0.99	[0.85, 1.09]	N/A	0.94	[0.56, 1.17]	N/A
	40	0.96	[0.63, 1.12]	N/A	0.92	[0.38, 1.32]	N/A
Cov. ratio	20	1.03	[0.57, 1.46]	0	0.87	[0.05, 2.07]	276/8000
	40	0.90	[0.34, 1.49]	334/8000	0.79	[0.016, 2.88]	332/8000

the times CGS led to the selection of those records that had non-zero values for the variable Bed to be moved to another stratum. In those cases the Outpatient stratum received records with low counts for Bed from another stratum, but for those exceptional 3 cases, the incoming records had all zeros for Bed, resulting in a non-estimable coefficient. However, with the exception of those three cases, when CGS was used, the original and masked confidence intervals for Bed had larger overlap and the means and covariance matrix were better preserved for stratum Outpatient. In fact, CGS led to the choice of records with low counts for Bed which fits well the description of the Outpatient hospital stratum, while RGS on several occasions moved records with large values for Beds, which is inconsistent for Outpatient stratum.

4   Concluding Discussion and Future Work

In this paper we presented a Conditional Group Swapping method suitable for categorical variables which define strata or subpopulations. This swapping method is designed with the goal to reduce the damage incurred by the disclosure limitation to the relationships between the variables within the strata and in the overall data. Our experimental results showed that the method has the potential to better preserve inferential properties, such as confidence intervals for the regression coefficients specific to particular strata and for the overall data, than

Random Swapping. For numerical variables the means and covariance matrices within the strata are less distorted as well.

We believe that in practice CGS should not be the only method that is applied to the data, especially if there are continuous variables in the data. Similar to other swapping approaches, CGS can be used together with other SDL methods. For example, one can apply Conditional Group Swapping to strata-defining variables and then add multivariate noise $\mathbf{E}$ to the continuous variables within each stratum s with strata-specific parameters:

$$\mathbf{X}_m^s = \mathbf{X}_o^s + \mathbf{E}^s \tag{3}$$

where $\mathbf{X}_o^s$ and $\mathbf{X}_m^s$ are the original and masked (continuous) data in stratum h, $\mathbf{E} \sim N(\mathbf{0}, c\Sigma_{orig^s})$, Σ_{orig^h} is the covariance matrix of the original data in stratum h, c is the parameter of the method. Such noise preserves the correlation structure within the strata. Conditional Group Swapping is designed with the same goal, so the combination of these two methods may work in synergy. Investigation of the best combinations of Conditional Group Swapping with other methods is one of the directions of our future research.

Another direction for future research is the investigation of the risk associated with the method. We believe that the risk assessment is more comprehensive and practically useful when done for the final version of the masked data, which, as we noted above, will result from the application of our Conditional Group Swapping together with other SDL methods. Indeed, if there are continuous variables in the data and they are not masked, then re-identification risk can be high regardless of the protection of categorical variables, because the values of continuous variables are virtually unique. CGS method is not suited for continuous variables, however, as mentioned above, it can be used in combination with additive noise. So, we carried out several experiments with this combination, in particular we applied it to both out data sets. The value $c = 0.15$ was used as a parameter of noise (see [13, 19] for recommendations for c). Changes in utility were insignificant, in particular the average confidence interval overlaps decreased by about 3 to 5 percent, and the range of variation was almost the same.

Next, we estimated the re-identification disclosure risk, defined as an average percentage of correctly identified records when record linkage techniques [7, 11] are used to match the original and masked data. Specifically, we assume that the intruder tries to link the masked file with an external database containing a subset of the attributes present in the original data (see [19]).

The re-identification disclosure risk for the Titanic data masked with multivariate noise and CGS was low: about 4 % of all records were correctly identified for $n_s = 20$ and about 3 % for $n_s = 40$. For SMHO data the risk was even lower, it was about 2 % for $n_s = 20$ and 1.5 % for $n_s = 40$.

As we mentioned above, these experiments do not represent a comprehensive risk analysis, however, they give an idea of the magnitude of risk. Thorough investigation of the disclosure risk for the combination of Conditional Group Swapping together with different SDL methods is the topic of our future research.

Also, in the future we plan to compare the presented approach with a wider range of SDL methods, for example data shuffling [18], latin hypercube sampling [3] and other methods which showed good utility and disclosure risk properties.

Acknowledgments. The authors would like to thank Alan Dorfman and Van Parsons for valuable suggestions and help during the preparation of the paper. The findings and conclusions in this paper are those of of the authors and do not necessarily represent the official position of the Centers for Disease Control and Prevention.

References

1. Brand, R.: Microdata protection through noise addition. In: Domingo-Ferrer, J. (ed.) Inference Control in Statistical Databases. LNCS, vol. 2316, pp. 97–116. Springer, Heidelberg (2002)
2. Dalenius, T., Reiss, S.P.: Data-swapping: A technique for disclosure control. J. Stat. Plann. Infer. **6**, 73–85 (1982)
3. Dandekar, R.A., Cohen, M., Kirkendall, N.: Sensitive micro data protection using latin hypercube sampling technique. In: Domingo-Ferrer, J. (ed.) Inference Control in Statistical Databases. LNCS, vol. 2316, pp. 117–125. Springer, Heidelberg (2002)
4. Defays, D., Anwar, N.: Micro-aggregation: a generic method. In: Proceedings of the 2nd International Symposium on Statistical Confidentiality, pp. 69–78. Office for Official Publications of the European Community, Luxembourg (1995)
5. Drechsler, J.: Synthetic Datasets for Statistical Disclosure Control: Theory and Implementation. Springer, New York (2011)
6. Elinder, M., Erixson, O.: Gender, social norms, and survival in maritime disasters. Proc Nat. Acad. Sci. USA **109**(33), 13220–13224 (2012)
7. Fellegi, I.P., Sunter, A.B.: A theory for record linkage. J. Am. Stat. Assoc. **64**, 1183–1210 (1969)
8. Gomatam, S., Karr, A.F., Chunhua, L., Sanil, A.: Data swapping: a risk-utility framework and web service implementation. Technical Report 134, National Institute of Statistical Sciences, Research Triangle Park, NC (2003)
9. Hundepool, A., Domingo-Ferrer, J., Franconi, L., Giessing, S., Lenz, R., Longhurst, J., Schulte-Nordholt, E., Seri, G., DeWolf, P.-P.: Handbook on Statistical Disclosure Control (version 1.2). ESSNET SDC project (2010). http://neon.vb.cbs.nl/casc
10. Hundepool, A., Domingo-Ferrer, J., Franconi, L., Giessing, S., Schulte Nordholt, E., Spicer, K., Wolf, P.-P.: Statistical Disclosure Control. Wiley, New York (2012)
11. Jaro, M.A.: Advances in record-linkage methodology as applied to matching the 1985 Census of Tampa, Florida. J. Am. Stat. Assoc. **84**, 414–420 (1989)
12. Kaggle. The Home of Data Science. http://www.kaggle.com
13. Karr, A.F., Kohnen, C.N., Oganian, A., Reiter, J.P., Sanil, A.P.: A framework for evaluating the utility of data altered to protect confidentiality. Am. Stat. **60**(3), 224–232 (2006)
14. Kim, J.J.: A method for limiting disclosure in microdata based on random noise and transformation. In: Proceedings of the ASA Section on Survey Research Methodology, pp. 303–308 (1986)
15. Lin, Y.-X.: Density approximant based on noise multiplied data. In: Domingo-Ferrer, J. (ed.) PSD 2014. LNCS, vol. 8744, pp. 89–104. Springer, Heidelberg (2014)

16. Mitra, R., Reiter, J.P.: Adjusting survey weights when altering identifying design variables via synthetic data. In: Domingo-Ferrer, J., Franconi, L. (eds.) PSD 2006. LNCS, vol. 4302, pp. 177–188. Springer, Heidelberg (2006)
17. Moor, R.: Controlled data swapping techniques for masking public use microdata sets. U.S. Census Bureau (1996)
18. Muralidhar, K., Sarathy, R.: Data shuffling: a new masking approach for numerical data. Manag. Sci. **52**(5), 658–670 (2006)
19. Oganian, A.: Security and Information Loss in Statistical Database Protection. Ph.D. thesis, Universitat Politecnica de Catalunya (2003)
20. Oganian, A., Karr, A.F.: Combinations of SDC methods for microdata protection. In: Domingo-Ferrer, J., Franconi, L. (eds.) PSD 2006. LNCS, vol. 4302, pp. 102–113. Springer, Heidelberg (2006)
21. Oganian, A., Karr, A.F.: Masking methods that preserve positivity constraints in microdata. J. Stat. Plann. Infer. **141**(1), 31–41 (2011)
22. Reiss, S.P., Post, M.J., Dalenius, T.: Non-reversible privacy transformations. In: Proceedings of the ACM Symposium on Principles of Database Systems, 29–31 March, pp. 139–146 (1982)
23. Rosenbaum, P.R., Rubin, D.B.: The Central Role of the propensity score in observational studies for Causal Effects. Biometrika **70**, 41–55 (1983)
24. Takemura, A.: Local recoding and record swapping by maximum weight matching for disclosure control of microdata sets. J. Offic. Stat. **18**, 275–289 (2002)
25. Templ, M.: Statistical disclosure control for microdata using the R-package sdcMicro. Trans. Data Priv. **1**(2), 67–85 (2008)
26. Torra, V.: Microaggregation for categorical variables: a median based approach. In: Domingo-Ferrer, J., Torra, V. (eds.) PSD 2004. LNCS, vol. 3050, pp. 162–174. Springer, Heidelberg (2004)
27. Valliant, R., Dever, J.A., Kreuter, F.: Package 'PracTools': Tools for Designing and Weighting Survey Samples (2015). https://cran.r-project.org/web/packages/PracTools/PracTools.pdf
28. Woo, M.-J., Reiter, J.P., Oganian, A., Karr, A.F.: Global measures of data utility for microdata masked for disclosure limitation. J. Priv. Confidentiality **1**(1), 111–124 (2009)

A Rule-Based Approach to Local Anonymization for Exclusivity Handling in Statistical Databases

Jens Albrecht[1], Marc Fiedler[2]([⊠]), and Tim Kiefer[3]

[1] Technische Hochschule Nürnberg, Fakultät Informatik, Nürnberg, Germany
jens.albrecht@th-nuernberg.de
[2] GfK SE, Data and Technology, Nürnberg, Germany
marc.fiedler@gfk.com
[3] Technische Universität Dresden, Database Systems Group, Dresden, Germany
tim.kiefer@tu-dresden.de

Abstract. Statistical databases in general and data warehouses in particular are used to analyze large amounts of business data in predefined as well as ad-hoc reports. Operators of statistical databases must ensure that individual sensitive data, e.g., personal data, medical data, or business-critical data, are not revealed to unprivileged users while making use of these data in aggregates. Business rules must be defined and enforced to prevent disclosure. The unsupervised nature of ad-hoc reports, defined by the user and unknown to the database operator upfront, adds to the complexity of keeping data secure. Storing sensitive data in statistical databases demands automated methods to prevent direct or indirect disclosure of such sensitive data.

This document describes a rule-based approach to local recoding of sensitive data. It introduces the notion of exclusivity to describe quasi-identifiers with local rules based on the multidimensional data model. It further defines options to treat exclusive entities that may disclose sensitive data. The local anonymization minimizes information loss and works well even with very large data sets.

1 Introduction

The protection of sensitive data is a cornerstone of trust, not only in public statistics but also in market research, online retail, healthcare, and many other business applications. Privacy protection is always about hiding the details, but details are often what's interesting. Therefore, methods for statistical disclosure control (SDC) seek an optimal balance between privacy and usability of the results [4].

In this paper, we present a new approach to privacy control in data warehouses based on the multidimensional data model. In terms of data modeling and data analysis, data warehouses and statistical databases share a lot of similarities [12]. Both focus on the analysis of multidimensional data and have to maintain privacy with minimal loss of information in reports. However, the data volume in data warehouses is generally quite large compared to, e.g., survey data

© Springer International Publishing Switzerland 2016
J. Domingo-Ferrer and M. Pejić-Bach (Eds.): PSD 2016, LNCS 9867, pp. 81–93, 2016.
DOI: 10.1007/978-3-319-45381-1_7

and the number of categories and observed objects can be very high (i.e., thousands of products with dozens of attributes). Hence, computationally expensive approaches to inference control often cannot be applied. Moreover, perturbation-based methods which modify the numerical data cannot be used, because business reporting in general needs exact, complete, and consistent figures.

The scenario motivating our approach is the retail data warehouse of GfK, a leading market research company [2]. GfK is a trusted source of relevant market and consumer information. Based on the world's largest sample of point-of-sales tracking data comprising 400.000 shops, 47 million distinct products, and 30 billion fact records, GfK runs an international point-of-sales panel to provide insights into what products are selling, when, where, and at what price, enabling manufacturers to offer the right products, and retailers to develop strategies that drive profitable sales. The raw data — which is the basis of all further analysis and reports — comes from various retailers worldwide. GfK clients have access to self-service OLAP-like reporting tools, where they can build ad-hoc reports based on microdata. Privacy protection in this setting means protecting the retailers, who do not want to disclose their identity or their sales volume in any product segment or region.

Attributes that directly identify the retailer, like shop or retailer name, can easily be hidden from the clients. However, products and brands which are the prime objects of market analysis can also identify retailers, e.g., if they are exclusively sold by a single retailer. Note, that exclusive products and tradebrands are a cornerstone of the product assortment strategy of each major retailer and often have a significant market share. Therefore, the sales values of these brands and products must be anonymized in reports.

Consider the example in Fig. 1 where products of brand `Okay` are exclusively sold by retailer `Jupiter`, because `Okay` is a tradebrand of that retailer. Thus, products of that brand and the brand itself are identifying the retailer and must be anonymized in reports. Furthermore, the example contains two products of brands `Panasonic` and `Samsung` which are exclusively sold by the retailers `Conwheel` and `MediaStore`. These products might also be anonymized, e.g., when the retailers demand protection, regardless that neither `Panasonic` nor `Samsung` is a tradebrand of any of those retailers. Note that exclusive sales only need to be protected, if external market knowledge like "Brand Okay is a tradebrand of retailer Jupiter" is potentially available to the report user.

Our approach to privacy protection in data warehouses uses an innovative combination of local suppression and generalization at cell level. It is based on exclusivity treatment rules specifying how exclusive data should be anonymized in a certain local context. The goal is to allow users to specify ad-hoc reports of the data and show the most detailed data while guaranteeing that given protected entities are neither revealed directly or indirectly via exclusivity.

Figure 2 shows how the application of treatment rules impacts a Hitlist report, which is a typical report in market research. The unmasked version of the report in Fig. 2a indirectly reveals exclusive sales of retailers and therefore cannot be distributed. The masked version of the report in Fig. 2b illustrates

Raw Data

ProductID	Product	Brand	TV Tuner	Technology	Conwheel	MediaStore	Jupiter	Sales Units
1	BLA42N18	Blaupunkt	Yes	LED	10	9	8	27
2	OKLED244	Okay	No	LED			8	8
3	OKLCD228	Okay	Yes	LCD			25	25
4	PAN29A65	Panasonic	No	LED		22		22
5	PAN42A78	Panasonic	Yes	LED	3	2	44	49
6	PAN40F65	Panasonic	Yes	LCD	3		9	12
7	PAN29B10	Panasonic	No	LED	14			14
8	SAM32F50	Samsung	Yes	LED		45		45
9	SAM46F61	Samsung	Yes	LED		23	18	41
	Total							243

Fig. 1. Sample of retail sales data containing exlusive cells (gray) which may disclose the retailer

Hitlist Report

Product ID	Product	Brand	TV Tuner	Technology	Sales Units
5	PAN42A78	Panasonic	Yes	LED	49
8	SAM32F50	Samsung	Yes	LED	45
9	SAM46F61	Samsung	Yes	LED	41
1	BLA42N18	Blaupunkt	Yes	LED	27
3	OKLCD228	Okay	Yes	LCD	25
4	PAN29A65	Panasonic	No	LED	22
7	PAN29B10	Panasonic	No	LED	14
6	PAN40F65	Panasonic	Yes	LCD	12
2	OKLED244	Okay	No	LED	8
	Total				243

Hitlist Report (anonymized)

Product ID	Product	Brand	TV Tuner	Technology	Sales Units
5	PAN42A78	Panasonic	Yes	LED	49
8	Suppressed	Suppressed	Yes	LED	45
9	SAM46F61	Samsung	Yes	LED	41
1	BLA42N18	Blaupunkt	Yes	LED	27
4	PAN29A65	Panasonic	No	LED	22
7	Suppressed	Panasonic	No	LED	14
6	PAN40F65	Panasonic	Yes	LCD	12
2, 3	Others	Others	Others	Others	33
	Total				243

(a) Hitlist report without treatment   (b) Hitlist report with treatment

Fig. 2. Hitlist report without and with anonymization

the result after the treatment rules have been applied. Exclusive sales will be suppressed or generalized, i.e., aggregated into Others, according to the privacy requirements of the retailer.

In short, our approach comprises several desirable properties:

- Applies context-specific local generalization and suppression.
- Preserves anonymity at the most detailed aggregate level that still protects the data source.
- Computes exact figures for all segments.
- Allows different treatment rules for each type of attribute.
- Easy to integrate in ETL process of a data warehouse.
- Works with very large data sets.

The remaining part of the paper is organized as follows: In Sect. 2, we give an overview of related work. Section 3 recaps the multidimensional data model. In Sect. 4, we introduce and formalize the concept of *exclusivity* and explain the local anonymization via exclusivity treatment rules. Section 5 outlines how exclusivity handling can be implemented in a real-world scenario. Section 6 concludes the paper.

2   Related Work

Three kinds of general protection scenarios can be distinguished: *Tabular data protection* deals with inference problems in statistical resulting tables [4,15]. *Microdata protection* or *privacy preserving data-publishing (PPDP)* methods try to modify microdata in order to make them publicly available but still prevent inferences to individuals [4,6]. In a *dynamic database protection (DDP)* scenario, users issue statistical queries to the database, as it is the case in interactive applications like OLAP [4]. Queries (resp. reports) are not known in advance and a system must dynamically determine which information must be hidden in a report to protect privacy. Our approach can be used for microdata as well as for dynamic database protection.

Privacy protection methods usually modify input data or query results. These modifications can be perturbative or non-perturbative [4,6]. Perturbation-based methods like data swapping or microaggregation are ideal for data mining [1,11], where detail data is needed, but exact figures are generally not required. In OLAP and reporting applications, however, exact figures are required, so only non-perturbative methods can be used. In this case, access to detail data must be restricted.

We classify restriction based methods into three groups: *Access control* methods limit the access to certain attributes (schema-level) and/or certain rows resp. cells in a data cube (instance or cell-level). *Query restriction* methods, e.g., [7,14], apply inference control techniques to a set of queries and dynamically prevent queries which might compromise security from being executed. *Anonymization-based* methods in contrast use generalization and suppression to include sensitive values in higher-level aggregates or mask details of quasi-identifying attributes.

[3,5] introduce notions for access control on multidimensional data. Primarily schema level security is addressed, i.e., hierarchy levels (attributes) like the *patient* can be labeled with a degree of privacy. Cell-level security can be achieved by so-called *sensitivity information assignment rules*. Only one treatment option is discussed here: to filter out the sensitive microdata in queries. Our approach is also rule-based, but works with anonymization instead of filters.

We focus on non-perturbative anonymization methods, which use a combination of generalization (also called recoding) and suppression. A well-known example for recoding has been introduced by Sweeney to achieve k-anonymity [13]. Overviews can be found in [6,16]. *Suppression* masks parts of quasi-identifying and/or sensitive attributes in a report. An example could be to suppress the day and month of the date of birth or the last digits of a telephone number. *Generalization* substitutes and aggregates the original values to more general values. Both, generalization and suppression can be done for all values of an attribute (global recoding), only for some instances (local recoding), or even instances in a certain context (cell-based recoding). For example, TV SAM32F50 by Samsung could be suppressed in data from stores in Germany, because it is locally sensitive there, but left unchanged anywhere else. Our approach supports scenarios like this one.

3 Data-Warehouse Data Model

OLAP databases are generally modeled according to the multidimensional data model, where attributes are classified into dimensions and facts. Figure 3a illustrates a simplified multidimensional model for retail sales data. The dimensions are Product, Shop, and Time. Cube cells contain the facts, i.e., the number and turnover of products sold in certain shops within certain time periods.

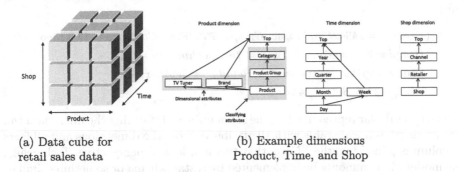

(a) Data cube for
retail sales data

(b) Example dimensions
Product, Time, and Shop

Fig. 3. Elements of the multidimensional data model

Formally, a multidimensional schema consists of a set of dimensional schemas $(D^1, \ldots, D^n)$ and a set of facts $(F^1, \ldots, F^m)$. The *schema of a dimension* D is a partially ordered set of attributes $(\{A_1, \ldots, A_k, Top\}; \rightarrow)$, where $\rightarrow$ denotes the functional dependency. Three examples for dimensional schemas are visualized as lattices in Fig. 3b.

A *path* P in a dimension is a fully ordered subset of the dimension's schema, e.g., the Shop dimension in Fig. 3b contains the path Shop $\rightarrow$ Retailer $\rightarrow$ Channel $\rightarrow$ Top. The *instance* of a path $A_1 \rightarrow A_2 \rightarrow \ldots \rightarrow A_m \rightarrow Top$ forms a generalization hierachy, or taxononmy, on the elements of the domains $dom(A_i)$ of the attributes with the generic node ALL at the top.

The *schema of a data cube* C consists of a set of dimensional schemas $\mathcal{D}$ and a set of facts $\mathcal{F}$, i.e.,

$$C = (\mathcal{D}, \mathcal{F}) = (\{D^1, \ldots, D^n\}, \{F^1, \ldots, F^m\}).$$

An instance of a cube C is a subset of the cross-product of the domains of all attributes and facts that contribute to the cube schema, i.e.,

$$C \subseteq \mathrm{dom}(D^1) \times \ldots \times \mathrm{dom}(D^n) \times \mathrm{dom}(F^1) \times \ldots \times \mathrm{dom}(F^m),$$

where $\mathrm{dom}(D^i) = \mathrm{dom}(A_1^i) \times \ldots \times \mathrm{dom}(A_{k_i}^i)$.

A *cube cell* c is an element of the cube instance:

$$c = (c_1, \ldots, c_k, f_1, \ldots f_m)$$

with $c_1 \in \mathrm{dom}(A_1^1), \ldots, c_k \in \mathrm{dom}(A_{kn}^n)$, and $f_i \in \mathrm{dom}(F^i)$. For the sake of simplicity and without loss of generality, we will work in the remaining part of the paper with a single fact f.

According to the definition, a cube cell includes not only the granular attributes in the dimensions, but also (logically denormalized) all higher level attributes. For example, if the product dimension is defined on the attributes {Product, Product Group, Category, Brand, TV Tuner, Technology}, the shop dimension by {Shop, Retailer} and the time dimension simply by {Year}, then a cube cell containing 44 sales of product PAN42A78 looks like this:

$$(Product = \text{PAN42A78}, Product\ Group = \text{PTV/Flat}, Category = \text{TV Video},$$
$$Brand = \text{Panasonic}, TV\ Tuner = \text{Yes}, Technology = \text{LED},$$
$$Shop = \text{Jupiter Store Munich}, Retailer = \text{Jupiter},$$
$$Year = 2015, Sales = 44).$$

Thus, in a tabular representation, the set of cube cells forming the microdata can be visualized as a wide table with all attributes from the dimensions and all facts as columns joined together. Note, that this is a logical perspective. Technically, the model will commonly be implemented by a star schema or something similar. The multidimensional data model adds knowledge about functional dependencies between the attributes and the distinction between dimensions and facts.

A *report* in the context of this paper consists of one or more tables or cross-tables similar to those in Fig. 2, which are usually derived from a microdata cube as in Fig. 1. Note, that in the unmasked version of the Hitlist report the retailer Jupiter is indirectly disclosed to everybody who knows, that Okay is a tradebrand of that retailer even though retailers are not shown separately.

4   Exclusivity in Multidimensional Data

In this section, the basic multidimensional model is extended to describe parts of the data that are sensitive and must be protected from unauthorized access. Furthermore, the concept of *exclusivity* is formalized. Exclusivity allows the deduction of sensitive data through other, non-sensitive data.

4.1   Protected Identifiers and Protected Cube

Identifiers in a privacy context generally denote attributes, that uniquely identify the individuals or organizations to be protected [4]. In the following, we call those attributes *protected identifiers* in contrast to identifiers for unprotected dimensional entities like products or countries.

Let *PID* be the set of protected identifiers and let $A_j \rightarrow A_i$ denote a functional dependency between two attributes, i.e., for each value of A_j there exists exactly one value of A_i. Then the following implication holds:

$$A_i \in PID \land A_j \rightarrow A_i \implies A_j \in PID. \tag{1}$$

In our example in Fig. 3b, *PID* is the set of attributes directly identifying the retailer of a given transaction in the data set, i.e., *Retailer* ∈ *PID*. Because there exists a functional dependency *Shop* → *Retailer*, the attribute *Shop* is also an identifier for *Retailer*, i.e., *PID* = {*Retailer, Shop*}. The whole path *Shop* → *Retailer* must be protected. A path of protected identifiers which are functionally related forms a *protected path*.

In order to ensure that sensitive data are never directly accessible, we define the *protected cube* as a data cube that omits protected identifiers in its schema.

Definition 1. *Let* $PID \subseteq D^i$. *The **protected cube** with regard to the protected identifiers PID is defined as*

$$PC = (\{D^1, \ldots, D^i \setminus PID, \ldots, D^n\}, \{F^1, \ldots, F^m\}).$$

Protected cubes ensure, that protected identifiers are never directly disclosed in reports. Note, that if a higher level attribute is removed from the schema of a cube, the whole protected path of attributes functionally determining A_j must be removed as well, because of implication (1).

We assume that all attributes in *PID* are part of a single dimension, because dimensions should be orthogonal and there should be no functional dependencies between attributes of different dimensions in a well-designed multidimensional schema [8, 10].

4.2 Cross-Dimensional Identification via Exclusivity

Quasi-identifiers disclose protected individuals with some degree of ambiguity [4]. Typical examples in the literature list attributes like date of birth, gender, address etc. which can be linked to persons if some external knowledge is available (e.g., [9, 11]). From a multidimensional perspective, all those attributes are coming from the same dimension.

However, identification of certain protected elements in one dimension might also be possible, if there is an exclusive relation to elements of another dimension. Consider the sales of the brand `Okay` in the sample data set in Fig. 1. In our example, `Okay` is a so-called tradebrand of the retailer `Jupiter`, i.e., `Jupiter` exclusively sells the products of this brand. Note, that exclusivity does not imply a functional dependency on schema-level. Instead, the concept of exclusivity is defined by implication rules on instance level (i.e., cell level) and is therefore an inherently local property of the data set.

Definition 2. *A cube cell* $c = (c_1, \ldots, c_i, \ldots, c_n, f)$ *is **exclusive** with respect to* c_i, *if a subset of the dimensional attribute values* $\{c_p, \ldots, c_q\}$ *(not containing* c_i*) imply the value of* c_i, *i.e.,*

$$\forall c' = (c'_1, \ldots, c'_i, \ldots, c'_n, f) : if \{c'_p, \ldots, c'_q\} = \{c_p, \ldots, c_q\}, then \ c'_i = c_i.$$

A rule $\{c_p, \ldots, c_q\} \to \{c_i\}$ *is called **exclusivity rule**.*

Consider the following exclusivity rules in the market research scenario:

$$\{Brand = \text{Okay}\} \rightarrow \{Retailer = \text{Jupiter}\},$$
$$\{Product = \text{SAM32F50}, Country = \text{DE}\} \rightarrow \{Retailer = \text{MediaStore}\}, \text{ and}$$
$$\{Product = \text{PAN29B10}, Year = 2015\} \rightarrow \{Retailer = \text{Conwheel}\}.$$

The first rule states that all cube cells with product sales from brand Okay imply retailer Jupiter. The second rule says that product SAM32F50 is exclusively sold by MediaStore in Germany. The third rule shows an example with a restriction to a certain time period. The rules illustrate real world examples and, if a report user has access to this kind of expert knowledge, the respective cube cells would disclose the retailers. This makes the attributes *Brand* and *Product* quasi-identifiers for *Retailer* within a certain local context. Global recoding of *Brand* and *Product* would result in unacceptable information loss, because information about all brands and products would be anonymized even though only very few brands and products are exclusively sold by a single retailer. Local recoding of affected brands and products for selected periods or countries instead minimizes the negative impact of anonymization.

Exclusivity rules are often available to statistical agencies or market researchers preparing the data. Basically, there are three sources for exclusivity rules:

1. Public knowledge: Some exclusivity rules can be gathered from public information sources, for example facts about tradebrands can easily be found in the internet.
2. Data provider: More specific rules may come from the organizations or individuals to be protected, like the rule that product SAM32F50 is in Germany exclusively sold by MediaStore.
3. Data inspection: The most expensive way is to analyze the data, e.g., to find products, brands, or even specific product segments which are exclusively sold by certain retailers. If the search space can be limited to a few attributes (e.g., Product, Brand per Country) it is possible to calculate the share of each retailer in that context. If the share is 100 %, an exclusivity rule was found.

4.3   Local Anonymization by Exclusivity Treatment Rules

As pointed out earlier, exclusivity is inherently local. Any kind of global generalization or suppression scheme would anonymize large amounts of uncritical data and result in inappropriate information loss. Therefore, we propose a combination of local cell generalization and suppression utilizing the knowledge about exlusivity rules. The idea is to specify for each exclusivity rule how the respective exclusive cube cells are to be anonymized.

As the approach is rule-based, anonymizations are not determined algorithmically but must be manually or semi-manually specified as treatment rules. This has the advantages, that (a) the anonymization scheme stays consistent over time, which is very important for periodical statistics, (b) processing is very fast even for very large amounts of data, and (c) only a minimal set of data is anonymized. However, k-anonymity [13] is not guaranteed.

To specify how exclusive data should be anonymized, the right hand side of an exclusivity rule is replaced by treatment options for one or all (local) quasi-identifiers. Thus, a treatment rule has the form $\{c_1, \ldots, c_r\} \rightarrow \{(a_1, t_1), \ldots, (a_k, t_k)\}$, where $a_1, \ldots, a_k$ are the quasi-identifiers and $t_1, \ldots, t_k$ specify how these attributes are to be anonymized. We propose the two treatment options suppress (local suppression) and aggregate (local generalization). Examples of treatment rules could be:

$\{Brand = \text{Okay}\} \rightarrow \{(Product, \text{aggregate}), (Brand, \text{aggregate})\}$,

$\{Product = \text{SAM32F50}, Country = \text{DE}\} \rightarrow \{(Product, \text{suppress}, (Brand, \text{suppress})\}$, and

$\{Product = \text{PAN29B10}, Year = 2015\} \rightarrow \{(Product, \text{suppress})\}$.

The application of the set $\mathcal{R}$ of all treatment rules to an instance of a protected cube PC transforms it to an anonymized cube instance AC. For this operation we introduce the anonymization operator $\Phi_{\mathcal{R}}$:

$$AC = \Phi_{\mathcal{R}}(PC).$$

Anonymization is done by *local recoding* of the dimensional values. LeFevre et al. [9] introduce an anonymization function ϕ to map quasi-identifying attributes globally to generalized values. In our scenario, the anoymization operation is applied locally, i.e., for each cube cell separately, according to the treatment rules.

Let PC be a protected cube. If a cube cell is matched by a rule in $\mathcal{R}$, the respective attributes are locally suppressed or generalized. Formally, the anonymization operator $\Phi_{\mathcal{R}}$ on the protected cube applies the local recoding $\phi_{\mathcal{R}}$ to each cube cell, i.e.,

$$\Phi_{\mathcal{R}}(PC) = \{\phi_{\mathcal{R}}(c) \mid c \in PC\}.$$

Figure 4 shows the anonymized data from Fig. 1 after application of the treatment rules $\mathcal{R}$ by $\Phi_{\mathcal{R}}$. For the sake of traceability we left the retailer information

Raw Data (anonymized)								
Product ID	**Product**	**Brand**	**TV Tuner**	**Technology**	Conwheel	MediaStore	Jupiter	**Sales Units**
1	BLA42N18	Blaupunkt	Yes	LED	10	9	8	27
2	Others	Others	No	LED			8	8
3	Others	Others	Yes	LCD			25	25
4	PAN29A65	Panasonic	No	LED		22		22
5	PAN42A78	Panasonic	Yes	LED	3	2	44	49
6	PAN40F65	Panasonic	Yes	LCD	3		9	12
7	Suppressed	Panasonic	No	LED	14			14
8	Suppressed	Suppressed	Yes	LED		45		45
9	SAM46F61	Samsung	Yes	LED		23	18	41
	Total							243

Fig. 4. Sample of retail sales data after application of anonymization $\Phi_{\mathcal{R}}$

marked gray, because the protected cube does not include protected identifiers by definition. Figure 2b shows how the Hitlist report shown in Fig. 2a changes when calculated based on this anonynmized cube.

Treatment rules in contrast to algorithmic approaches (e.g. [9,13]) are a very flexible mechanism to implement individual privacy requirements of the data suppliers. The practical definition of exclusivity rules can be stronger and/or weaker than the formal definition of exclusivity. On the one hand, for example, a product can already be declared as exclusive with respect to a retailer, if this retailer has a market share of more than 90 % for this product. On the other hand, a product with a small market share might be coincidentally sold only at a single retailer, but does not need treatment because the retailer does not consider this information confidential.

5 Applying Anonymization in Practise

Real-world data warehouses implemented by relational star- or snowflake schemas often comprise tera- or petabytes of data. In these scenarios, protected cubes can easily be built by omitting protected identifiers at schema level, e.g. via relational views or built-in security mechanisms of OLAP applications. However, building an anonymized cube by applying local treatment rules on instance level across the whole dataset demands elaborate strategies. One option is to apply the rules at query execution time in a specialized OLAP application. However, this is a complex and time consuming transformation and therefore not desirable.

A better option is the application of exclusivity rules in the ETL process, which is generally used to harmonize, cleanse, and load data into the data warehouse. Given that exclusivity is local in nature, exclusivity rules can be applied incrementally during the mostly periodical, e.g., daily, weekly, or monthly data loads as indicated in Fig. 5. The whole data cube is composed of incrementally appended subcubes, each of which is defined by a certain combination of dimensional attribute values, e.g., product category, time period, and country. In GfK's data warehouse, these subcubes contain on average 15 million cube cells per day, which is a quite manageable data volume compared to the size of the whole data cube with more than 30 billion cells.

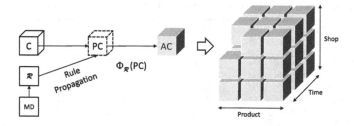

Fig. 5. Application of treatment rules as part of data warehouse ETL process

The treatment rules in GfK's data warehouse are stored as part of the master data (MD) of the product dimension, identifying which products and/or brands are exclusively sold by a particular retailer, possibly limited to certain time periods and countries. The treatment rules $\mathcal{R}$ are propagated to the protected cube as part of the data load process. Since this annotated protected cube is an intermediate data structure, it is illustrated with dotted lines. Figure 6a illustrates the relational representation of such an annotated protected cube.

The implementation of the anonymization operator $\Phi_{\mathcal{R}}$ takes the propagated treatment rules as input and replaces the ids (foreign keys) of products needing anonymization by artificially generated products. Figure 7 shows a subset of the extended product dimension and its relational representation. The product ids 2 and 3 of Okay products are substitued in the fact table by the product id 0 representing the general Others product. Others is serving as an accumulator for all products that need to be aggregated within a product group. For the

Fact Table (Protected Cube)

Product ID	Product Treatment Rule	Brand Treatment Rule	Year	Retailer	Sales Units
7		suppress	2015	Conwheel	4
8	suppress	suppress	2015	MediaStore	13
9			2015	MediaStore	23
9			2015	Jupiter	18
2	aggregate	aggregate	2015	Jupiter	7
3	aggregate	aggregate	2015	Jupiter	10
7			2016	Conwheel	13
7			2016	MediaStore	9
7			2016	Jupiter	24
8	suppress	suppress	2016	Jupiter	15
9			2016	Conwheel	5
9			2016	MediaStore	14
9			2016	Jupiter	12
2	aggregate	aggregate	2016	Jupiter	10
3	aggregate	aggregate	2016	Jupiter	11
Total					**188**

Fact Table (Anonymized Cube)

Treated Product ID	Year	Sales Units
701	2015	4
811	2015	13
9	2015	23
9	2015	18
0	2015	7
0	2015	10
7	2016	13
7	2016	9
7	2016	24
811	2016	15
9	2016	5
9	2016	14
9	2016	12
0	2016	10
0	2016	11
Total		**188**

(a) Protected cube with annotations for treatments   (b) Anonymized cube

Fig. 6. Protected and anonymized cube illustrated as relational fact tables

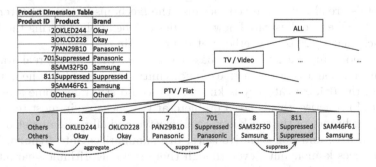

Fig. 7. Instance of treated product dimension and respective dimension table

two suppressed products 7 and 8, indivual surrogates are introduced. Figure 6a shows the resulting fact table.

In GfK, mainly product and brand identifiers are suppressed, and in some productgroups additional quasi-identifiers for products or brands. Applying suppression on product- and brand level produces at maximum three additional substitutes per product: product attributes suppress, brand attributess suppress, or both. Similarly, there are three variants of Others per product group. However, these new elements are only necessary for products or brands in exclusive contexts and constitute in reality for just 8 % of all products. The size of the about 700 times larger fact table remains unchanged, as only product identifiers are replaced.

Note, that neither the protected nor the anonymized cube in Fig. 6a is aggregated, both have the same size as the original data set. In market research this is necessary to enable the calculation of specific facts, e.g., quantity-weighted prices and sales distributions. In scenarios where only summarizable facts are of interest, the anonymized cube could be aggregated and the resulting microdata could be freely published.

As shown by these simple examples, exclusivity treatment can be done as part of the data warehouse load process, by applying an identifier mapping to the fact table and by merging newly created artificial elements into the dimension tables. This introduces negligible overhead in the ETL process, and requires neither additional processing power nor special logic at report runtime.

6   Summary and Conclusion

Motivated by a use case from point-of-sales reporting at GfK, we introduced a non-perturbative rule-based mechanism to avoid disclosure of the sales performance of GfK's data suppliers. We showed that removing protected identifiers on schema-level does not avoid disclosure of retailers due to exclusivity. Additionally, an exclusivity treatment must be applied on cross-dimensionally related attributes. We proposed two treatment options for local suppression and generalization and showed how the application of these rules anonymizes OLAP reports sufficiently with very little information loss. Whereas suppression leaves sales on detail level, generalization aggregates exclusive sales into Others in reports. Totals are always preserved, because the mechanism does not filter out exclusive sales. We also illustrated how the rules can be applied during the ETL process, such that there is no overhead at query runtime.

As the whole approach is purely based on rules, there is no real way to measure information loss or the degree of anonymity. For example, if the number of brands in the Others category is known, one could still infer some sensitive information. In practise, this is not a problem at GfK because in most reports a few interesting brands or the top-k items are selected and all remaining elements, including sensitive and non-sensitive sales, are summarized into Others. This defacto preserves k-anonymity, even though there is no algorithmic guarantee.

The process described in the last chapter is a simplified version of what is really happening in the GfK data warehouse. In reality, there are several more treatment options to fulfill special privacy requirements. Every day about 15 million fact records are anonymized using the rule-based approach for exclusivity handling, and hundreds of privacy-protected reports are produced.

References

1. Agrawal, R., Srikant, R.: Privacy-preserving data mining. In: Proceedings of the 2000 ACM SIGMOD International Conference on Management of Data - SIGMOD 2000, vol. 29(2), pp. 439–450 (2000)
2. Albrecht, J., Lehner, W., Teschke, M., Kirsche, T.: Building a real data warehouse for market research. In: Proceedings 8th International Conference on Database and Expert Systems Applications, DEXA 1997, pp. 651–656 (1997)
3. Blanco, C., De Guzmán, I.G.R., Fernández-Medina, E., Trujillo, J., Piattini, M.: Applying an MDA-based approach to consider security rules in the development of secure DWs. In: Proceedings of the International Conference on Availability, Reliability and Security, ARES 2009, pp. 528–533 (2009)
4. Domingo-Ferrer, J.: A survey of inference control methods for privacy-preserving datamining. In: Aggarwal, C.C., Yu, P.S. (eds.) Privacy-Preserving Data Mining, vol. 34, pp. 53–80. Springer, Boston (2008)
5. Fernández-Medina, E., Trujillo, J., Villarroel, R., Piattini, M.: Developing secure data warehouses with a UML extension. Inf. Syst. 32(6), 826–856 (2007)
6. Fung, B.C.M., Wang, K., Chen, R., Yu, P.S.: Privacy-preserving data publishing. ACM Comput. Surv. 42(4), 1–53 (2010)
7. Hua, M., Zhang, S., Wang, W., Zhou, H., Shi, B.-L.: FMC: an approach for privacy preserving OLAP. In: Tjoa, A.M., Trujillo, J. (eds.) DaWaK 2005. LNCS, vol. 3589, pp. 408–417. Springer, Heidelberg (2005)
8. Lechtenbörger, J., Vossen, G.: Multidimensional normal forms for data warehouse design. Inf. Syst. 28(5), 415–434 (2003)
9. LeFevre, K., DeWitt, D.J., Ramakrishnan, R.: Mondrian multidimensional K-anonymity. In: Proceedings International Conference on Data Engineering, vol. 2006, p. 25 (2006)
10. Lehner, W., Albrecht, J., Wedekind, H.: Normal forms for multidimensional databases. In: Proceedings Tenth International Conference on Scientific and Statistical Database Management (Cat. No. 98TB100243), pp. 63–72. IEEE Computer Society, july 1998
11. Malik, M.B., Ghazi, M.A., Ali, R.: Techniques, privacy preserving data mining : current scenario and future prospects. In: 2012 Third International Conference on Computer and Communication Technology, pp. 26–32 (2012)
12. Shoshani, A.: OLAP and statistical databases: similarities and differences. In: Proceedings of the Sixteenth ACMSIGACT-SIGMOD-SIGART Symposium on Principles Of Database Systems PODS 1997, pp. 185–196 (1997)
13. Sweeney, L.: Achieving k-anonymity privacy protection using generalization and suppression. Int. J. Uncertainty Fuzziness Knowl. Based Syst. 10(5), 1–18 (2002)
14. Wang, L., Wijesekera, D., Jajodia, S.: Cardinality-based inference control in data cubes. J. Comput. Secur. 12, 655–692 (2004)
15. Willenborg, L., Waal, T.D.: Elements of Statistical Disclosure Control. Springer, New York (2001)
16. Xu, Y., Ma, T., Tang, M., Tian, W.: A survey of privacy preserving data publishing using generalization and suppression. Appl. Math. Inf. Sci. 8(3), 1103–1116 (2014)

Perturbative Data Protection of Multivariate Nominal Datasets

Mercedes Rodriguez-Garcia[1], David Sánchez[1(✉)],
and Montserrat Batet[2]

[1] UNESCO Chair in Data Privacy, Department of Computer Engineering
and Maths, Universitat Rovira i Virgili, Tarragona, Spain
mercedes.rodriguez@uca.es, david.sanchez@urv.cat
[2] Internet Interdisciplinary Institute (IN3),
Universitat Oberta de Catalunya, Castelldefels, Spain
montserrat.batet@urv.cat

Abstract. Many of the potentially sensitive personal data produced and compiled in electronic sources are nominal and multi-attribute (e.g., personal interests, healthcare diagnoses, commercial transactions, etc.). For such data, which are discrete, finite and non-ordinal, privacy-protection methods should mask original values to prevent disclosure while preserving the underlying semantics of nominal attributes and the (potential) correlation between them. In this paper we tackle this challenge by proposing a semantically-grounded version of numerical correlated noise addition that, by relying on structured knowledge sources (ontologies), is capable of perturbing/masking multivariate nominal attributes while reasonably preserving their semantics and correlations.

Keywords: Privacy · Data protection · Semantics · Ontologies

1 Introduction

Personal data are crucial both for business and research purposes. *Microdata* (multivariate records, each one detailing the attributes of an individual) are of especial interest because, unlike *macrodata*, they do not restrict the type and granularity of data analyses. However, publishing *microdata* for secondary use may compromise the privacy of the individuals. To protect privacy while ensuring that data are analytically useful, data protection methods [1, 2] aim at balancing the trade-off between disclosure risk and data utility preservation.

Even though most protection methods focus on numerical data, nowadays, *nominal* microdata (categorical values or textual responses) account a significant amount of the personal data available from social networks, polls, healthcare records or web browsing [3]. Unlike numbers, nominal data are finite, discrete and, in general, non-ordinal. Thus, it is not possible to apply the arithmetic and statistical operators required for data masking. Moreover, whereas the utility of numerical data is a function of their statistical properties, nominal data utility depends on the preservation of semantics [4]. In this paper, we tackle this challenging scenario and specifically focus on multivariate nominal datasets, which are especially relevant for research.

© Springer International Publishing Switzerland 2016
J. Domingo-Ferrer and M. Pejić-Bach (Eds.): PSD 2016, LNCS 9867, pp. 94–106, 2016.
DOI: 10.1007/978-3-319-45381-1_8

1.1 Related Work

Many data protection methods have been proposed within the Statistical Disclosure Control [1] and Privacy Preserving Data Publishing [2] disciplines. For nominal data, non-perturbative methods based on replacing values by generalizations are the most widespread ones. These, however, suffer from several drawbacks that may hamper data utility [5]. First, generalization produces a loss of granularity because input values are replaced by a reduced set of categories/generalizations. Moreover, if outlier values are present in the sample, the need to generalize them may result in very coarse generalizations and, thus, a high loss of information/utility.

Perturbative methods such as data microaggregation, rank swapping, data shuffling or noise addition are free from these drawbacks since they mask data while maintaining their granularity. From these, only microaggregation has been adapted to semantically mask nominal data [6, 7]. Microaggregation is usually employed to enforce k-anonymity [8] by building clusters of k indistinguishable records and replacing original values with a representative value of the cluster (e.g., the mean). To minimize the information loss associated to the replacement, microaggregation requires static and homogenous data.

By contrast, noise addition is able to mask records individually, which is useful when data are generated as a stream and should be protected on the fly [9]. Moreover, unlike methods based on a homogenous data aggregation, noise addition allows defining per-individual perturbation/masking levels, which is useful to accommodate heterogeneous privacy needs. Finally, noise addition has recently gained more relevance due to the ε-differential privacy model [10], whose enforcement relies on Laplacian noise addition.

Because of its mathematical roots, noise addition has been rarely applied to categorical/nominal data. In [11] the authors propose changing the values of each attribute according to a probability distribution. In [12], it is suggested dividing the categorical attributes into sub-attributes that present natural orders, which are needed for noise addition. In the context of differential privacy, some mechanisms have been proposed to perturb discrete data: the geometric mechanism (a discrete probability distribution alternative to the Laplace one) [13] and the exponential mechanism (which probabilistically chooses the output of a discrete function) [14]. However, both of them rely on the data distribution rather than on semantics; this fact makes them more suitable for discrete numerical values rather than nominal data. In [15] the use of "semantic noise" is suggested to mask text, but its calculation is not specified.

1.2 Contributions

In [16] we presented the notion of *semantic noise* as a semantically-grounded version of numerical uncorrelated noise addition. With this method we were able of masking individual nominal attributes while preserving their semantics better than with perturbation mechanisms based only on data distributions.

In this paper we extend this work to support correlated noise, which is needed to protect non-independent attributes in multivariate datasets while preserving their correlations. As in [16], we exploit ontologies to capture and manage the underlying semantics of nominal values and employ semantic operators alternative to the arithmetical ones used for numerical noise addition (i.e., distance, mean, variance and covariance). Finally, we propose a semantic algorithm to add user-defined correlated noise to pairs of nominal attributes.

The rest of the paper is organized as follows. Section 2 provides background on (numerical) noise addition. Section 3 describes the semantic management of nominal values, and details our semantic correlated noise addition method. Section 4 reports the results of several empirical experiments with nominal multivariate datasets. Section 5 depicts the conclusions and provides some lines of future research.

2   Background

Uncorrelated noise [17] perturbs individual attributes by adding normally distributed random noise. Let us have a dataset X, consisting on a $(n \times p)$ matrix with p attributes of n records, where $X_j = \{x_{1j}, ..., x_{ij}, ..., x_{nj}\}$ is the jth attribute and x_{ij} is the value of the attribute j corresponding to the individual/record i. To perturb the attribute X_j, each value x_{ij} is replaced by a noisy version z_{ij}:

$$Z_j = X_j + \varepsilon_j, \tag{1}$$

where $\varepsilon_j = \{\epsilon_{1j}, ..., \epsilon_{ij}, ..., \epsilon_{nj}\}$ is the noise sequence. Being $X_j \sim N(\mu_{Xj}, \sigma^2_{Xj})$ a vector with mean μ_{Xj} and variance σ^2_{Xj}, $\varepsilon_j \sim N(0, \sigma^2_{Xj})$ is a vector of normally distributed random errors with mean zero and variance σ^2_{Xj}. The error variance σ^2_{Xj} is proportional to the original attribute variance as follows:

$$\sigma^2_{\varepsilon_j} = \alpha\sigma^2_{Xj}, \quad \alpha > 0 \tag{2}$$

The factor α determines the amount of noise, whose value usually ranges between 0.1 and 0.5 [18]. This method preserves the mean of the attributes and keeps the variance proportional in a factor of $1 + \alpha$; but, since noise is independently applied to each attribute, the covariance between noise-added attributes is zero and, thus, the attribute correlations are not preserved.

To solve this issue, which is crucial for research because most datasets are multivariate and correlated and many data analyses are performed over records rather than attributes, correlated noise addition has been proposed [19]. The noise applied to multiple attributes of a dataset X follows matrix notation:

$$Z = X + \varepsilon, \tag{3}$$

where the $(n \times p)$ data matrix $X \sim N(\mu_X, \Sigma_X)$ follows a multivariate normal distribution with mean the p-dimensional vector μ_X and covariance matrix the $(p \times p)$ matrix Σ_X. The diagonal elements of Σ_X are the variances of the attributes and the off-diagonal elements are the covariances between the attributes. Similarly $\varepsilon \sim N(0, \Sigma_\varepsilon)$ is a $(n \times p)$ matrix with p noise sequences of n error values. The noise matrix ε also follows a multivariate normal distribution with mean the p-dimensional vector 0 and covariance matrix the $(p \times p)$ matrix Σ_ε. The covariance matrix of errors is proportional to the covariance of the data:

$$\sum\nolimits_\varepsilon = \alpha \sum\nolimits_X, \quad \alpha > 0 \tag{4}$$

Correlated noise addition preserves the mean of each attribute, keeps the covariance matrix proportional in a factor $1 + \alpha$ to the covariance matrix of the original data and maintains the correlation between the attributes.

3 Nominal Data Perturbation with Semantic Correlated Noise

We propose a (perturbative) correlated noise addition method for masking multivariate nominal datasets that exploits ontologies to capture and manage data semantics. The idea is to replace original values by other semantically similar concepts, whose similarity is proportional to the desired magnitude of noise (which states the level of masking to be applied to the data).

3.1 Semantic Management of Nominal Data

To manage the semantics of nominal data we first define the *semantic domain* of the attributes according to the concepts modelled in an ontology τ, which we exploit as knowledge base. An ontology can be viewed as a tree/graph whose nodes depict the concepts from an area of knowledge and the edges detail the semantic relationships between them. Thus, our method assumes the availability of an appropriate ontology for the domain of interest.

Let $A = \{a_1, ..., a_i, ..., a_n\}$ an attribute of a dataset X of n records, where a_i is the value of the individual i mapped to a concept in τ. The *semantic domain* of an attribute A is defined as the set of concepts belonging to the category of A (e.g. if A contains diseases, $D(A)$ is the set of all diseases).

$$D(A) = \{c \in Category(A)\} \tag{5}$$

Then, we extract the *taxonomy of the semantic domain* $D(A)$ from the ontology τ, as the hierarchy $\tau(D(A))$ including of all concepts that are taxonomic specializations of the Least Common Subsumer of $D(A)$, $LCS(D(A))$; that is, the most specific ancestor in τ that subsumes all concepts in $D(A)$).

$$\tau(D(A)) = \bigcup_{c_i \in \tau} \{c_i \mid LCS(D(A)) \geq c_i\} \tag{6}$$

Contrary to the numerical domain, $D(A)$ is finite, discrete and, in general, non-ordinal. To manage it, in [16] we defined semantic versions of the arithmetic operators needed for uncorrelated noise addition (i.e. distance, mean and variance). Because they are also used for correlated noise addition, we briefly introduce them here.

The most basic operator is the one that measures the *distance* between two values. In the semantic domain, in which nominal values should be compared according to the semantics of the concepts to which they refer, the dissimilarity between the meaning of two concepts c_i and c_j can be computed through their semantic distance, $sd(c_i,c_j)$. In [16] the distance version of the well-known Wu and Palmer (W&P) semantic similarity measure [20] was selected to compute $sd(c_i,c_j)$ because it provides values normalized in the range [0..1], which is coherent with a Normal noise distribution, it provides a non-logarithmic and non-exponential assessment of the distance, which avoids the concentration of values in the high or low zones of the output range, and it is computationally efficient.

By relying on the semantic distance $sd(\cdot,\cdot)$, we can define the *semantic mean of a nominal attribute A* as the concept c form $\tau(D(A))$ that minimizes the sum of the semantic distances respect to all a_i in A.

$$s\mu(A) = \arg \min_{c \in \tau(D(A))} \left(\sum_{a_i \in A} sd(c, a_i) \right) \qquad (7)$$

On the other hand, the *semantic variance of a nominal attribute A* can be defined as the average of squared semantic distances between each concept a_i in A and the semantic mean $s\mu(A)$.

$$s\sigma^2(A) = \frac{\sum\limits_{a_i \in A} (sd(a_i, s\mu(A)))^2}{n} \qquad (8)$$

As stated in Sect. 2, for correlated noise addition, we also require two additional operators measuring the linear dependence between attribute pairs: *covariance* and *correlation*. In the numerical domain, the standard arithmetical covariance is positive when the greater values of one attribute mainly correspond with the greater values of the other attribute and the same holds for the lesser values. To distinguish great values from small values a total order over the attribute domains must exist, as it is the case of numbers. However, the domains of nominal attributes are usually non-ordinal and, thus, we need a measure of statistical dependence that does not rely on total orders.

In [21], Székely defined the *distance covariance* and the *distance correlation*; they measure the statistical dependence between variables by relying on the pairwise distance (or dissimilarity) between the values of each variable. In this paper, we adapt these measures to the nominal domain by relying on the above mentioned semantic distance $sd(\cdot,\cdot)$.

Let $A = \{a_1,..., a_n\}$ and $B = \{b_1,..., b_n\}$ be two nominal attributes of a sample of n individuals, where a_i and b_i represent the values for individual i. According to the definition of *distance covariance*, we must first compute the n by n distance matrices of each attribute. To do so, we must compute all pairwise distances of each attribute that, in the semantic domain, correspond to the semantic distances sd between nominal value pairs. Distance matrices are used to obtain the n by n double centered distance matrices that, for attribute A, is computed as follows:

$$\left(\delta_{ij}^A\right)_{i,j=1}^n = \left(sd(a_i,a_j) - \delta_{i.}^A - \delta_{j}^A + \delta_{..}^A\right)_{i,j=1}^n, \tag{9}$$

where $sd(a_i,a_j)$ is the value of the $(i,j)^{\text{th}}$ element of the distance matrix of attribute A, $\delta_{i.}^A$ is the mean of ith row of the distance matrix, δ_j^A is the mean of jth column of the distance matrix and $\delta_{..}^A$ is the mean of all values of the distance matrix (i.e. the grand mean). The notation is similar for attribute B.

Then, the *semantic distance covariance* between two nominal attributes A and B is the square root of the arithmetic mean of the products $\delta_{ij}^A\delta_{ij}^B$. It satisfies $sdCov(A, B) \geq 0$ and it is 0 if and only if A and B are independent [21].

$$sdCov(A,B) = \frac{1}{n}\sqrt{\sum_{i,j=1}^n \delta_{ij}^A\delta_{ij}^B} \tag{10}$$

As for the numerical correlation, the *semantic distance correlation* of two nominal attributes A and B can be computed as the nonnegative number obtained by dividing their distance covariance, $sdCov(A,B)$, by the product of their distance standard deviations.

$$sdCor(A,B) = \begin{cases} \frac{sdCov(A,B)}{\sqrt{sdVar(A)\times sdVar(B)}}, & sdVar(A) \times sdVar(B) > 0 \\ 0, & sdVar(A) \times sdVar(B) = 0 \end{cases}, \tag{11}$$

where $sdVar(A)$ and $sdVar(B)$ are the *semantic distance variances* of attributes A and B, which are a special case of distance covariance when the two attributes are identical (i.e., $sdVar(A) = sdCov(A,A)$). The *semantic distance correlation* results are bounded in the $[0..1]$ range, and it is 0 if and only if A and B are independent. As in the numerical case, values close to zero suggest a weak semantic association between attributes, while larger values suggest a stronger association.

3.2   Semantic Noise for Multivariate Datasets

In this section we adapt the standard correlated noise addition method to the semantic domain of nominal data. The goal of our method is twofold: (i) to mask original nominal values by replacing them with terms within a semantic distance coherent with the desired level of protection, and (ii) to preserve, as much as possible the semantic features of the data. Regarding the latter, we specifically aim at preserving, as much as possible, the semantic mean of each attribute; obtaining a per-attribute dispersion proportional to the variance of the original data and the noise magnitude; obtaining a pairwise attribute dispersion proportional to the covariance the original data and the noise magnitude; and preserving, as much as possible, the semantic correlation between the attributes.

Unlike uncorrelated noise, the sequences of correlated noise must consider the degree of correlation between attributes in order to preserve their level of association (see Sect. 2). For simplicity of the following explanation, let us assume that the dataset X has only two nominal attributes A and B with n records (for datasets with more than two attributes, the process depicted below would be applied for attribute pairs). By relying on the operators detailed above, the generated *semantic correlated noise* consists on a $(n \times 2)$ matrix of random numbers $\varepsilon_{A,B} = \{(\epsilon_{a1}, \epsilon_{b1}), \ldots, (\epsilon_{ai}, \epsilon_{bi}), \ldots, (\epsilon_{an}, \epsilon_{bn})\}$ that follows a multivariate normal distribution $\varepsilon_{A,B} \sim N(0, \alpha\Sigma_{A,B})$, with mean the vector 0 and covariance the matrix $\alpha\Sigma_{A,B}$, being α the desired level of semantic noise.

$$\sum_{A,B} = \begin{pmatrix} sdVar(A) & sdCov(A,B) \\ sdCov(B,A) & sdVar(B) \end{pmatrix} \tag{12}$$

In the numerical domain, the noise magnitude can be directly and coherently added/subtracted to/from the original values. Specifically, if a positive noise is added to an original value greater (lower) than any other value in the attribute sample, the masked value will get away from (closer to) the other values in the sample in the same magnitude; on the contrary, if the noise is negative, the masked value will get closer to (away from) the other values. Since the error is normally distributed around zero, the total added/subtracted magnitudes are compensated so that the aggregated relative differences are maintained in the masked outcome, thus preserving the statistical features of the data like the mean.

In the semantic domain, we propose replacing original nominal values by other concepts in the underlying taxonomy $\tau(D(\cdot))$ whose semantic distance is, ideally, equal to the noise magnitude. However, since nominal data are not ordinal, if we replace an original value by another concept at a certain distance, we cannot guarantee that the new concept is also closer to or farther from the other values at the same distance.

To solve this issue we propose a heuristic that guides the replacement of values according to the noise sign and that aims at improving the preservation of the correlation between attributes. In general, to better preserve the correlation between the two attributes A and B, if a positive noise is added to an original value a_i greater (lower) than its pair b_i, the new value should get away from (closer to) b_i at the same magnitude; on the contrary, if the error is negative, the new value should get closer to (away from) b_i. Thus, the magnitude of the accumulated additions and subtractions between the pair will compensate each other.

In order to balance the number of "movements" between the attribute value pairs and preserve the attribute correlation, the heuristic we propose interprets the noise sign with respect to a reference point defined by the pair corresponding to the value that is being replaced: if the noise is positive, the concept c in $\tau(D(A))$ that will replace a_i must be farther from its pair b_i than a_i, that is, $sd(c, b_i) > sd(a_i, b_i)$, and vice versa; if the noise is negative, the concept c must be closer to b_i than a_i, that is, $sd(c, b_i) < sd(a_i, b_i)$, and vice versa. Understandably, both attributes must belong to the same semantic domain, that is, $\tau(D(A)) = \tau(D(B))$.

Algorithm 1 formalizes the semantic correlated noise addition process according to the proposed heuristic. First, the taxonomies $\tau(D(A))$ and $\tau(D(B))$ associated to the domains of attributes A and B are obtained as detailed in Sect. 3.1. Then, the values of each attribute are mapped to concepts of $\tau(D(A))$ and $\tau(D(B))$. In lines 4 and 5, the (2×2) covariance matrix $\Sigma_{A,B}$ is built and the $(n \times 2)$ noise matrix $\varepsilon_{A,B} \sim N(0, \alpha\Sigma_{A,B})$ is generated. Finally, in lines 6 to 14, attribute A is masked according to the desired magnitude of correlated noise and by following the proposed heuristic. The same process is applied for attribute B.

It is important to note that, to provide a priori privacy guarantees, our method should be applied in the context of a noise-based privacy model, such as ε-differential privacy [10]. Through a specially tailored noise distribution, this model guarantees that the outputs are insensitive (up to a factor dependent on ε) to modifications of one input record. In this way, the participation of one individual in a survey (or any specifics that he has contributed to the survey) will not be jeopardized by more than a $1 + \varepsilon$ factor.

Algorithm 1. Semantic correlated noise addition for two attributes A and B.

Input : A, B : nominal attributes of n records; τ: taxonomy; α: semantic noise level
Output : A^*, B^* : noise-added masked nominal attributes
1: $\tau(D(A)) \leftarrow$ obtain_taxonomy$(D(A), \tau)$
2: $A \leftarrow$ map$(A, \tau(D(A)))$
3: Apply lines 1-2 to the attribute B
4: $\Sigma_{A,B} \leftarrow$ compute_CovarianceMatrix(A, B)
5: $\varepsilon_{A,B} \leftarrow$ generate_NoiseMatrix$(\alpha, \Sigma_{A,B})$ // $\varepsilon_{A,B} = \left\{ \left(\epsilon_{a_1}, \epsilon_{b_1}\right), ..., \left(\epsilon_{a_n}, \epsilon_{b_n}\right) \right\} \sim N(0, \alpha\Sigma_{A,B})$
6: **for all** a_i in A **do**
7:    **if** $\epsilon_{a_i} = 0$ **then**
8:      $a_i^* \leftarrow a_i$
9:    **else if** ϵ_{a_i} is positive **then**
10:      $a_i^* \leftarrow \underset{c \in \tau(D(A))}{\arg\min} \left\{ sd(c, a_i) \mid sd(c, a_i) \geq \left| \epsilon_{a_i} \right| \wedge sd(c, b_i) > sd(a_i, b_i) \right\}$
11:    **else if** ϵ_{a_i} is negative **then**
12:      $a_i^* \leftarrow \underset{c \in \tau(D(A))}{\arg\min} \left\{ sd(c, a_i) \mid sd(c, a_i) \geq \left| \epsilon_{a_i} \right| \wedge sd(c, b_i) < sd(a_i, b_i) \right\}$
13:    **end if**
14: **end for**
15: Apply lines 6-14 to attribute B
16: **return** A^*, B^*

4 Experiments

In this section we evaluate the algorithm proposed in Sect. 3 with two nominal datasets with significantly different degrees of attribute correlation, and compare them with the semantic uncorrelated noise addition method we presented in [16]. As evaluation data, we use a patient discharge dataset provided by the California Office of Statewide Health Planning and Development, where each record describes a patient and contains two -correlated- nominal attributes: the principal diagnosis and the secondary diagnosis. SNOMED-CT [22], a healthcare knowledge base modeling all the diagnoses in the dataset, has been used as the ontology that provides the data semantics.

In the first experiment we have used a sample of 1,350 records with a high *semantic distance correlation* (0.94) between the two nominal attributes (i.e., $A = principal$ *diagnosis* and $B = secondary$ *diagnosis*). The dataset has been masked with the algorithm we propose in Sect. 3 and also with the uncorrelated noise addition method presented in [16], for the usual values of the noise parameter $\alpha = [0.1..0.5]$. Figure 1 shows the semantic distance correlation (*sdCor*, Eq. (11)) of the noise-added attributes for the different methods and values of α. As expected, the results show that *sdCor* is preserved by the correlated noise addition algorithm better than by the uncorrelated method [16], a difference that is more noticeable for large values of α. Notice that the semantic correlation (and, as we discuss below, any other feature of the data) cannot be perfectly preserved when adding noise to nominal data because of two reasons. First, in the semantic domain, value replacements are discrete and noise magnitudes rarely match to exact concept distances in the taxonomy τ (especially for coarse grained taxonomies); thus, some approximation errors will be accumulated during the noise-based replacement of values. Second, because the size of the taxonomy $\tau(D(A))$ is limited, there will be cases in which we cannot find a replacement concept that is as far as stated by the magnitude of the noise; in these cases, the farthest concept will be used as replacement, thus truncating the noise magnitude to the distance of that most distant concept.

In addition to the attribute correlation, noise addition methods should also preserve other features of the data. In [16] it was shown that the uncorrelated noise method was able to reasonably preserve the semantic mean ($s\mu$, Eq. (7)) of individual attributes and to maintain a data dispersion ($s\sigma^2$, Eq. (8)) proportional. The semantic correlated noise addition method we present here should also preserve these features, in addition to maintain a proportional semantic distance covariance (*sdCov*, Eq. 10).

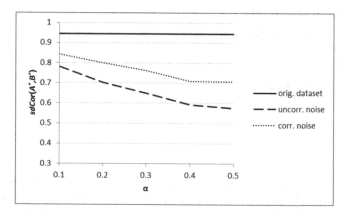

Fig. 1. Semantic distance correlation (*sdCor*) for a sample of 1,350 records with strongly correlated attributes (0.94).

For the evaluation dataset, the mean of attribute A is $s\mu(A)$ = *Furuncle of chest wall* and of attribute B is $s\mu(B)$ = *Viral hepatitis with hepatic coma* and the variances are $s\sigma^2(A) = 0.22$ and $s\sigma^2(B) = 0.24$. Finally the covariance between A and B is $sdCov(A,B) = 0.2553$. In Table 1 we depict how these features have been preserved in the noise-added attributes for the different methods and values of α. Specifically, we measured the semantic distance between the mean concept of each noise-added attribute and that of the original sample, and the absolute difference between the actual variances/covariance of the masked attributes and the expected variances/covariance after adding noise with $\alpha = \{0.2, 0.4\}$. In all cases a value of 0 states that the means, variances or covariance have been perfectly controlled. In addition, we also measured the actual root mean square error (RMSE), computed as the root average square semantic distance between original and masked value pairs; this measures the overall loss of semantics in the masked dataset, which should be similar to the target RMSE associated to the noise sequences to be added for each α.

Table 1. Evaluation metrics for noise-added attributes with the semantic uncorrelated method from [16] and the semantic correlated method proposed here.

Metric	Semantic uncorrelated noise addition		Semantic correlated noise addition	
	$\alpha = 0.2$	$\alpha = 0.4$	$\alpha = 0.2$	$\alpha = 0.4$
$sd(s\mu(A), s\mu(A^*))$	0.20	0.20	0	0.20
$\|s\sigma^2(A^*) - (1 + \alpha)\, s\sigma^2(A)\|$	0.02	0.07	0.01	0.04
$Actual\_RMSE(A)$	0.28	0.36	0.30	0.39
$Target\_RMSE(A)$	0.21	0.31	0.23	0.34
$sd(s\mu(B), s\mu(B^*))$	0	0	0.18	0.18
$\|s\sigma^2(B^*) - (1 + \alpha)\, s\sigma^2(B)\|$	0.05	0.09	0.05	0.08
$Actual\_RMSE(B)$	0.24	0.34	0.28	0.37
$Target\_RMSE(B)$	0.21	0.31	0.24	0.34
$\|sdCov(A^*,B^*) - (1 + \alpha)\, sdCov(A,B)\|$	0.18	0.26	0.17	0.25

Because α determines the magnitude of the added noise, the larger the α, the larger the RMSE (and the stricter the masking). Table 1 shows that the actual RMSE is similar to the target RMSE (i.e., the methods are able to controllably perturb data), even though not identical. As discussed above, the differences between target and actual RMSEs are due to the need to discretize and truncate noise magnitudes to the concepts in the taxonomy. The mean of the masked attributes is preserved up to a similar degree for the two methods, and we cannot observe significant differences between values of α. Note that, although the correlated noise addition algorithm is more focused in preserving the correlation, it is still capable of preserving the mean up to a similar or even better degree than the uncorrelated method. Table 1 also shows that the absolute difference between the variances and covariance of the masked sample and the expected ones are nearly 0 for the two methods. This means that the correlated noise

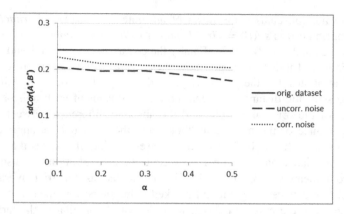

Fig. 2. Semantic distance correlation (*sdCor*) for a sample of 1,049 records with weakly correlated attributes (0.24).

addition algorithm, in addition to better preserve correlations, also maintain dispersions as controlled as the uncorrelated method.

In a second experiment we extracted a sample of 1,049 records with weakly correlated diagnoses (*sdCor(A,B)* = 0.24). As above, Fig. 2 shows the *semantic distance correlation* for the different methods and values of α. In this case, the differences between methods are much smaller. This suggests that, for weakly correlated attributes, the correlated method may not be worth it.

5   Conclusions

In this paper, we have presented a semantically-grounded version of the numerical correlated noise addition method, which is capable of masking multivariate nominal datasets while reasonably preserving their semantics and attribute correlations. Unlike other perturbative data protection methods based on clustering records, our proposal is able to protect records individually, which is useful when protecting dynamic data or data streams, or when we need to accommodate per-individual privacy needs. Moreover, unlike the non-perturbative methods usually employed to mask nominal values (i.e., based on generalizations), our approach does not coarsen the granularity of the data. Finally, our proposal to noise addition is general and can support the usual noise distributions (e.g., Normal, Laplace) employed in noise-based data protection schemas and privacy models (such as ε-differential privacy).

As future work, we plan define other heuristics to assist the value replacement process so that we are able to preserve better a particular feature of the data, in case the posterior data analysis strongly depends on that. Moreover, we also plan to compare our approach with non-semantic ones, such as the discrete but distributional-oriented geometric and exponential mechanisms employed by ε-differential privacy for categorical data.

Acknowledgements. This work was supported by the EU Commission under the H2020 project "CLARUS", by the Spanish Government through projects TIN2014-57364-C2-R "SmartGlacis", TIN2011-27076-C03-01 "Co-Privacy" and TIN2015-70054-REDC "Red de excelencia Consolider ARES" and by the Government of Catalonia under grant 2014 SGR 537. M. Batet is supported by a Postdoctoral grant from Ministry of Economy and Competitiveness (MINECO) (FPDI-2013-16589).

References

1. Hundepool, A., Domingo-Ferrer, J., Franconi, L., Giessing, S., Nordholt, E.S., Spicer, K., Wolf, P.-P.: Microdata. In: Statistical Disclosure Control, pp. 23–130. Wiley (2012)
2. Domingo-Ferrer, J., Sánchez, D., Soria-Comas, J.: Database Anonymization: Privacy Models, Data Utility, and Microaggregation-based Inter-model Connections. Morgan & Claypool Publishers (2016)
3. Ramirez, E., Brill, J., Ohlhausen, M., Wright, J., Mc-Sweeny, T.: Data brokers: a call for transparency and accountability. Federal Trade Commission, Technical Report, May 2014
4. Sánchez, D., Batet, M.: C-sanitized: a privacy model for document redaction and sanitization. J. Assoc. Inf. Sci. Technol. **67**, 148–163 (2016)
5. Soria-Comas, J., Domingo-Ferrer, J., Sánchez, D., Martínez, S.: t-closeness through microaggregation: strict privacy with enhanced utility preservation. IEEE Trans. Knowl. Data Eng. **27**, 3098–3110 (2015)
6. Martínez, S., Sánchez, D., Valls, A.: Semantic adaptive microaggregation of categorical microdata. Comput. Secur. **31**, 653–672 (2012)
7. Batet, M., Erola, A., Sánchez, D., Castellà-Roca, J.: Utility preserving query log anonymization via semantic microaggregation. Inf. Sci. **242**, 49–63 (2013)
8. Samarati, P., Sweeney, L.: Protecting privacy when disclosing information: k-anonymity and its enforcement through generalization and suppression. Computer Science Laboratory, SRI International (1998)
9. Krempl, G., Zliobaite, I., Brzezinski, D., Hüllermeier, E., Last, M., Lemaire, V., Noack, T., Shaker, A., Sievi, S., Spiliopoulou, M., Stefanowski, J.: Open challenges for data stream mining research. ACM SIGKDD Explor. Newslett. **16**, 1–10 (2014)
10. Dwork, C.: Differential privacy. Automata Lang. Programm. **4052**, 1–2 (2006)
11. Kooiman, P., Willenborg, L., Gouweleeuw, J.: Pram: a method for disclosure limitation of microdata. Research Paper 9705, Statistics Netharlands, P.O. Box 4000, 2270 JM Voorburg, The Netharlands (1997)
12. Giggins, H., Brankovic, L.: Protecting privacy in genetic databases. In: Proceeding of the 6th Engineering Mathematics and Applications Conference (EMAC 2003), vol. 2, Sydney, Australia, pp. 73–78 (2003)
13. Ghosh, A., Roughgarden, T., Sundararajan, M.: Universally utility-maximizing privacy mechanisms. In: Proceeding of the ACM Symposium on Theory of Computing (STOC 2009), pp. 351–360 (2009)
14. McSherry, F., Talwar, K.: Mechanism design via differential privacy. In: Proceeding of Annual IEEE Symposium on Foundations of Computer Science (FOCS 2007), pp. 94–103 (2007)
15. Abril, D., Navarro-Arribas, G., Torra, V.: On the declassification of confidential documents. In: Torra, V., Narakawa, Y., Yin, J., Long, J. (eds.) MDAI 2011. LNCS, vol. 6820, pp. 235–246. Springer, Heidelberg (2011)

16. Rodriguez-Garcia, M., Batet, M., Sánchez, D.: Semantic noise: privacy-protection of nominal microdata through uncorrelated noise addition. In: Proceeding of the 27th IEEE International Conference on Tools with Artificial Intelligence, ICTAI 2015, Vietri sul Mare, Italy, pp. 1106–1113 (2015)
17. Conway, R., Strip, D.: Selective partial access to a database. Cornell University, Technical Report (1976)
18. Tendick, P.: Optimal noise addition for preserving confidentiality in multivariate data. J. Stat. Plann. Infer. **27**, 341–353 (1991)
19. Kim, J.: A method for limiting disclosure in microdata based on random noise and transformation. In: Proceeding of the ASA Section on Survey Research Methods, pp. 370–374 (1986)
20. Wu, Z., Palmer, M.: Verbs semantics and lexical selection. In: Proceeding of the Annual Meeting of the Association for Computational Linguistics, pp. 133–139 (1994)
21. Székely, G.J., Rizzo, M.L., Bakirov, N.K.: Measuring and testing dependence by correlation of distances. Ann. Stat. **35**, 2769–2794 (2007)
22. Spackman, K.A.: SNOMED CT milestones: endorsements are added to already-impressive standards credentials. Healthcare Inf. Bus. Mag. Inf. Commun. Syst. **21**, 54–56 (2004)

Spatial Smoothing and Statistical Disclosure Control

Edwin de Jonge and Peter-Paul de Wolf[✉]

Statistics Netherlands, The Hague, The Netherlands
pp.dewolf@cbs.nl

Abstract. Cartographic maps have many practical uses and can be an attractive alternative for disseminating detailed frequency tables. However, a detailed map may disclose private data of individual units of a population. We will describe some smoothing algorithms to display spatial distribution patterns. In certain situations, the disclosure risk of a spatial distribution pattern, can be formulated in terms of a frequency table disclosure problem. In this paper we will explore the effects of spatial smoothing related to statistical disclosure control.

1 Introduction

Users of statistical data are often interested in spatial distribution patterns. A policy-maker may be interested in the distribution of income over the neighborhoods of a city, a health care professional may want to now where to find incidences of infections and a social worker may want to focus on locations that are at high risk for social problems. While these are all examples of valid and relevant needs, they are at odds with confidentiality. When a location is too detailed, e.g. an address, or when its neighborhood has few inhabitants, e.g. one isolated household, the displayed density or estimate at that location is very revealing. An useful 'safe' map is a trade-off between utility and disclosure risk: we want to maximize the utility of a map, while minimizing the disclosure risk for the persons or organizations it concerns.

In this paper we will explore the possibilities and challenges of using maps to disseminate statistical output. We will discuss possible smoothing techniques, their function as estimators of spatial distributions and the implications of these smoothing techniques on statistical confidentiality.

1.1 Maps for Dissemination

Statistical data are traditionally disseminated as tabular data, which have been aggregated, such as frequency tables or data cubes. Geographical data of this

---

The views expressed in this paper are those of the authors and do not necessarily reflect the policy of Statistics Netherlands.

The authors like to thank Ton de Waal for reviewing an earlier version of this paper.

J. Domingo-Ferrer and M. Pejić-Bach (Eds.): PSD 2016, LNCS 9867, pp. 107–117, 2016.
DOI: 10.1007/978-3-319-45381-1_9

kind, is typically aggregated for a specific administrative region such as a neighborhood, municipality, county, province or state. These data are of great value to assess and analyze the properties of these regions, but have a large drawback: spatial patterns in areal or regional data are distorted by the shape and size of the regions. Regions have irregular shapes and differ wildly in size. Often administrative regions are constructed in a specific way, e.g., to display some 'fixed' property (fire brigade areas), some 'fixed' number of addresses (postal code areas) or some 'fixed' ground area (provinces). This implies that different regions often have different densities concerning the units of interest, which affects the size and shape of those regions. E.g., to obtain a region with a 'fixed' number of people, densely populated regions tend to be small, whereas scarcely populated regions tend to be large. Since figures that NSIs publish are aggregations of all members of a population in a region, the size of the region has a large impact on density estimates. These properties of (administrative) regions make aggregated regional data and maps less suited for displaying and analyzing detailed spatial patterns.

Note that the use of administrative areas usually gives rise to the so called differencing problem. In case two types of areas are not properly nested, the intersection of these regions is a smaller (possibly non-administrative) region. Differencing the aggregated values belonging to the intersecting regions, may thus lead to information on that smaller region.

An alternative to aggregated data per region is publishing individual records with detailed locations. Publication of such so-called microdata is accurate but introduces other problems: issues concerning confidentiality and issues concerning spatial density estimation. Indeed, publishing records of individuals with confidential data and their detailed locations, poses a disclosure risk. To reduce that risk, the output will have to be modified somehow. On the other hand, detailed microdata does not reveal spatial patterns by itself. Plotting the locations on the map may reveal such patterns, but the resulting map may be too coarse, when the number of locations is small, or too dense, when the number of records is high. To derive and visualize the spatial distribution, it is preferred to estimate the density function itself.

A recent popular other dissemination method for geographic data is to publish spatial grid data. The geography is rastered in rectangular cells (a grid) and for each cell an aggregate is calculated. Gridding data can be seen as a discrete form of density estimation, since it is essentially a two dimensional histogram. By its nature, it can also be regarded as a form of data protection: individual data is being aggregated into larger groups thereby reducing the risk of disclosing individual information, although it might still lead to so-called group disclosure. Group disclosure is considered to be a problem when some sensitive information can be disclosed on a group of respondents, where that group can easily be identified (as a group).

Gridding data is a simple procedure, is very fast and thus seems a good candidate for estimating spatial distributions and at the same time protecting confidential data of individual units. However in practice it is difficult, if not impossible, to find a grid cell size that serves both goals: a detailed cell size discloses individuals, but a larger cell size removes spatial patterns. Another disadvantage of gridding is that for larger cell sizes it results in 'pixelized' maps, which amplifies the discrete nature of its density estimation procedure. These pixelized maps can be perceived by users as being 'broken'. For small cell sizes a gridded map quickly becomes noisy, also hiding 'larger' spatial patterns in the data.

We pose that a cartographic map displaying a spatial density is a dissemination medium in its own right. The analytic focus of a map is how density depends on location. A published map is neither a frequency table nor a set of records of individuals. Instead it shows a density per location, so the procedure to reduce disclosure risk needs to operate on locations instead of data cube cells or microdata records. A disclosure risk arises when the map is combined with a microdata set containing locations: the map may reveal an attribute of certain (groups of) individuals in the microdata set.

Note that our cartographic map will show a spatial density on the complete area the map is referring to. The map is not partitioned in different (sub)-areas. Hence, the problem of differencing is not apparent in our situation. Moreover, in case our map is combined with other maps, obtaining sensitive information by differencing distributions is much harder compared to differencing aggregated information on non-nested (sub)-areas.

1.2   Related Work

Displaying statistical information on a map, starts with the underlying microdata. One approach might be to protect the underlying microdata with the usual (non-spatial) methods and then plot the results on a map. Chapter 3 of [3] deals with disclosure control methods in case of general microdata sets. Standard methods to reduce the risk of disclosure in microdata sets include data swapping, noise addition, coarsening and top/bottom coding.

As stated earlier, geographical data often is first aggregated over an area, be it an administrative area or a grid square. That way the data can be viewed as aggregated (tabulated) data. Depending on the variable that is aggregated, this may constitute either frequency count tables or quantitative tables. Hence another approach might be to protect the tabulated data and then plot the results on a map. Standard methods to reduce the risk of disclosure in publishing tabulated data may be found in Chaps. 4 and 5 of [3] and include coarsening (redesigning the table), noise addition, rounding and suppression.

It thus may be tempting to first protect either the microdata or the tabulated data and then plot the results on a map. However, the standard methods usually do not take the spatial character of the data into account, which is suboptimal when the main utility is in that spatial distribution. In Chap. 3 of [4], a general framework for disclosure control of geographic population data is given. The proposed method is targeted at publishing information on a predefined spatial configuration, like a grid or a set of administrative areas.

Privacy related to geography is often seen in studies of health and crime data. See e.g., [1] for a nice overview of spatial methods related to health data. The methods mentioned in that paper are targeted at producing masked microdata sets to researchers, where the individual data can no longer be linked with identifiable persons.

In the world of official statistics, an attempt to relate geographically important census data to disclosure risk was described in [8].

In [5] a method is proposed to deal with spatial data that may include a mixture of points, lines and/or areas. The ideas mentioned in that paper make use of gridding: aggregating data over a grid and then representing the resulting data using a heatmap approach. The disclosure control part of the problem is dealt with at the level of the grid cells: the values of those grid cells are considered to be tabular data and sensitivity measures for tabular data are used to define 'unsafe' grid cells. Whenever a grid cell is considered 'unsafe', it may be suppressed or combined with neighboring grid cells to form a (locally) larger grid cell. This can be seen as a rudimentary way of smoothing.

In [9] another approach to asses the disclosure risk is mentioned. In that paper the original values in the underlying microdata are, per record, replaced by the values that result from the smoothing. Essentially, the value in each record is replaced by the value of the smoothed spatial estimate at the geographical location of the individual of the record in question. The effect on the disclosure risk of this smoothing is then measured using the microdata set with those smoothed values.

Smoothing aggregates neighboring values to a new value. It can be seen as a combination of coarsening (aggregation) and data collapsing (borrowing information from other records). It should be noted that above actions typically are executed upon individual records.

Most of the methods referred to above, either aim at producing a protected microdata set or at producing a plot on a (predefined) set of areas, like a grid or a set of administrative areas. In contrast this paper sees a cartographic density map as a publication in its own right. Therefore we try to protect the output directly, not the microdata that serves as input. Moreover, our cartographic density map is an attempt to estimate a spatially continuous version of the density, covering the complete map and not restricted to a specific partitioning of the total map-area.

2   Method

The goal of the method is to publish a spatial estimate of the relative frequencies of a variable with at least one sensitive category. Dichotomous variables (having only two categories) often have asymmetric sensitivity: being unemployed is considered sensitive, being employed is not; having social benefits is considered sensitive, not having social benefits is not. We will call a variable that has at least one sensitive category a *sensitive variable*.

When looking at the spatial distribution of a sensitive variable, a certain location could reveal sensitive information about individuals that are somehow 'linked' to it. Examples are the income of a person living in a certain house or the ethnicity of a group of people that live in a certain neighborhood or the emission of toxic gases by a certain factory. We will define a *sensitive location* to be a location where sensitive information can be derived about individual units that can be linked to that location.

A plot of a spatial distribution on its own usually does not easily reveal information on individual units. The minimum additional information needed to use spatial distributions for disclosing information about individual units, is that an attacker knows which units are linked to a particular location. We thus consider the attacker scenario where the attacker has information on the identities of units linked to specific locations. The attacker may then derive the probability of occurrence of a certain attribute for identified individuals.

In the remainder of this section we describe the procedure for dichotomous variables with one sensitive category. The method however can be extended to categorical variables with multiple sensitive categories.

As mentioned in the discussion above and in [5] as well, smoothing the data is an estimation approach (estimating a spatial distribution) that has some disclosure protection build in. Our smooth spatial density estimate in case of a dichotomous variable with one sensitive category, can be considered to represent relative frequencies at spatial locations. We will thus make use of some suggested rules related to frequency count tables (see e.g., Chap. 5 of [3]):

- There should be at least k_0 observations related to an identifiable group of individuals
- The frequency on the sensitive category, relative to the size of the identifiable group of individuals, should not be larger than f_{max}

The first rule is related to the fact that when a small group of individuals (say a single household, or a group of two people) can be identified at a location, not many different relative frequencies are possible. It thus becomes easy to derive information on a small group of individuals. The threshold is usually taken to be an absolute number, like in k-anonymity.

The second rule is related to so called group disclosure. In that case an attribute can be derived that holds for a group of individuals at a certain location. The size of the group of individuals should not constitute a too large portion of the total number of individuals at that location, even though that total number may be large. For example knowing that a certain location has 100 % unemployment reveals information on each individual at that location, without actually identifying any individual. Group disclosure can be considered to be similar to the type of attack l-diversity tries to prevent.

Taking those rules into account, we end up with the following approach:

1. Determine a minimum number of observations k_0. Locations with less then k_0 nearby neighbors are called *rare sensitive* locations.
2. Determine a maximum allowed relative frequency $f_{max} \leq 1$. Locations where $\hat{f}(x, y) > f_{max}$ are called *group sensitive* locations.

3. For sensitive variable v, construct a spatial density estimate $\hat{f}(x, y)$ of the relative frequency. Note that the estimation may involve one or more tuning parameters α_i and includes a density estimation $\hat{d}(x, y)$ of the total population to be able to get a *relative* frequency.

4. Discretize $\hat{f}(x, y)$ in at most 5 levels that will correspond to different colors on the map. The maximum of 5 levels is a generally accepted convention in data-visualization. The levels should partition the interval $[0, f_{\max}]$. Assign group sensitive locations with $\hat{f}(x, y) > f_{\max}$ to the top level. Assign rare sensitive locations to the bottom level.

5. Repeat 3 and 4, for different choices of estimation parameters α_i and choose the parameter that fits well with the data to be displayed.

We discuss the steps of the approach further in the following sections.

2.1   Spatial Estimation

This paper makes no assumptions on the spatial distribution function and describes two non-parametric techniques to estimate $\hat{f}(x, y)$. However if more detailed knowledge on d and v is available their distributions could alternatively be modeled using a parametric method.

Kernel density estimation. A commonly used non-parametric method is Kernel Density Estimation (KDE), which is often applied in bivariate density estimation. In KDE each data point (x_i, y_i) is 'smeared' out over a local neighborhood with a kernel K_α which size depends on tuning parameter α. The distribution function $\hat{g}_\alpha(x, y)$ is then the sum of all smoothed data points:

$$\hat{g}_\alpha(x, y) = \sum_{i=1}^{n} K_\alpha\left(x, y, x_i, y_i\right)$$

with $\hat{g}_\alpha(x, y)$ the estimated spatial function.

In a fixed kernel density estimator α is the *bandwidth* h, which defines the size of the local neighborhood for each point (x_i, y_i)

$$\hat{g}_h(x, y) = \sum_{i=1}^{n} K\left(\frac{x - x_i}{h}, \frac{y - y_i}{h}\right)$$

Typical kernel functions used are Gaussian and Epanechnikov, but can be any non-negative function that integrates to 1 and has mean of 0. In many cases different kernels will produce very similar estimates for equivalent bandwidths. The more influential parameter is the bandwidth h. For simplicity we assume that h is equal in x and y direction. A practical advantage of spatial KDE is that its bandwidth is a physical dimension: it blurs sensitive locations, e.g. a bandwidth of 500 m, makes locating sensitive variables within 100 m difficult.

Estimating the relative frequency $\hat{f}_h(x,y)$, involves two KDE estimates: the spatial distribution $\hat{v}_h(x,y)$ of sensitive category c and the total population distribution $\hat{d}_h(x,y)$. Since the units for which $v = c$ are a subset of total population d and both estimations use the same kernel and bandwidth, it holds that $\hat{v}_h(x,y) \leq \hat{d}_h(x,y)$.

The estimated spatial fraction is defined as:

$$\hat{f}_h(x,y) = \begin{cases} \frac{\hat{v}_h(x,y)}{\hat{d}_h(x,y)} \in [0,1] \\ \text{undefined} \qquad\qquad \text{if } \hat{d}_h(x,y) = 0 \end{cases}$$

Locations for which $\hat{f}_h(x,y)$ is undefined, will be displayed transparently in the resulting map, since a fraction of an empty population has no practical meaning.

The disclosure risk for the fixed bandwidth kernel estimator depends on h in the following different aspects:

– A small bandwidth may produce many group sensitive locations (x_g, y_g) with $\hat{f}_h(x_g, y_g) > f_{\max}$, disclosing groups of individuals.
– A small bandwidth h may create rare sensitive locations (x_r, y_r) for which $\hat{d}_h(x_r, y_r)$ is small, meaning that the neighborhood h contains less then k_0 observations. These locations may disclose individuals, even for $0 < \hat{f}_h(x_r, y_r) < f_{\max}$. From a spatial estimation point of view, the estimation error in these locations is large.

Concerning disclosure risk, it might seem wise to choose a large bandwidth but this will decrease the utility of the map because spatial patterns may be smoothed out too much. It should be noted that a (too) small h increases disclosure risk, but may decrease utility as well. A (too) small bandwidth leads to noisy pictures in which the spatial pattern may not be too apparent. Choosing an appropriate bandwidth h thus reflects the utility-risk trade-off, but not in a linear way.

A good bandwidth strongly depends on the true population distribution $d(x,y)$. It is therefore advisable to try various increasing values of bandwidths. This is comparable to [2] who calculates the masking at multiple scales.

Alternatively, one could use an adaptive bandwidth: a bandwidth depending on the underlying spatial distribution of the population. This could e.g., lead to setting h larger in scarcely populated areas and smaller in more densely populated areas. An implementation of an adaptive bandwidth estimator needs further investigation.

Nearest neighbor estimation. Instead of using a fixed bandwidth density estimator and first calculate $\hat{v}(x,y)$ and $\hat{d}(x,y)$, we can estimate the relative frequency $\hat{f}_k(x,y)$ directly by calculating the fraction of k nearest neighbors (k-nn) which belong to sensitive category c:

$$\hat{f}_k(x,y) = \frac{1}{k} \sum_{i \in \text{k-nn}(x,y)} \delta_{v_i,c}$$

in which $\delta_{v_i,c}$ is the Kronecker delta function.

Note that choosing $k > k_0$ fulfills the rareness condition automatically. The problem of finding a good bandwidth is now changed into finding a good k, the number of neighbors. When chosen too small, densely populated locations are very pronounced in the data set. More importantly a small k may give many group sensitive locations with $\hat{f}_k(x, y) > f_{max}$. A large k will remove spatial patterns, especially for low density locations, because their neighbors are far away. It is advised to test the map for various values of k. [6,7] use a similar technique (equipop) to estimate detailed census block or neighborhood statistics. The resulting spatial estimation can be seen as a special case of an adaptive bandwidth estimator. Each location (x, y) has a different sized neighborhood based on the local spatial density. Due to the nearest neighbor procedure, the resulting spatial density contains discrete transitions: further investigation is needed to evaluate the usability of this procedure and its relation to both risk and utility assessment.

2.2 Discretization

To display the estimated spatial distribution $\hat{f}(x, y)$, we suggest to make use of a discretized version of that estimator. This serves several goals. Firstly it connects each interval or level to a color in the resulting map. A small set of gradual colors helps users to focus on important locations and sudden changes in relative frequency. Secondly, it naturally creates a top and bottom level which can be used for top and bottom coding sensitive locations. Without discretization a separate procedure would be needed to create a top and bottom threshold. While smoothing may reduce the relative frequency of a sensitive location (such that $\hat{f}(x, y)$ is below f_{max}), the resulting map may still suggest where the exact sensitive location was, because the relative frequency is smeared out gradually in the neighborhood and the color gradient points to the exact location. Discretization blurs the exact location, by grouping high frequencies in one level.

The discretization is on the interval $[0, f_{max}]$, so *not* on $[0, 1]$, to assure that the top level does not consist of only group sensitive locations. Otherwise, the map would highlight the sensitive locations being the opposite of what we try to achieve.

Note that if $\max(\hat{f}(x, y)) < f_{max}$, the discretization interval can be further reduced to $[0, \max(\hat{f}(x, y))]$.

2.3 Example

Figure 1 shows two maps of a synthetic spatial dataset using a fixed kernel density estimator with two different bandwidths: $h = 50$ m and $h = 100$ m. Moreover, we used $f_{max} = 0.9$ and the number of levels $L = 5$. Clearly visible is the difference in granularity: with a bandwidth of 50 m, the map shows many local modes and seems noisy. With a bandwidth of 100 m the patterns are more smooth. Note that since $\max(\hat{f}_{100m}(x, y)) < \max(\hat{f}_{50m}(x, y))$ the (color) levels in both maps do not represent the same intervals.

Fig. 1. KDE estimates of relative frequency for $h = 50\,\mathrm{m}$ (upper) and $h = 100\,\mathrm{m}$ (lower)

3 Discussion

Smoothing has several attractive features: it is a well known technique for univariate and bivariate data. Smoothed maps may show distinct but smooth spatial patterns, which can be controlled with one or more parameters. Its utility is in showing spatial patterns, which are of interest to (local) policy makers. In this paper we have concentrated on smooth estimation of relative frequencies. We think that such estimators are especially of interest to policy makers.

Since smoothing has some intrinsic capacity of statistical disclosure control, we have taken the view that smoothing may be a good way to balance the availability of spatial information while keeping the disclosure risks at an acceptable level.

The current paper has only an exploratory character and hence we end with several issues that may be interesting for further research. We have grouped those issues into three categories: utility related, disclosure related and anomalies.

▶ **Utility related**

- Smoothing can sometimes 'over-smooth' and hide or remove a spatial pattern unwantedly. This obviously depends on the true data distribution and the size of the smoothing parameter. For the moment, choosing the right value that properly reveals a pattern remains a human task. Automatic bandwidth selection for spatial data would however be an interesting path to investigate.

- A restriction parameter for the nearest neighbor estimator on the radius h_{max} in which neighbors are to be found, may prevent locations having long-distance nearest neighbors. Without the restriction, the map could show positive relative frequencies in areas where there exists no population.

▶ **Disclosure related**

- In this paper, we have borrowed some basic disclosure risk rules from the theory concerning frequency count tables to define the disclosure risk for our smooth estimate of the relative frequency. Additional research is needed to find a more general approach to assess the disclosure risk when disseminating spatial plots.

- When disseminating two maps of the relative frequencies of category c and c' (being the opposite of having c), it becomes obvious that group disclosure as defined by $\hat{f}(x,y) > f_{max}$ in case of category c is connected with a small value of $\hat{f}(x,y)$ for category c'. This is comparable to the situation in frequency count tables where group disclosure corresponds to too much concentration in category c and hence to too little observations in category c'. For frequency count tables it is common practice to apply a disclosure control method to both cells. Hence, we would need to take action in both released maps as well.

▶ **Anomalies**

- Spatial estimation smears out points to neighboring locations. This may introduce *false density*: locations which in reality have no density, but are near a dense location, now can have a probability mass assigned larger than 0. This is most apparent when the leakage is into areas like rivers, parks, lakes and woods, i.e., areas that are known to be unpopulated. In those cases, the map should be accompanied with a short description on what the map displays, so the user will not start to doubt the overall utility of the map. Alternatively the method could exclude those locations: e.g. in one-dimensional kernel smoothing sometimes boundary kernels are used. If and how that can be used in case of spatial smoothing, needs further investigation: what would be the impact on the smoothing algorithm and what new artifacts may be introduced?

- Smoothing may introduce *false sensitive locations*. This is a special case of the previous remark: a location which is not sensitive for a small bandwidth may have many sensitive neighboring locations. Combining that information by smoothing with a (larger) bandwidth would make the non-sensitive location seem sensitive. It has to be investigated how to deal with this.

- Spatial estimation may introduce *dislocated modes*: two modes that are located closely to each other can be blended into a single mode at an in-between location. The resulting map suggests that the in-between location is important, while in reality two (or more) nearby locations are of more interest. The blending of modes depends on the smoothing parameter. The larger the windows of smoothing, the likelier it is to happen. Research is needed how this influences the usefulness of such maps.

The preliminary results for producing smooth maps that protect sensitive variables are promising: a map with a sensitive variable can be made safe, while showing spatial patterns. However, further research is needed to improve the results. Summarizing and adding to the above mentioned issues, interesting tasks include the derivation of a disclosure risk function for smooth maps, producing maps with sensitive continuous variables, automatic bandwidth selection methods and deriving utility measures for spatial distribution plots.

References

1. Armstrong, M.P., Rushton, G., Zimmerman, D.L.: Geographically masking health data to preserve confidentiality. Stat. Med. **18**, 497–525 (1999)
2. Clarke, K.C.: A multiscale masking method for point geographic data. Int. J. Geogr. Inf. Sci. **30**(2), 300–315 (2016)
3. Hundepool, A., Domingo-Ferrer, J., Franconi, L., Giessing, S., Schulte Nordholt, E., Spicer, K., de Wolf, P.P.: Statistical Disclosure Control. Wiley Series in Survey Methodology. John Wiley & Sons, Ltd., Hoboken (2012). ISBN: 978-1-119-97815-2
4. Markkula, J.: Geographic personal data, its privacy protection and prospects in a location-based service environment. Ph.d. thesis, University of Jyväskylä (2003). https://jyx.jyu.fi/dspace/handle/123456789/13227
5. O'Keefe, C.M.: Confidentialising maps of mixed point and diffuse spatial data. In: Domingo-Ferrer, J., Tinnirello, I. (eds.) PSD 2012. LNCS, vol. 7556, pp. 226–240. Springer, Heidelberg (2012)
6. Östh, J., Malmberg, B., Anderson, E.: Introducing equipop. In: 6th International Conference on Population Geographies (2011)
7. Östh, J., Clark, W.A., Malmberg, B.: Measuring the scale of segregation using k-nearest neighbor aggregates. Geogr. Anal. **47**(1), 34–49 (2015)
8. Young, C., Martin, D., Skinner, C.: Geographically intelligent disclosure control for flexible aggregation of census data. Int. J. Geogr. Inf. Sci. **23**(4), 457–482 (2009)
9. Zhou, Y., Dominici, F., Louis, T.A.: A smoothing approach for masking spatial data. Ann. Appl. Stat. **4**(3), 1451–1475 (2010). doi:10.1214/09-AOAS325

Protection Using Privacy Models

On-Average KL-Privacy and Its Equivalence to Generalization for Max-Entropy Mechanisms

Yu-Xiang Wang[1,2]([✉]), Jing Lei[2], and Stephen E. Fienberg[1,2,3]

[1] Machine Learning Department, Carnegie Mellon University,
Pittsburgh, PA 15213, USA
[2] Department of Statistics, Carnegie Mellon University, Pittsburgh, PA 15213, USA
yuxiangw@cs.cmu.edu
[3] Heinz College, Carnegie Mellon University, Pittsburgh, PA 15213, USA

Abstract. We define On-Average KL-Privacy and present its properties and connections to differential privacy, generalization and information-theoretic quantities including max-information and mutual information. The new definition significantly weakens differential privacy, while preserving its minimal design features such as composition over small group and multiple queries as well as closeness to post-processing. Moreover, we show that On-Average KL-Privacy is *equivalent* to generalization for a large class of commonly-used tools in statistics and machine learning that samples from Gibbs distributions—a class of distributions that arises naturally from the maximum entropy principle. In addition, a byproduct of our analysis yields a lower bound for generalization error in terms of mutual information which reveals an interesting interplay with known upper bounds that use the same quantity.

Keywords: Differential privacy · Generalization · Stability · Information theory · Maximum entropy · Statistical learning theory

1 Introduction

Increasing privacy concerns have become a major obstacle for collecting, analyzing and sharing data, as well as communicating results of a data analysis in sensitive domains. For example, the second Netflix Prize competition was canceled in response to a lawsuit and Federal Trade Commission privacy concerns, and the National Institute of Health decided in August 2008 to remove aggregate Genome-Wide Association Studies (GWAS) data from the public website, after learning about a potential privacy risk. These concerns are well-grounded in the context of the Big-Data era as stories about privacy breaches from improperly-handled data set appear very regularly (e.g., medical records [3], Netflix [27], NYC Taxi [37]). These incidences highlight the need for formal methods that provably protect the privacy of individual-level data points while allowing similar database level of utility comparing to the non-private counterpart.

There is a long history of attempts to address these problems and the risk-utility tradeoff in statistical agencies [8,9,19] but most of the methods developed

© Springer International Publishing Switzerland 2016
J. Domingo-Ferrer and M. Pejić-Bach (Eds.): PSD 2016, LNCS 9867, pp. 121–134, 2016.
DOI: 10.1007/978-3-319-45381-1_10

do not provide clear and quantifiable privacy guarantees. Differential privacy [10,14] succeeds in the first task. While it allows a clear quantification of the privacy loss, it provides a worst-case guarantee and in practice it often requires adding noise with a very large magnitude (if finite at all), hence resulting in unsatisfactory utility, cf., [16,32,38].

A growing literature focuses on weakening the notion of differential privacy to make it applicable and for a more favorable privacy-utility trade-off. Popular attempts include (ϵ, δ)-approximate differential privacy [13], personalized differential privacy [15,22], random differential privacy [17] and so on. They each have pros and cons and are useful in their specific contexts. There is a related literature addressing the folklore observation that "differential privacy implies generalization" [5,11,12,18,30,35].

The implication of generalization is a minimal property that we feel any notion of privacy should have. This brings us to the natural question:

– Is there a weak notion of privacy that is equivalent to generalization?

In this paper, we provide a partial answer to this question. Specifically, we define On-Average Kullback-Leibler(KL)-Privacy and show that it characterizes On-Average Generalization[1] for algorithms that draw sample from an important class of maximum entropy/Gibbs distributions, i.e., distributions with probability/density proportional to

$$\exp(-\mathcal{L}(\text{Output}, \text{Data}))\pi(\text{Output})$$

for a loss function $\mathcal{L}$ and (possibly improper) prior distribution π.

We argue that this is a fundamental class of algorithms that covers a big portion of tools in modern data analysis including Bayesian inference, empirical risk minimization in statistical learning as well as the private releases of database queries through Laplace and Gaussian noise adding. From here onwards, we will refer this class of distributions "MaxEnt distributions" and the algorithm that output a sample from a MaxEnt distribution "posterior sampling".

Related Work: This work is closely related to the various notions of algorithmic stability in learning theory [7,21,26,29]. In fact, we can treat differential privacy as a very strong notion of stability. Thus On-average KL-privacy may well be called On-average KL-stability. Stability implies generalization in many settings but they are often only sufficient conditions. Exceptions include [26,29] who show that notions of stability are also necessary for the consistency of empirical risk minimization and distribution-free learnability of any algorithms. Our specific stability definition, its equivalence to generalization and its properties as a privacy measure has not been studied before. KL-Privacy first appears in [4] and is shown to imply generalization in [5]. On-Average KL-privacy further weakens KL-privacy. A high-level connection can be made to leave-one-out cross validation which is often used as a (slightly biased) empirical measure of generalization, e.g., see [25].

---

[1] We will formally define these quantities.

2 Symbols and Notation

We will use the standard statistical learning terminology where $z \in \mathcal{Z}$ is a data point, $h \in \mathcal{H}$ is a hypothesis and $\ell : \mathcal{Z} \times \mathcal{H} \to \mathbb{R}$ is the loss function. One can think of the negative loss function as a measure of utility of h on data point z. Lastly, $\mathcal{A}$ is a possibly randomized algorithm that maps a data set $Z \in \mathcal{Z}^n$ to some hypothesis $h \in \mathcal{H}$. For example, if $\mathcal{A}$ is the empirical risk minimization (ERM), then $\mathcal{A}$ chooses $h^* = \mathrm{argmin}_{h \in \mathcal{H}} \sum_{i=1}^{n} \ell(z_i, h)$.

Just to point out that many data analysis tasks can be cast into this form, e.g., in linear regression, $z_i = (x_i, y_i) \in \mathbb{R}^d \times \mathbb{R}$, h is the coefficient vector and ℓ is just $\|y_i - x_i^T h\|^2$; in k-means clustering, $z \in \mathbb{R}^d$ is just the feature vector, $h = \{h_1, ..., h_k\}$ is the collection of k-cluster centers and $\ell(z, h) = \min_j \|z - h_j\|^2$. Simple calculations of statistical quantities can often be represented in this form too, e.g., calculating the mean is equivalent to linear regression with identity design, and calculating the median is the same as ERM with loss function $|z - h|$.

We also consider cases when the loss function is defined over the whole data set $Z \in \mathcal{Z}$, in this case the loss function is also evaluated on the whole data set by the structured loss $\mathcal{L} : h \times \mathcal{Z} \to \mathbb{R}$. Generally speaking, Z could be a string of text, a news article, a sequence of transactions of a credit card user, or rather just the entire data set of n iid samples. We will revisit the structured loss variant with more concrete examples later. However we would like to point out that this is equivalent to the above case when we only have one (much more complicated) data point and the algorithm $\mathcal{A}$ is applied to only one sample.

3 Main Results

We first describe differential privacy and then it will become very intuitive where KL-privacy and On-Average KL-privacy come from. Roughly speaking, differential privacy requires that for any data sets Z and Z' that differs by only one data point, the algorithm $\mathcal{A}(Z)$ and $\mathcal{A}(Z')$ samples output h from two distributions that are very similar to each other. Define "Hamming distance"

$$d(Z, Z') := \#\{i = 1, ..., n : z_i \neq z'_i\}. \tag{1}$$

Definition 1 (ϵ-Differential Privacy [10]**).** *We call an algorithm $\mathcal{A}$ ϵ-differentially private (or in short ϵ-DP), if*

$$\mathbb{P}(\mathcal{A}(Z) \in H) \leq \exp(\epsilon)\mathbb{P}(\mathcal{A}(Z') \in H)$$

for $\forall$ Z, Z' obeying $d(Z, Z') = 1$ and any measurable subset $H \subseteq \mathcal{H}$.

More transparently, assuming the range of $\mathcal{A}$ is the whole space $\mathcal{H}$, and also assume $\mathcal{A}(Z)$ defines a density on $\mathcal{H}$ with respect to a base measure on $\mathcal{H}^2$, then ϵ-Differential Privacy requires

---

[2] These assumptions are only for presentation simplicity. The notion of On-Average KL-privacy can naturally handle mixture of densities and point masses.

$$\sup_{Z,Z':d(Z,Z')\leq 1}\sup_{h\in\mathcal{H}}\log\frac{p_{\mathcal{A}(Z)}(h)}{p_{\mathcal{A}(Z')}(h)}\leq\epsilon.$$

Replacing the second supremum with an expectation over $h\sim\mathcal{A}(Z)$ we get the maximum KL-divergence over the output from two adjacent data sets. This is KL-Privacy as defined in Barber and Duchi [4], and by replacing both suprema with expectations we get what we call On-Average KL-Privacy. For $Z\in\mathcal{Z}^n$ and $z\in\mathcal{Z}$, denote $[Z_{-1},z]\in\mathcal{Z}^n$ the data set obtained from replacing the first entry of Z by z. Also recall that the KL-divergence between two distributions F and G is $D_{\mathrm{KL}}(F\|G)=\mathbb{E}_F\frac{dF}{dG}$.

Definition 2 (On-Average KL-Privacy). *We say $\mathcal{A}$ obeys ϵ-On-Average KL-privacy for some distribution $\mathcal{D}$ if*

$$\mathbb{E}_{Z\sim\mathcal{D}^n,z\sim\mathcal{D}}D_{\mathrm{KL}}(\mathcal{A}(Z)\|\mathcal{A}([Z_{-1},z]))\leq\epsilon.$$

Note that by the property of KL-divergence, the On-Average KL-Privacy is always nonnegative and is 0 if and only if the two distributions are the same almost everywhere. In the above case, it happens when $z=z'$.

Unlike differential privacy that provides a uniform privacy guarantee for any users in $\mathcal{Z}$, on-average KL-Privacy is a distribution-specific quantity that measures the amount of average privacy loss of an average data point $z\sim\mathcal{D}$ suffer from running data analysis $\mathcal{A}$ on a data set Z drawn iid from the same distribution $\mathcal{D}$.

We argue that this kind of average privacy guarantee is practically useful because it is specified for each data distribution separately, therefore is able to adapt to easy problems by revealing the smaller privacy loss. It is also much less sensitive to outliers. After all, when differential privacy fails to provide a sufficiently small ϵ (e.g., $\epsilon<1$) due to peculiar data sets that exist in $\mathcal{Z}^n$ but rarely appear in practice, we would still be interested to gauge how a randomized algorithm protects a typical user's privacy.

Now we define what we mean by *generalization*. Let the empirical risk $\hat{R}(h,Z)=\frac{1}{n}\sum_{i=1}^n\ell(h,z_i)$ and the actual risk be $R(h)=\mathbb{E}_{z\sim\mathcal{D}}\ell(h,z)$.

Definition 3 (On-Average Generalization). *We say an algorithm $\mathcal{A}$ has on-average generalization error ϵ if $\left|\mathbb{E}R(\mathcal{A}(Z))-\mathbb{E}\hat{R}(\mathcal{A}(Z),Z)\right|\leq\epsilon$.*

This is weaker than the standard notion of generalization in machine learning which requires $\mathbb{E}|R(\mathcal{A}(Z))-\hat{R}(\mathcal{A}(Z),Z)|\leq\epsilon$. Nevertheless, on-average generalization is sufficient for the purpose of proving consistency for methods that approximately minimize the empirical risk.

3.1 The Equivalence to Generalization

It turns out that when $\mathcal{A}$ assumes a special form, that is, sampling from a Gibbs distribution, we can completely characterize generalization of $\mathcal{A}$ using On-Average KL-Privacy. This class of algorithms include the most general mechanism for differential privacy — exponential mechanism [23], which casts many other noise adding procedures as special cases. We will discuss a more compelling reason why restricting our attention to this class is not limiting in Sect. 3.3.

Theorem 4 (On-Average KL-Privacy $\Leftrightarrow$ Generalization). *Given an arbitrary loss function $\ell(z, h)$ (for example, it can be $-\log q(z|h)$ for some probabilistic model q parameterized by h) and let*

$$\mathcal{A}(Z) : h \sim p(h|Z) \propto \exp\left(-\sum_{i=1}^{n} \ell(z_i, h) - r(h) \right).$$

If in addition $\mathcal{A}(Z)$ obeys that for every Z, the distribution $p(h|Z)$ is well-defined (in that the normalization constant is finite), then $\mathcal{A}$ satisfies ϵ-On-Average KL-Privacy if and only if $\mathcal{A}$ has on-average generalization error ϵ.

The proof, given in the Appendix, uses a ghost sample trick and the fact that the expected normalization constants of the sampling distribution over Z and Z' are the same.

Remark 1 (Structural Loss). Take $n = 1$, and loss function be $\mathcal{L}$. Then for an algorithm $\mathcal{A}$ that samples with probability proportional to $\exp(-\mathcal{L}(h, Z) - r(h))$: ϵ-On-Average KL-Privacy is equivalent to ϵ-generalization of the structural loss, i.e., it implies that $|\mathbb{E}\mathcal{L}(Z, \mathcal{A}(Z)) - \mathbb{E}\mathcal{L}(Z', \mathcal{A}(Z))| \leq \epsilon$.

Remark 2 (Dispersion parameter γ). The case when $\mathcal{A} \propto \exp(-\gamma[\mathcal{L}(h, Z) - r(h)])$ for a constant γ can be handled by redefining $\mathcal{L}' = \gamma\mathcal{L}$. In that case, ϵ_γ-On-Average KL-Privacy with respect to $\mathcal{L}'$ implies ϵ_γ/γ generalization with respect to $\mathcal{L}$. For this reason, larger γ may not imply strictly better generalization.

Remark 3 (Comparing to differential privacy). Note that here we do not require ℓ to be uniformly bounded, but if we do, i.e. $\sup_{z \in \mathcal{Z}, h \in \mathcal{H}} |\ell(z, h)| \leq B$, then the same algorithm $\mathcal{A}$ above obeys $4B\gamma$-Differential Privacy [23,34] and it implies $O(B\gamma)$-generalization. This, however, could be much larger than the actual generalization error (see our examples in Sect. 6).

3.2 Preservation of Other Properties of DP

We now show that despite being much weaker than DP, On-Average KL-privacy inherits some major properties of differential privacy (under mild additional assumptions in some cases).

Lemma 5 (Closeness to Post-processing). *Let f be any (possibly randomized) measurable function from $\mathcal{H}$ to another domain $\mathcal{H}'$, then for any Z, Z'*

$$D_{\mathrm{KL}}(f(\mathcal{A}(Z))\|f(\mathcal{A}(Z'))) \leq D_{\mathrm{KL}}(\mathcal{A}(Z)\|\mathcal{A}(Z')).$$

Proof. This directly follows from the data processing inequality for the Rényi divergence in Van Erven and Harremoës [33, Theorem 1].

Lemma 6 (Small group privacy). *Let $k \leq n$. Assume $\mathcal{A}$ is posterior sampling as in Theorem 4. Then for any $k = 1, ..., n$, we have*

$$\mathbb{E}_{[Z, z_{1:k}] \sim \mathcal{D}^{n+k}} D_{\mathrm{KL}} \left(\mathcal{A}(Z) \| \mathcal{A}([Z_{-1:k}, z_{1:k}]) \right) = k \mathbb{E}_{[Z, z] \sim \mathcal{D}^{n+1}} D_{\mathrm{KL}} \left(\mathcal{A}(Z) \| \mathcal{A}([Z_{-1}, z]) \right).$$

Lemma 7 (Adaptive Composition Theorem). *Let $\mathcal{A}$ be ϵ_1-(On-Average) KL-Privacy and $\mathcal{B}(\cdot, h)$ be ϵ_2-(On-Average) KL-Privacy for every $h \in \Omega_{\mathcal{A}}$ where the support of random function $\mathcal{A}$ is $\Omega_{\mathcal{A}}$. Then $(\mathcal{A}, \mathcal{B})$ jointly is $(\epsilon_1 + \epsilon_2)$-(On-Average) KL-Privacy.*

Due to space constraint, the proofs of Lemmas 6 and 7 (as well as other unproven results) are presented in the full version of this paper [36].

3.3   Posterior Sampling as Max-Entropy Solutions

In this section, we give a few theoretical justifications why restricting to posterior sampling is not limiting the applicability of Theorem 4 much. First of all, Laplace, Gaussian and Exponential Mechanism in the Differential Privacy literature are special cases of this class. Secondly, among all distributions to sample from, the Gibbs distribution is the variational solution that simultaneously maximizes the conditional entropy and utility. To be more precise on the claim, we first define conditional entropy.

Definition 8 (Conditional Entropy). *Conditional entropy*

$$H(\mathcal{A}(Z)|Z) = -\mathbb{E}_Z \mathbb{E}_{h \sim \mathcal{A}(Z)} \log p(h|Z)$$

where $\mathcal{A}(Z) \sim p(h|Z)$.

Theorem 9. *Let $Z \sim \mathcal{D}^n$ for any distribution $\mathcal{D}$. A variational solution to the following convex optimization problem*

$$\min_{\mathcal{A}} \quad -\frac{1}{\gamma} \mathbb{E}_{Z \sim \mathcal{D}^n} H(\mathcal{A}(Z)|Z) + \mathbb{E}_{Z \sim \mathcal{D}^n} \mathbb{E}_{h \sim \mathcal{A}(Z)} \sum_{i=1}^{n} \ell_i(h, z_i) \tag{2}$$

is $\mathcal{A}$ that outputs h with distribution $p(h|Z) \propto \exp\left(-\gamma \sum_{i=1}^{n} \ell_i(h, z_i)\right)$.

Proof. This is an instance of Theorem 3 in Mir [24] (first appeared in Tishby et al. [31]) by taking the distortion function to be the empirical risk. Note that this is a simple convex optimization over the functions and the proof involves substituting the solution into the optimality condition with a specific Lagrange multiplier chosen to appropriately adjust the normalization constant.   □

The above theorem is distribution-free, and in fact works for every instance of Z separately. Condition on each Z, the variational optimization finds the distribution with maximum information entropy among all distributions that satisfies a set of utility constraints. This corresponds to the well-known principle of maximum entropy (MaxEnt) [20]. Many philosophical justifications of this principle has been proposed, but we would like to focus on the statistical perspective and treat it as a form of regularization that penalizes the complexity of the chosen distribution (akin to Akaike Information Criterion [1]), hence avoiding overfitting to the data. For more information, we refer the readers to references on MaxEnt's connections to thermal dynamics [20], to Bayesian inference and convex duality [2] as well as its modern use in modeling natural languages [6].

Note that the above characterization also allows for any form of prior $\pi(h)$ to be assumed. Denote prior entropy $\tilde{H}(h) = -\mathbb{E}_{h \sim \pi(h)} \log(\pi(h))$, and define information gain $\tilde{I}(h; Z) = \tilde{H}(h) - H(h|Z)$. The variational solution of $p(h|Z)$ that minimizes $\tilde{I}(h; Z) + \gamma \mathbb{E}_Z \mathbb{E}_{h|Z} \mathcal{L}(Z, h)$ is proportional to $\exp(-\mathcal{L}(Z, h))\pi(h)$. This provides an alternative way of seeing the class of algorithms $\mathcal{A}$ that we consider. $\tilde{I}(h; Z)$ can be thought of as a information-theoretic quantification of privacy loss, as described in [24,39]. As a result, we can think of the class of $\mathcal{A}$ that samples from MaxEnt distributions as the most private algorithm among all algorithms that achieves a given utility constraint.

4   Max-Information and Mutual Information

Recently, Dwork et al. [11] defined approximate max-information and used it as a tool to prove generalization (with high probability). Russo and Zou [28] showed that the weaker mutual information implies on-average generalization under a distribution assumption of the entire space of $\{\mathcal{L}(h, Z)|h \in \mathcal{H}\}$ induced by distribution of Z. In this section, we compare On-Average KL-Privacy with these two notions. Note that we will use Z and Z' to denote two completely different data sets rather than adjacent ones as we had in differential privacy.

Definition 10 (Max-Information, Definition 11 in [11]). *We say $\mathcal{A}$ has an β-approximate max-information of k if for every distribution $\mathcal{D}$,*

$$I_\infty^\beta(Z; \mathcal{A}(Z)) = \max_{(H,Z) \subset \mathcal{H} \times \mathcal{Z}^n : \mathbb{P}(h \in H, Z \in \tilde{Z}) > \beta} \log \frac{\mathbb{P}(h \in H, Z \in \tilde{Z}) - \beta}{\mathbb{P}(h \in H)p(Z \in \tilde{Z})} \leq k.$$

This is alternatively denoted by $I_\infty^\beta(\mathcal{A}, n) \leq k$. We say $\mathcal{A}$ has a pure max-information of k if $\beta = 0$.

It is shown that differential privacy and short description length imply bounds on max-information [11], hence generalization. Here we show that the pure max-information implies a very strong On-Average-KL-Privacy for any distribution $\mathcal{D}$ when we take $\mathcal{A}$ to be a posterior sampling mechanism.

Lemma 11 (Relationship to max-information). *If $\mathcal{A}$ is a posterior sampling mechanism as described in Theorem 4, then $I_\infty(\mathcal{A}, n) \leq k$ implies that $\mathcal{A}$ obeys k/n-On-Average-KL-Privacy for any data generating distribution $\mathcal{D}$.*

An immediate corollary of the above connection is that we can now significantly simplify the proof for "max-information $\Rightarrow$ generalization" for posterior sampling algorithms.

Corollary 12 *Let $\mathcal{A}$ be a posterior sampling algorithm. $I_\infty(\mathcal{A}, n) \leq k$ implies that $\mathcal{A}$ generalizes with rate k/n.*

We now compare to mutual information and draw connections to [28].

Definition 13 (Mutual Information). *The mutual information*

$$I(\mathcal{A}(Z); Z) = \mathbb{E}_Z \mathbb{E}_{h \sim \mathcal{A}(Z)} \log \frac{p(h, Z)}{p(h)p(Z)}$$

where $\mathcal{A}(Z) \sim p(h|Z)$, $p(Z) = \mathcal{D}^n$ and $p(h) = \int p(h|Z)p(Z)dZ$.

Lemma 14 (Relationship to Mutual Information). *For any randomized algorithm $\mathcal{A}$, let $\mathcal{A}(Z)$ be an RV, and Z, Z' be two data sets of size n. We have*

$$I(\mathcal{A}(Z); Z) = D_{\mathrm{KL}}(\mathcal{A}(Z) \| \mathcal{A}(Z')) + \mathbb{E}_Z \mathbb{E}_{h \sim \mathcal{A}(Z)} \left[\mathbb{E}_{Z'} \log p(\mathcal{A}(Z')) - \log \mathbb{E}_{Z'} p(\mathcal{A}(Z')) \right],$$

which by Jensen's inequality implies $I(\mathcal{A}(Z), Z) \leq D_{\mathrm{KL}}(\mathcal{A}(Z) \| \mathcal{A}(Z'))$.

A natural observation is that for MaxEnt $\mathcal{A}$ defined with $\mathcal{L}$, mutual information lower bounds its generalization error. On the other hand, Proposition 1 in Russo and Zou [28] states that under the assumption that $\mathcal{L}(h, Z)$ is σ^2-subgaussian for every h, then the on-average generalization error is always smaller than $\sigma\sqrt{2I(\mathcal{A}(Z); \mathcal{L}(\cdot, Z))}$. Similar results hold for sub-exponential $\mathcal{L}(h, Z)$ [28, Proposition 3].

Note that in their bounds, $I(\mathcal{A}(Z); \mathcal{L}(\cdot, Z))$ is the mutual information between the choice of hypothesis h and the loss function for which we are defining generalization on. By data processing inequality, we have $I(\mathcal{A}(Z); \mathcal{L}(\cdot, Z)) \leq I(\mathcal{A}(Z); Z)$. Further, when $\mathcal{A}$ is posterior distribution, it only depends on Z through $\mathcal{L}(\cdot, Z)$, namely $\mathcal{L}(\cdot, Z)$ is a sufficient statistic for $\mathcal{A}$. As a result $\mathcal{A} \perp Z | \mathcal{L}(\cdot, Z)$. Therefore, we know $I(\mathcal{A}(Z); \mathcal{L}(\cdot, Z)) = I(\mathcal{A}(Z); Z)$. Combine this observation with Lemma 14 and Theorem 4, we get the following characterization of generalization through mutual information.

Corollary 15 (Mutual information and generalization). *Let $\mathcal{A}$ be an algorithm that samples $\propto \exp(-\gamma \mathcal{L}(h, Z))$, and $\mathcal{L}(h, Z) - R(h)$ is σ^2-subgaussian for any $h \in \mathcal{H}$, then*

$$\frac{1}{\gamma} I(\mathcal{A}(Z); Z) \leq \left| \mathbb{E}_Z \mathbb{E}_{h \sim \mathcal{A}(Z)}[\mathcal{L}(h, Z) - R(h)] \right| \leq \sigma\sqrt{2I(\mathcal{A}(Z); Z)}.$$

If $\mathcal{L}(h, Z) - R(h)$ is σ^2-subexponential with parameter (σ, b) instead, then we have a weaker upper bound $bI(\mathcal{A}(Z); Z) + \sigma^2/(2b)$.

The corollary implies that for each γ we have an intriguing bound that says $I(\mathcal{A}; Z) \leq 2\gamma^2\sigma^2$ for any distribution of Z, $\mathcal{H}$ and $\mathcal{L}$ such that $\mathcal{L}(\cdot, Z)$ is σ^2-subgaussian. One interesting case is when $\gamma = 1/\sigma$. This gives

$$\sigma I(\mathcal{A}(Z); Z) \leq \left| \mathbb{E}_Z \mathbb{E}_{h \sim \mathcal{A}(Z)}[\mathcal{L}(h, Z) - R(h)] \right| \leq \sigma\sqrt{2I(\mathcal{A}(Z); Z)}.$$

The lower bound is therefore sharp up to a multiplicative factor of $\sqrt{I(\mathcal{A}(Z); Z)}$.

5 Connections to Other Attempts to Weaken DP

We compare and contrast the On-Average KL-Privacy with other notions of privacy that are designed to weaken the original DP. The (certainly incomplete) list includes (ϵ, δ)-approximate differential privacy (Approx-DP) [13], random differential privacy (Rand-DP) [17], Personalized Differential Privacy (Personal-DP) [15,22] and Total-Variation-Privacy (TV-Privacy) [4,5]. Table 1 summarizes and compares of these definitions.

Table 1. Summary of different privacy definitions.

Privacy definition	Z	z	Distance (pseudo)metric
Pure DP	$\sup_{Z \in \mathcal{Z}^n}$	$\sup_{z \in \mathcal{Z}}$	$D_\infty(P\|Q)$
Approx-DP	$\sup_{Z \in \mathcal{Z}^n}$	$\sup_{z \in \mathcal{Z}}$	$D_\infty^\delta(P\|Q)$
Personal-DP	$\sup_{Z \in \mathcal{Z}^n}$	for each z	$D_\infty(P\|Q)$ or $D_\infty^\delta(P\|Q)$
KL-Privacy	$\sup_{Z \in \mathcal{Z}^n}$	$\sup_{z \in \mathcal{Z}}$	$D_{\mathrm{KL}}(P\|Q)$
TV-Privacy	$\sup_{Z \in \mathcal{Z}^n}$	$\sup_{z \in \mathcal{Z}}$	$\|P - Q\|_{TV}$
Rand-Privacy	$1 - \delta_1$ any $\mathcal{D}^n$	$1 - \delta_1$ any $\mathcal{D}$	$D_\infty^{\delta_2}(P\|Q)$
On-Avg KL-Privacy	$\mathbb{E}_{Z \sim \mathcal{D}^n}$ for each $\mathcal{D}$	$\mathbb{E}_{Z \sim \mathcal{D}}$ for each $\mathcal{D}$	$D_{\mathrm{KL}}(P\|Q)$

A key difference of On-Average KL-Privacy from almost all other previous definitions of privacy, is that the probability is defined only over the random coins of private algorithms. For this reason, even if we convert our bound into the high probability form, the meaning of the small probability δ would be very different from that in Approx-DP. The only exception in the list is Rand-DP, which assumes, like we do, the $n + 1$ data points in adjacent data sets Z and Z' are draw iid from a distribution. Ours is weaker than Rand-DP in that ours is a distribution-specific quantity.

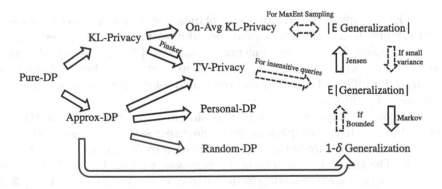

Fig. 1. Relationship of different privacy definitions and generalization.

Among these notions of privacy, Pure-DP and Approx-DP have been shown to imply generalization with high probability [5,12]; and TV-privacy was more shown to imply generalization (in expectation) for a restricted class of queries (loss functions) [5]. The relationship between our proposal and these known results are clearly illustrated in Fig. 1. To the best of our knowledge, our result is the first of its kind that crisply characterizes generalization.

Lastly, we would like to point out that while each of these definitions retains some properties of differential privacy, they might not possess all of them simultaneously and satisfactorily. For example, (ϵ, δ)-approx-DP does not have a satisfactory group privacy guarantee as δ grows exponentially with the group size.

6 Experiments

In this section, we validate our theoretical results through numerical simulation. Specifically, we use two simple examples to compare the ϵ of differential privacy, ϵ of on-average KL-privacy, the generalization error, as well as the utility, measured in terms of the excess population risk.

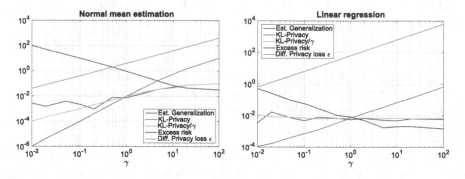

Fig. 2. Comparison of On-Avg KL-Privacy and Differential Privacy on two examples.

The first example is the private release of mean, we consider Z to be the mean of 100 samples from standard normal distribution truncated between $[-2, 2]$. Hypothesis space $\mathcal{H} = \mathbb{R}$, loss function $\mathcal{L}(Z, h) = |Z - h|$. $\mathcal{A}$ samples with probability proportional to $\exp(-\gamma|Z - h|)$. Note that this is the simple Laplace mechanism for differential privacy and the global sensitivity is 4, as a result this algorithm is 4γ-differentially private.

The second example we consider is a simple linear regression in 1D. We generate the data from a simple univariate linear regression model $y = xh + \text{noise}$, where x and the noise are both sampled iid from a uniform distribution defined on $[-1, 1]$. The true h is chosen to be 1. Moreover, we use the standard square loss $\ell(z_i, h) = (y_i - x_i h)^2$. Clearly, the data domain $\mathcal{Z} = \mathcal{Y} \times \mathcal{X} = [-1, 1] \times [-2, 2]$ and if we constrain $\mathcal{H}$ to be within a bounded set $[-2, 2]$, $\sup_{x,y,\beta}(y - x\beta)^2 \leq 16$ and the posterior sampling with parameter γ obeys 64γ-DP.

Figure 2 plots the results over an exponential grid of parameter γ. In these two examples, we calculate on-Average KL-Privacy using known formula of the KL-divergence of Laplace and Gaussian distributions. Then we stochastically estimate the expectation over data. We estimate the generalization error in the direct formula by evaluating on fresh samples. As we can see, appropriately scaled On-Average KL-Privacy characterizes the generalization error precisely as the theory predicts. On the other hand, if we just compare the privacy losses, the average ϵ from a random data set given by On-Avg KL-Privacy is smaller than that for the worst case in DP by orders of magnitudes.

7   Conclusion

We presented On-Average KL-privacy as a new notion of privacy (or stability) on average. We showed that this new definition preserves properties of differential privacy including closedness to post-processing, small group privacy and adaptive composition. Moreover, we showed that On-Average KL-privacy/stability characterizes a weak form of generalization for a large class of sampling distributions that simultaneously maximize entropy and utility. This equivalence and connections to certain information-theoretic quantities allowed us to provide the first lower bound of generalization using mutual information. Lastly, we conduct numerical simulations which confirm our theory and demonstrate the substantially more favorable privacy-utility trade-off.

A   Proofs

We now prove Theorem 4 and Lemma 14. Due to space limit, proof of Lemmas 6, 7 and 11 are given in the technical report [36].

Proof of Theorem 4. We start by a ghost sample argument.

$$
\mathbb{E}_{z \sim \mathcal{D}, Z \sim \mathcal{D}^n} \mathbb{E}_{h \sim \mathcal{A}(Z)} \left[\ell(z, h) - \frac{1}{n} \sum_{i=1}^{n} \ell(z_i, h) \right]
$$

$$
= \mathbb{E}_{Z' \sim \mathcal{D}^n, Z \sim \mathcal{D}^n} \mathbb{E}_{h \sim \mathcal{A}(Z)} \left[\frac{1}{n} \sum_{i=1}^{n} \ell(z_i', h) - \frac{1}{n} \sum_{i=1}^{n} \ell(z_i, h) \right]
$$

$$
= \frac{1}{n} \sum_{i=1}^{n} \mathbb{E}_{z_i' \sim \mathcal{D}, Z \sim \mathcal{D}^n} \mathbb{E}_{h \sim \mathcal{A}(Z)} \left[\ell(z_i', h) - \ell(z_i, h) \right]
$$

$$
= \frac{1}{n} \sum_{i=1}^{n} \mathbb{E}_{z_i' \sim \mathcal{D}, Z \sim \mathcal{D}^n} \mathbb{E}_{h \sim \mathcal{A}(Z)} \left[\ell(z_i', h) + \sum_{j \neq i} \ell(z_j, h) + r(h) - \ell(z_i, h) - \sum_{j \neq i} \ell(z_j, h) - r(h) \right]
$$

$$
= \frac{1}{n} \sum_{i=1}^{n} \mathbb{E}_{z_i' \sim \mathcal{D}, Z \sim \mathcal{D}^n} \mathbb{E}_{\mathcal{A}(Z)} \left[-\log p_{\mathcal{A}([Z_{-i}, z_i'])}(h) + \log p_{\mathcal{A}(Z)(h)}(h) + \log K_i - \log K_i' \right]
$$

$$
= \frac{1}{n} \sum_{i=1}^{n} \mathbb{E}_{z_i' \sim \mathcal{D}, Z \sim \mathcal{D}^n} \mathbb{E}_{\mathcal{A}(Z)} \left[\log p_{\mathcal{A}(Z)}(h) - \log p_{\mathcal{A}([Z_{-i}, z_i'])}(h) \right]
$$

$$
= \mathbb{E}_{z \sim \mathcal{D}, Z \sim \mathcal{D}^n} \mathbb{E}_{h \sim \mathcal{A}(Z)} \left[\log p_{\mathcal{A}(Z)}(h) - \log p_{\mathcal{A}([Z_{-1}, z])}(h) \right].
$$

The K_i and K_i' are partition functions of $p_{\mathcal{A}(Z)}(h)$ and $p_{\mathcal{A}([Z_{-i},z_i'])}(h)$ respectively. Since $z_i \sim z_i'$, we know $\mathbb{E}K_i - \mathbb{E}K_i' = 0$. The proof is complete by noting the non-negativity of On-Average KL-Privacy, which allows us to take absolute value without changing the equivalence.   □

Proof of Lemma 14. Denote $p(\mathcal{A}(Z)) = p(h|Z)$. $p(h, Z) = p(h|Z)p(Z)$. The marginal distribution of h is therefore $p(h) = \int_Z p(h, Z)dZ = \mathbb{E}_Z p(\mathcal{A}(Z))$. By definition,

$$
\begin{aligned}
I(\mathcal{A}(Z); Z) &= \mathbb{E}_Z \mathbb{E}_{h|Z} \log \frac{p(h|Z)p(Z)}{p(h)p(Z)} \\
&= \mathbb{E}_Z \mathbb{E}_{h|Z} \log p(h|Z) - \mathbb{E}_Z \mathbb{E}_{h|Z} \log \mathbb{E}_{Z'} p(h|Z') \\
&= \mathbb{E}_Z \mathbb{E}_{h|Z} \log p(h|Z) - \mathbb{E}_{Z,Z'} \mathbb{E}_{h|Z} \log p(h|Z') \\
&\quad + \mathbb{E}_{Z,Z'} \mathbb{E}_{h|Z} \log p(h|Z') - \mathbb{E}_Z \mathbb{E}_{h|Z} \log \mathbb{E}_{Z'} p(h|Z') \\
&= \mathbb{E}_{Z,Z'} \mathbb{E}_{h|Z} \log \frac{p(h|Z)}{p(h|Z')} + \mathbb{E}_{Z,Z'} \mathbb{E}_{h|Z} \log p(h|Z') - \mathbb{E}_Z \mathbb{E}_{h|Z} \log \mathbb{E}_{Z'} p(h|Z') \\
&= D_{\mathrm{KL}}(\mathcal{A}(Z), \mathcal{A}(Z')) + \mathbb{E}_Z \mathbb{E}_{h|Z} [\mathbb{E}_{Z'} \log p(\mathcal{A}(Z')) - \log \mathbb{E}_{Z'} p(\mathcal{A}(Z'))]
\end{aligned}
$$

The inequality remark in the last line follows from Jensen's inequality.   □

References

1. Akaike, H.: Likelihood of a model and information criteria. J. Econometrics **16**(1), 3–14 (1981)
2. Altun, Y., Smola, A.J.: Unifying divergence minimization and statistical inference via convex duality. In: Lugosi, G., Simon, H.U. (eds.) COLT 2006. LNCS (LNAI), vol. 4005, pp. 139–153. Springer, Heidelberg (2006)
3. Anderson, N.: "anonymized" data really isn't and here's why not (2009). http://arstechnica.com/tech-policy/2009/09/your-secrets-live-online-in-databases-of-ruin/
4. Barber, R.F., Duchi, J.C.: Privacy and statistical risk: Formalisms and minimax bounds. arXiv preprint arXiv:1412.4451 (2014)
5. Bassily, R., Nissim, K., Smith, A., Steinke, T., Stemmer, U., Ullman, J.: Algorithmic stability for adaptive data analysis. arXiv preprint arXiv:1511.02513 (2015)
6. Berger, A.L., Pietra, V.J.D., Pietra, S.A.D.: A maximum entropy approach to natural language processing. Comput. Linguist. **22**(1), 39–71 (1996)
7. Bousquet, O., Elisseeff, A.: Stability and generalization. J. Mach. Learn. Res. **2**, 499–526 (2002)
8. Duncan, G.T., Elliot, M., Salazar-González, J.J.: Statistical Confidentiality: Principle and Practice. Springer, New York (2011)
9. Duncan, G.T., Fienberg, S.E., Krishnan, R., Padman, R., Roehrig, S.F.: Disclosure limitation methods and information loss for tabular data. In: Confidentiality, Disclosure and Data Access: Theory and Practical Applications for Statistical Agencies, pp. 135–166 (2001)
10. Dwork, C.: Differential privacy. In: Bugliesi, M., Preneel, B., Sassone, V., Wegener, I. (eds.) ICALP 2006. LNCS, vol. 4052, pp. 1–12. Springer, Heidelberg (2006)

11. Dwork, C., Feldman, V., Hardt, M., Pitassi, T., Reingold, O., Roth, A.: Generalization in adaptive data analysis and holdout reuse. In: Advances in Neural Information Processing Systems (NIPS 2015), pp. 2341–2349 (2015)
12. Dwork, C., Feldman, V., Hardt, M., Pitassi, T., Reingold, O., Roth, A.L.: Preserving statistical validity in adaptive data analysis. In: ACM on Symposium on Theory of Computing (STOC 2015), pp. 117–126. ACM (2015)
13. Dwork, C., Kenthapadi, K., McSherry, F., Mironov, I., Naor, M.: Our data, ourselves: privacy via distributed noise generation. In: Vaudenay, S. (ed.) EUROCRYPT 2006. LNCS, vol. 4004, pp. 486–503. Springer, Heidelberg (2006)
14. Dwork, C., McSherry, F., Nissim, K., Smith, A.: Calibrating noise to sensitivity in private data analysis. In: Halevi, S., Rabin, T. (eds.) TCC 2006. LNCS, vol. 3876, pp. 265–284. Springer, Heidelberg (2006)
15. Ebadi, H., Sands, D., Schneider, G.: Differential privacy: now it's getting personal. In: ACM Symposium on Principles of Programming Languages, pp. 69–81. ACM (2015)
16. Fienberg, S.E., Rinaldo, A., Yang, X.: Differential privacy and the risk-utility tradeoff for multi-dimensional contingency tables. In: Domingo-Ferrer, J., Magkos, E. (eds.) PSD 2010. LNCS, vol. 6344, pp. 187–199. Springer, Heidelberg (2010)
17. Hall, R., Rinaldo, A., Wasserman, L.: Random differential privacy. arXiv preprint arXiv:1112.2680 (2011)
18. Hardt, M., Ullman, J.: Preventing false discovery in interactive data analysis is hard. In: IEEE Symposium on Foundations of Computer Science (FOCS 2014), pp. 454–463. IEEE (2014)
19. Hundepool, A., Domingo-Ferrer, J., Franconi, L., Giessing, S., Nordholt, E.S., Spicer, K., De Wolf, P.P.: Statistical Disclosure Control. Wiley (2012)
20. Jaynes, E.T.: Information theory and statistical mechanics. Phys. Rev. **106**(4), 620 (1957)
21. Kearns, M., Ron, D.: Algorithmic stability and sanity-check bounds for leave-one-out cross-validation. Neural Comput. **11**(6), 1427–1453 (1999)
22. Liu, Z., Wang, Y.X., Smola, A.: Fast differentially private matrix factorization. In: ACM Conference on Recommender Systems (RecSys 2015), pp. 171–178. ACM (2015)
23. McSherry, F., Talwar, K.: Mechanism design via differential privacy. In: IEEE Symposium on Foundations of Computer Science (FOCS 2007), pp. 94–103 (2007)
24. Mir, D.J.: Information-theoretic foundations of differential privacy. In: Garcia-Alfaro, J., Cuppens, F., Cuppens-Boulahia, N., Miri, A., Tawbi, N. (eds.) FPS 2012. LNCS, vol. 7743, pp. 374–381. Springer, Heidelberg (2013)
25. Mosteller, F., Tukey, J.W.: Data analysis, including statistics (1968)
26. Mukherjee, S., Niyogi, P., Poggio, T., Rifkin, R.: Learning theory: stability is sufficient for generalization and necessary and sufficient for consistency of empirical risk minimization. Adv. Comput. Math. **25**(1–3), 161–193 (2006)
27. Narayanan, A., Shmatikov, V.: Robust de-anonymization of large sparse datasets. In: IEEE Symposium on Security and Privacy, pp. 111–125. IEEE, September 2008
28. Russo, D., Zou, J.: Controlling bias in adaptive data analysis using information theory. In: International Conference on Artificial Intelligence and Statistics (AISTATS 2016) (2016)
29. Shalev-Shwartz, S., Shamir, O., Srebro, N., Sridharan, K.: Learnability, stability and uniform convergence. J. Mach. Learn. Res. **11**, 2635–2670 (2010)
30. Steinke, T., Ullman, J.: Interactive fingerprinting codes and the hardness of preventing false discovery. arXiv preprint arXiv:1410.1228 (2014)

31. Tishby, N., Pereira, F.C., Bialek, W.: The information bottleneck method. arXiv preprint arXiv:physics/0004057 (2000)
32. Uhlerop, C., Slavković, A., Fienberg, S.E.: Privacy-preserving data sharing for genome-wide association studies. J. Priv. Confidentiality **5**(1), 137 (2013)
33. Van Erven, T., Harremoës, P.: Rényi divergence and kullback-leibler divergence. IEEE Trans. Inf. Theor. **60**(7), 3797–3820 (2014)
34. Wang, Y.X., Fienberg, S.E., Smola, A.: Privacy for free: posterior sampling and stochastic gradient monte carlo. In: International Conference on Machine Learning (ICML 2015) (2015)
35. Wang, Y.X., Lei, J., Fienberg, S.E.: Learning with differential privacy: stability, learnability and the sufficiency and necessity of erm principle. J. Mach. Learn. Res. (to appear, 2016)
36. Wang, Y.X., Lei, J., Fienberg, S.E.: On-average kl-privacy and its equivalence to generalization for max-entropy mechanisms (2016). preprint http://www.cs.cmu.edu/~yuxiangw/publications.html
37. Yau, N.: Lessons from improperly anonymized taxi logs (2014). http://flowingdata.com/2014/06/23/lessons-from-improperly-anonymized-taxi-logs/
38. Yu, F., Fienberg, S.E., Slavković, A.B., Uhler, C.: Scalable privacy-preserving data sharing methodology for genome-wide association studies. J. Biomed. Inform. **50**, 133–141 (2014)
39. Zhou, S., Lafferty, J., Wasserman, L.: Compressed and privacy-sensitive sparse regression. IEEE Trans. Inf. Theor. **55**(2), 846–866 (2009)

Correcting Finite Sampling Issues in Entropy *l*-diversity

Sebastian Stammler, Stefan Katzenbeisser, and Kay Hamacher(✉)

Technische Universität Darmstadt, 64287 Darmstadt, Germany
{stammler,hamacher}@cbs.tu-darmstadt.de
http://www.kay-hamacher.de

Abstract. In statistical disclosure control (SDC) anonymized versions of a database table are obtained via generalization and suppression to reduce de-anonymization attacks, ideally with minimal utility loss. This amounts to an optimization problem in which a measure of remaining diversity needs to be improved. The feasible solutions are those that fulfill some privacy criteria, e.g., the entropy *l*-diversity. In the statistics it is known that the naive computation of an entropy via the Shannon formula systematically underestimates the (real) entropy and thus influences the resulting equivalence classes. In this contribution we implement an asymptotically unbiased estimator for the Shannon entropy and apply it to three test databases. Our results show previously performed systematic miscalculations; we show that by an unbiased estimator one can increase the utility of the data without compromising privacy.

Keywords: Anonymity · l-diversity · Finite sampling · Statistics · Information theory

1 Introduction

In order to enable the statistical analysis of private data sets, database tables are anonymized (e.g., by suppression or generalization). The process of anonymizing a table amounts to finding a trade-off between maintaining the utility of a dataset and the potential (maximal) threat to privacy of those persons whose data is stored in the data-set, mainly in form of one or several tables. These tables frequently contain a sensitive attribute that might be correlated with personally identifiable information (PII) in other columns of the table(s). While there exist threats beyond de-identification and revelation of sensitive information on individuals [18,20], e.g., record linkage [26], we follow here the rationale that protecting sensitive information is a necessary condition on top of which additional protection requirements can be formulated.

To (quantitatively) assess the privacy implications, researchers have proposed to use various privacy criteria. Based upon those criteria one can then use operations such as generalization and suppression for statistical disclosure control (SDC) to modify the table in question such as to contain only information that

© Springer International Publishing Switzerland 2016
J. Domingo-Ferrer and M. Pejić-Bach (Eds.): PSD 2016, LNCS 9867, pp. 135–146, 2016.
DOI: 10.1007/978-3-319-45381-1_11

does not reveal more information on any individual than a given criterion with their privacy parameters allows.

Different metrics have been proposed in the literature; from the seminal work by Sweeney on k-anonymity [27,28] to t-closeness [15], δ-disclosure privacy [3] and δ-presence [19]. One measure that protects datasets against attribute disclosure is the l-diversity criterion [16] which is the basis of our work and will be discussed in detail in the subsequent section.

2   The Entropy l-diversity

By generalization of the quasi-identifiers or, additionally, suppressions of records, we convert a sensitive table T_S of raw data into an anonymized table T. We will call the set of records that have the same quasi-identifiers an *equivalence class*. Now, we can define the entropy l-diversity measure [16] as follows:[1] Let E be an equivalence class and $S(E)$ the domain of sensitive attributes within this equivalence class. Then define the naive Shannon entropy of an equivalence class as

$$H_E := - \sum_{s \in S(E)} p(s) \cdot \log p(s) \tag{1}$$

where $p(s)$ is the *probability* of the sensitive attribute s in $S(E)$. In information theory it is always the *probabilities* that gives rise to entropies; these *probabilities*, however, are not necessarily known; in particular they are not equal to *frequencies* and counts. If one uses for $p(s)$ the frequencies $\frac{n(s)}{\sum_\sigma n(\sigma)}$ we will call the resulting entropies the naive Shannon estimator H_S.

Independent of our choice of estimating the (real) entropy, the anonymized table T is said to fulfill the *entropy l-diversity* criterion, if for all equivalence classes we have

$$\forall E : H_E \geq \log (l) \tag{2}$$

Among all anonymizations that fulfill the criterion, one then chooses the optimal due to some information loss criterion (cmp. Sect. 5.1). Note, that the set of equivalence classes fulfilling Eq. (2) implicitly define a feasible region for the optimization problem of minimal information loss.

2.1   Entropy and Relation to Channel Theorem

There is a close relation with the demand implied via the threshold $\log (l)$ in Eq. (2) and the information communicated or revealed: we can regard the generalization and suppression procedure to meet the requirement in Eq. (2) as a communication channel between the table curator with full access and the analyst

---

[1] Note, that typically the logarithm to base two is used. However, any base will do as we can simply rescale all equations and inequalities. We will use the natural logarithm in the subsequent parts as the estimator we use is more easily derived in nats rather than bits.

who is supposed to only see a "protected version". Then, the channel capacity is $C = \sup_{p_X(x)} I(X;Y)$ where $I(X;Y)$ is the mutual information between the original data X (represented by a random variable) and the output Y. This, in turn, can be written [17] as an entropy difference $I(X;Y) = H(X) - H(X|Y)$, where $H(X)$ is the entropy in the processed data and $H(X|Y)$ is the one in a generalized and/or suppressed set of variables, namely H_S. But the last term is bounded from below via Eq. (2) and thus limits the information revealed $I(X;Y)$. Therefore, deriving a generalization and suppression scheme that fulfills Eq. (2) with the smallest margin would be most helpful. The solution to this optimization problem depends drastically on the number and topology of feasible solutions to Eq. (2). Any shortcomings and conceptual problems in the correct evaluation of H_S poses the risk that we work on a smaller set of feasible solutions that are eventually available.

Now, the straight-forward way to evaluate the entropy H_E in Eq. (1) is to plug into the probabilities p an estimator thereof[2]. The naive estimators of probabilities are their respective frequencies[3]. Between those two, there exists, however, a subtle difference: frequencies $\hat{n}(s)$ are maximum likelihood estimators of the probabilities. Now, the entropy is a non-linear function of the probabilities and therefore the order of the two operations – maximum likelihood estimation and function evaluation – *cannot* be exchanged.

Thus, we actually *must not* evaluate an estimator for $\hat{p}(s)$ first and plug it into Eq. (1) to obtain H_E. Rather, we have to estimate the entropy itself. This subtle differences is subject to a long history of published work in probability theory, statistics, and theoretical physics.

Herein lies our contribution: we formulate a proposition (see below) on the trade-off between privacy and utility, how an unbiased estimator minimizes the margin towards the $\log(l)$ threshold without compromising the desired degree of privacy. To this end, we have evaluated different, published estimators $\hat{H}$ of the Shannon entropy based on the characteristics typical for statistical databases. It turns out that the one proposed by Grassberger (see below) is conceptually the only capable to meet the requirements of a good estimator for H_S, in particular it is asymptotically unbiased. This choice was then applied to a number of test data bases to empirically show the benefits of using a consistent estimator for H_S.

3   Better Estimators

Within the statistics and physics community it is well-known that better estimators than H_S of the entropy exist. Corrections can be either computational, e.g., based on bootstrap-procedures [10, 29] or analytic estimators. Obviously the latter provide more efficiency as one typically has to evaluate some few (special)

---

[2] often written $\hat{p}(s)$ to indicate that these are not the "real" probabilities, but rather stem from a statistical model or concrete data.

[3] frequencies are just the counts divided by the total number of observations, namely $\hat{p}(s) = \frac{\hat{n}(s)}{\sum_\sigma \hat{n}(\sigma)}$.

functions and can avoid both thousand of iterations and the problems of small fluctuations due to the inherent stochasticity of bootstrap procedures.

Simple estimators were proposed quite some time ago. The simplest being the one by Harris [23]

$$H_H := H_S - \frac{M-1}{2N} + \frac{1}{12N^2} \left(1 - \sum_{i=1}^{M} \frac{1}{\hat{p}_i} \right) + \mathcal{O}\left(N^{-3}\right)$$

where N is the total number of observations (entries in one equivalence cluster in our setting) and M the number of different buckets, that is, different instances within an equivalence class. Clearly, a leading correction $\sim M/N$ is not necessarily negligible for typical databases and the strict demands on privacy. This and other improved estimators form a continuously growing literature – see [2, 22–24] for some background.

3.1   (Finite-)Size Scaling

More importantly, we learn from the Harris estimator H_H that any correction to the naive one H_S will be of leading order $\sim M/N$. Now, one would immediately assume that the size of the data set would influence the difference between any improved estimator and the naive one of Eq. (1). But quiet to the contrary! The M and N apply to *each* entropy computation for *each* equivalence class E in the requirement of Eq. (2). Therefore, increasing, e.g., the database size not necessarily leads to larger M and N in the correction as the number of entries under a modified generalization $E \rightarrow E'$ typically reduces these values when splitting an equivalence class E due to more data points available in a larger database.

3.2   Choice of an Appropriate Estimator

Grassberger [8,9] has proposed different improvements on an *unbiased* estimator of entropies. In [9] Grassberger first reported an estimator for a generalized entropy, the Rényi entropy.

While an important contribution in itself, the Rényi entropy is of little use in the setting of privacy research as it is not extensive in physics parlance – that is, it does not scale with the overall size of the dataset. This, however, we consider to be an important property for a diversity index and the threat assessment: if we, for example, double the size of the database, we still want a threshold to have the same meaning. Therefore, we restrict our discussion to the only extensive entropy, the Shannon entropy.

After intensive literature review we chose Grassberger's entropy estimator $\hat{H}_G$ [8] as it is asymptotically unbiased and can be evaluated with reasonable computational demands. For a set of size $N = |E|$ with $M = |S(E)|$ buckets, let $n(s)$ denote the number of elements in the bucket representing s – we stick to the nomenclature of Eq. (1). Then[4]

---

[4] Note that $G(0)$ is never used but marks the start of the inductive definition. $\gamma \approx 0.577215\ldots$ is Euler's constant.

$$\hat{H}_G := G(N) - \frac{1}{N} \sum_{s \in S(E)} n(s) \cdot G\left(n(s)\right) \quad \text{with} \tag{3}$$

$$G(0) := -\gamma - \ln 2,$$
$$G(2n+1) := G(2n) \quad \text{and}$$
$$G(2n+2) := G(2n) + \frac{2}{2n+1} \tag{4}$$

We have implemented this variant as an additional l-diversity criterion using the ARX anonymization-framework [12,21], see below. To efficiently compute the $G(n)$, we have expressed them as a series in $1/n$, see Appendix A.

4 Proposition

> We propose that a better estimator, which becomes asymptotically unbiased[5], generally increases the number of feasible equivalence classes and thus can only improve the solution to the optimization problem of reducing information loss, that is utility in the SDC setting under sufficiently large privacy demands ($l \gtrsim 2$).

We therefore solve a shortcoming formulated by Machanavajjhala *et al.* [16] in their seminar work proposing l-diversity: *"Thus entropy l-diversity may sometimes be too restrictive."* With our improvement the entropy l-diversity becomes more relaxed without comprising the PPI beyond the $\log(l)$ limit.

In Sect. 3 we have already discussed the statistical background and argued that indeed Grassberger's estimator will increase the number of feasible anonymization schemes. What remains to be shown is that the so larger space of anonymizations results (almost always) in SDCs with smaller loss, i.e., larger utility. Sections 5 and 6 are devoted to this.

5 Experimental Setup

To test our proposition, we used the Java ARX-framework to run several anonymizations over a multi-dimensional parameter space. We anonymized different databases (see Table 1) and subsets of these. Furthermore, we also varied the privacy parameter l, the maximum allowed outliers and the metric due to which the optimal anonymization is chosen. We then compared the minimal information losses resulting from using the naive Shannon and Grassberger entropy estimators in criterion (2). The following section describes the details of our experiments.

---

[5] With growing counts the systematic bias tends to vanish in the statistical sense.

5.1   Measuring Information Loss – The Optimization Criterion

The two major aspects of interest to both practitioners and research in SDC are (a) the utility of the data after generalization and suppression and (b) the privacy guarantees or the information leaked.

For the latter we can – based on the theoretical work by Grassberger and others – easily evaluate our approach: as the naive estimator systematically underestimates the entropy, we generally increase the entropy estimation using the better estimator of Eq. (3) and hence generally arrive at a lesser anonymized table. But since the *entropy* is better estimated, we still accomplish conceptually the same level of privacy in the information-theoretic sense.

What remains, then, to be evaluated is the utility of the processed data table. Conceptually this is the opposite of the information lost in the process. ARX implements several metrics, of which we have used the following four, namely

Average equivalence class size. This single-dimensional (i.e., based on eq. classes) metric measures data utility based on the sizes of the equivalence classes [14]. The values of the quasi-identifiers are not taken into account.

KL-Divergence. The KL-Divergence is a relative entropy and describes how much information is need to obtain one distribution given another [17]. As a metric in ARX, it is also single-dimensional.

Loss [11]. This metric is multi-dimensional, i.e., it takes into account the loss of each attribute. The Loss measure for each quasi-identifier is multiplicative in nature (and in particular between 0 and 1), so we used the geometric mean as aggregation method. It is also the recommended default measure by ARX.

Non-Uniform Entropy. This multi-dimensional metric measures the loss of entropy [6]. Aggregation was done by summing, since entropy is additive in nature.

5.2   Evaluation on Three Databases

We used three reference databases whose details are listed in Table 1. They were kindly provided by the ARX-Team and are three of the five databases they used in their recent paper on the connection between data-suppression and utility [13]. They cover most of the interesting ranges that we frequently encounter in SDC:

Table 1. Details of the test data employed in our study.

DB name	no. rows	no. quasi-identifiers	sensitive attribute	m (#subsamples)[a]
adult	30,163	8	occupation	10,000
fars	100,938	7	US-state	1,000
ihis	1,193,505	8	education	1,000

[a]For different sizes, we took m random subsamples from the full data set

might it be census data (millions of records) down to the challenges in genomic privacy (a few hundred up to a few thousand[6]).

For each data base, we took m samples of different sizes, starting at size 100. We increased the size by factors of 3 until we reached the full data base size. Then, for each sample, we iterated over several parameters to get a full picture of the impact of Grassberger's correction on them. The parameters were the maximum allowed outliers (0; also 4 % for *fars* and 100 % for *adult*), the l of the l-diversity criterion[7], the four metrics described in Sect. 5.1 and finally applied the uncorrected entropy-l-diversity criterion and the Grassberger corrected one. Additionally, for each metric, we calculated the minimum and maximum information loss to be able to calculate a *relative information loss*.

We then also aggregated results over the same sample size for each database by geometric averaging of the *relative change in information loss*. The result can be seen in Fig. 2.

6 Empirical Results

6.1 Information Loss Improves

In Fig. 1 we illustrate, for a representative parameter setting, the systematic impact the unbiased estimator of Eq. (3) has vs. the naive estimator H_S of Eq. (1). Clearly, we see that our proposition holds. A few outliers in the left part of Fig. 1 occur. This is due a few buckets that are almost empty and the recursion on $G(n)$ in Eq. (3) shows very small, negative values. Interestingly, the removal of max. 4 % outliers (right subfigure) leads to a distinct split between the results for various l values. This, however, is easily understandable as ARX's removal of outliers naturally reduces the range of the empirical distribution for the attributes, thus the overall entropy is reduced by outlier removal. Then, the maximal l has a more pronounced effect on the remaining data entries.

Also note the very different scalings of the two plots. Allowing only 4 % outliers drastically improves data utility, strongly supporting the proposition in [13].

In Fig. 2 we show a comprehensive overview of our results. Clearly, the pattern of improvement by the Grassberger correction is – given a particular database – the same for all of the four information loss variants. As already argued in Sect. 3.1 we do not observe systematic influences of the *overall* size of the table.

More importantly, we observe that the information loss is for sufficiently large privacy demands ($l \gtrsim 2$) always smaller for the asymptotically unbiased estimator by Grassberger.

For $l < 2$, however, the naive estimator tends to show smaller information losses. This occurs as the number M of instances within an equivalence class becomes rather large for the implied many different equivalence classes and thus the number of entries within a single bucket very small (it might even vanish). For even arguments, $G(2n)$ is very slightly larger than $\log(2n)$, however quickly

---

[6] cmp. the 1000 genome project, http://www.1000genomes.org.

[7] The used l's and sizes can be seen in Fig. 2.

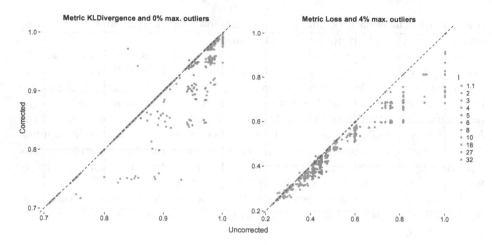

Fig. 1. Comparison of information loss (IL) for two different setups (left: no data suppression allowed and metric KL-Divergence; right: 4% outliers allowed and Loss metric • coloured by parameter l) on the *fars* data set with size 1000 subsampling (only 100 random samples shown per parameter set). The x-axis is the IL for the naive estimator of Eq. (1), the y-axis the one for Grassberger's correction in Eq. (3). All points below the diagonal are cases where the information loss via Grassberger's formula is smaller than for the naive estimator.

converging as n grows[8]. So if there are many buckets of small even sizes in just one equivalence class, it *can* happen, that the criterion (2) $\hat{H}_G \geq \log(l)$ is harder to fulfill for every eq. class, leading to fewer possible anonymizations and thus an optimum with lesser utility. Now, such schemes with less than 2 bits in "protection" can hardly be called privacy regimes, but rather maintains the raw data. We still have to acknowledge that here, our proposition would not hold[9].

6.2   Statistical Significance

We performed a one-sided, paired Wilcoxon signed rank test [25] on all of our results. The null hypothesis for $l \geq 2$ is that the information loss increases by the use of the Grassberger estimator, for $l < 2$ that it decreases. As we used several evaluation criteria (e.g. information loss metrics, sampling with partial overlap etc.) to the same feasible solution created by the criterion of Eq. (2) we added the conservative Bonferroni correction (cmp. [7]) to account for multiple testing. All results are statistically significant with p most of the time even smaller than the numerical accuracy of IEEE double precision. We therefore can safely reject the hypothesis that the usage of this unbiased entropy estimator leads to the same SDC scheme under generalization and suppression.

---

[8] $G(2)/\log(2) \approx 1.053, G(4)/\log(4) \approx 1.0072, G(6)/\log(6) \approx 1.0025$.

[9] We therefore recommend to carefully analyze what one wants to achieve: with such a small privacy guarantee ($l < 2$) on could consequentially just release the raw data.

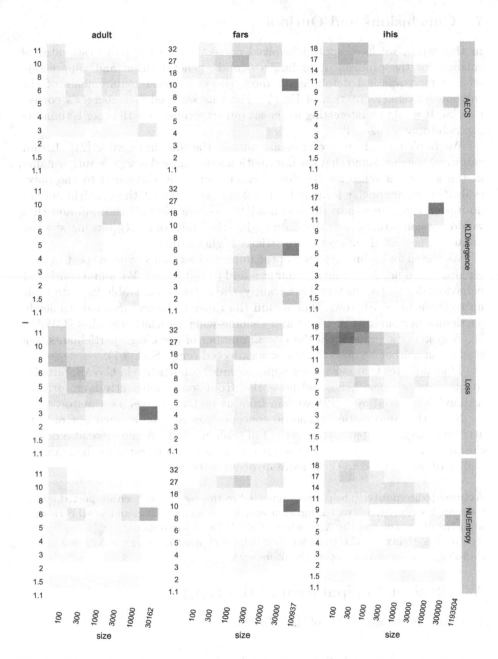

Fig. 2. Color-coding of geometrical average change in information loss under the four metrics of Sect. 5.1 for the three databases; averaged over sufficiently many samples for various sizes as laid out in Table 1. Blue means the IL decreased, red it increased. The *x*-axis shows the size (no. of rows) of a sample from the database(s). The *y*-axis the *l*-threshold chosen in Eq. (2). (Color figure online)

7   Conclusions and Outlook

In this study, we have applied involved corrections to the naive estimator of entropy for the problem of finding a "good" generalization and suppression scheme for statistical databases to meet the conceptual requirements of the entropy l-diversity criterion of Eq. (2). For this we used Grassberger's correction [8]. It would be interesting to repeat our experiments with other estimators like Schürmann's [23, 24].

We implemented our experiments within the ARX framework [21]. In our empirical tests we found that the information loss imposed is significantly smaller in the statistical sense for the Grassberger correction compared to the naive evaluation of Shannon's formula for $l \geq 2$. We argued that the definition of the limitation $\log(l)$ relates to the information theoretic channel theorem and thus we do not compromise sensitive data within the limits of $\log(l)$ [bits/nats] while obtaining protected data sets with strictly higher utility.

We tested for scaling behavior of the improved estimator with respect to data set size – read number of table columns – and to the given l. We found consistent behaviour for all parameters. Importantly, the effect is observable for data table up to some $10^4 - 10^6$ rows, thus within the range of typical medical studies in particular personalized medicine and genome-wide association studies (GWAS) [4]. A systematic influence of the *overall* number of rows, e.g., participants in a study, can neither be observed nor was expected (cmp. Sect. 3.1).

This indicates that our results apply to small cohorts like in GWAS, but also to big data settings, like social networks: from some sufficiently large privacy demand (expressed by $l \gtrsim 2$) we gain from using Grassberger's estimator.

Note, that information theoretic concepts can also be applied not only to attributes in tables, but rather to the full table itself [1]. Again, here also occurs the problem of system bias. In the future, we want to investigate the transferability of our method to those more involved settings.

Acknowledgements. The research reported in this paper has been supported by the German Federal Ministry of Education and Research (BMBF) [and by the Hessian Ministry of Science and the Arts] within CRISP (www.crisp-da.de).

We also thank the ARX-Team for their helpful support in using the API and understanding some inner workings of the framework.

A   Efficient Computation of the $G(n)$

The inductive definition (4) of the $G(n)$ can be rewritten as

$$G(2n) = -\gamma - \ln 2 + \sum_{k=0}^{n-1} \frac{2}{2k+1}$$

Remember that $G(2n + 1) := G(2n)$. This equation can simply be rewritten in terms of the Harmonic numbers $H_n := \sum_{k=1}^{n} 1/k$ as

$$G(2n) = -\gamma - \ln 2 + 2H_{2n-1} - H_{n-1}$$

The Harmonic numbers can also be expressed in terms of the Digamma function ψ as $H_{n-1} = \psi(n) + \gamma$. The Digamma function has the asymptotic series

$$\psi(x) = \ln x - \frac{1}{2x} - \sum_{k=1}^{\infty} \frac{B_{2k}}{2k\, x^{2k}},$$

using the Bernoulli numbers B_n. Combining the last two equations yields the asymptotic expansion for $G(2n)$:

$$G(2n) = \ln(2n) + \frac{1}{24\, n^2} - \frac{7}{960\, n^4} + \frac{31}{8064\, n^6} + \mathcal{O}\left(\frac{1}{n^8}\right)$$

This expansion is accurate to double presicion for $n \geq 50$. For smaller values we used a precomputed table. Note that, beside the logarithm, it suffices to calculate a single division $1/n^2$ and then proceed with nested multiplication.

References

1. Antal, L., Shlomo, N., Elliot, M.: Measuring disclosure risk with entropy in population based frequency tables. In: Domingo-Ferrer [5], pp. 62–78
2. Batu, T., Dasgupta, S., Kumar, R., Rubinfeld, R.: The complexity of approximating the entropy. SIAM J. Comput. **35**(1), 132–150 (2005)
3. Brickell, J., Shmatikov, V.: The cost of privacy: destruction of data-mining utility in anonymized data publishing. In: Proceedings of the 14th ACM SIGKDD International Conference on Knowledge Discovery and Data Mining, pp. 70–78. ACM (2008)
4. Craig, D.W., Goor, R.M., Wang, Z., Paschall, J., Ostell, J., Feolo, M., Sherry, S.T., Manolio, T.A.: Assessing and managing risk when sharing aggregate genetic variant data. Nat. Rev. Genet. **12**(10), 730–736 (2011). http://dx.doi.org/10.1038/nrg3067
5. Domingo-Ferrer, J. (ed.): PSD 2014. LNCS, vol. 8744. Springer, Heidelberg (2014)
6. Gionis, A., Tassa, T.: k-anonymization with minimal loss of information. IEEE Trans. Knowl. Data Eng. **21**(2), 206–219 (2009)
7. Goeman, J.J., Solari, A.: Multiple hypothesis testing in genomics. Stat. Med. **33**(11), 1946–1978 (2014)
8. Grassberger, P.: Entropy estimates from insufficient samplings arXiv:physics/0307138 (2008)
9. Grassberger, P.: Finite sample corrections to entropy and dimension estimates. Phys. Lett. A **128**(6), 369–373 (1988)
10. Hamacher, K.: Using lisp macro-facilities for transferable statistical tests. In: 9th European Lisp Symposium (accepted, 2016)
11. Iyengar, V.S.: Transforming data to satisfy privacy constraints. In: Proceedings of the Eighth ACM SIGKDD International Conference on Knowledge Discovery and Data Mining, pp. 279–288. ACM (2002)
12. Kohlmayer, F., Prasser, F., Eckert, C., Kemper, A., Kuhn, K.: Flash: efficient, stable and optimal k-anonymity. In: Privacy, Security, Risk and Trust (PASSAT), 2012 International Conference on and 2012 International Confernece on Social Computing (SocialCom), pp. 708–717, September 2012

13. Kohlmayer, F., Prasser, F., Kuhn, K.A.: The cost of quality: implementing generalization and suppression for anonymizing biomedical data with minimal information loss. J. Biomed. Inform. **58**, 37–48 (2015)
14. LeFevre, K., DeWitt, D.J., Ramakrishnan, R.: Mondrian multidimensional k-anonymity. In: Proceedings of the 22nd International Conference on Data Engineering, ICDE 2006, pp. 25–25. IEEE (2006)
15. Li, N., Li, T., Venkatasubramanian, S.: t-closeness: privacy beyond k-anonymity and l-diversity. In: IEEE 23rd International Conference on Data Engineering, ICDE 2007, pp. 106–115. IEEE (2007)
16. Machanavajjhala, A., Kifer, D., Gehrke, J., Venkitasubramaniam, M.: l-diversity: privacy beyond k-anonymity. ACM Trans. Knowl. Discov. Data (TKDD) **1**(1), 3 (2007)
17. MacKay, D.: Information Theory, Inference, and Learning Algorithms, 2nd edn. Cambridge University Press, Cambridge (2004)
18. Narayanan, A., Shmatikov, V.: Myths and fallacies of "personally identifiable information". Commun. ACM **53**(6), 24–26 (2010). http://doi.acm.org/10.1145/1743546.1743558
19. Nergiz, M.E., Atzori, M., Clifton, C.: Hiding the presence of individuals from shared databases. In: Proceedings of the 2007 ACM SIGMOD International Conference on Management of Data, SIGMOD 2007, pp. 665–676. ACM, New York (2007). http://doi.acm.org/10.1145/1247480.1247554
20. Ohm, P.: Broken promises of privacy: responding to the surprising failure of anonymization. UCLA Law Rev. **57**, 1701 (2009)
21. Prasser, F., Kohlmayer, F., Lautenschläger, R., Kuhn, K.A.: ARX - a comprehensive tool for anonymizing biomedical data. In: Proceedings of the AMIA 2014 Annual Symposium, Washington D.C., USA, November 2014
22. Roldán, É.: Estimating the Kullback-Leibler divergence. In: Irreversibility and Dissipation in Microscopic Systems, pp. 61–85. Springer International Publishing, Cham (2014)
23. Schürmann, T.: Bias analysis in entropy estimation. J. Phys. A: Math. Gen. **37**(27), L295 (2004)
24. Schürmann, T.: A note on entropy estimation. Neural Comput. **27**(10), 2097–2106 (2015)
25. Siegel, S.: Non-parametric Statistics for the Behavioral Sciences. McGraw-Hill, New York (1956)
26. Steorts, R.C., Ventura, S.L., Sadinle, M., Fienberg, S.E.: A comparison of blocking methods for record linkage. In: Domingo-Ferrer [5], pp. 253–268
27. Sweeney, L.: Achieving k-anonymity privacy protection using generalization and suppression. Int. J. Uncertainty, Fuzziness Knowl. Based Syst. **10**(5), 571–588 (2002)
28. Sweeney, L.: k-anonymity: a model for protecting privacy. Int. J. Uncertainty, Fuzziness Knowl. Based Syst. **10**(5), 557–570 (2002)
29. Weil, P., Hoffgaard, F., Hamacher, K.: Estimating sufficient statistics in co-evolutionary analysis by mutual information. Comput. Biol. Chem. **33**(6), 440–444 (2009)

Synthetic Data

Creating an 'Academic Use File' Based on Descriptive Statistics: Synthetic Microdata from the Perspective of Distribution Type

Kiyomi Shirakawa[1(✉)], Yutaka Abe[2], and Shinsuke Ito[3]

[1] National Statistics Center, Hitotsubashi University,
2-1 Naka, Kunitachi-shi, Tokyo 186-8603, Japan
kshirakawa@ier.hit-u.ac.jp
[2] Hitotsubashi University,
2-1 Naka, Kunitachi-shi, Tokyo 186-8603, Japan
y-abe@ier.hit-u.ac.jp
[3] Chuo University, 742-1 Higashinakano,
Hachioji-shi, Tokyo 192-0393, Japan
ssitoh@tamacc.chuo-u.ac.jp

Abstract. When creating synthetic microdata in Japan, the values from result tables are used to remove links to individual data. The result tables of conventional official statistics do not allow the generation of random numbers for reproducing individual data. Therefore, the National Statistics Center has created pseudo-individual data on a trial basis using the 2004 National Survey of Family Income and Expenditure. Although mean, variance, and correlation coefficient in the original data were reproduced in the synthetic microdata created, the trial did not include the creation of completely synthetic microdata from the result tables, and the reproduction of the distribution was not taken into account. In this study, a method for generating random numbers with a distribution close to that of the original data was tested, and new type of synthetic microdata called an 'Academic Use File' was created. Random numbers were generated completely from the values contained in the result tables. In addition, this test took into account the Anscombe's quartet as well as the sensitivity rule. As a result, based on the numerical values of the result tables, it was possible to determine the approach that best approximates the distribution type of the original data.

Keywords: Higher moment · Hyper-multidimensional cross-tabulation data · Microaggregation · Reproduce · Multivariate normal random numbers · Skewness · Kurtosis

1 Introduction

In statistics education and training, synthetic microdata that can be freely used are important. When creating synthetic microdata in Japan, the values from result tables are used in order to remove links to individual data in order to comply with Japanese legal requirements. Therefore, the National Statistics Center has created synthetic microdata

© Springer International Publishing Switzerland 2016
J. Domingo-Ferrer and M. Pejić-Bach (Eds.): PSD 2016, LNCS 9867, pp. 149–162, 2016.
DOI: 10.1007/978-3-319-45381-1_12

on a trial basis using the 2004 National Survey of Family Income and Expenditure, where the mean, variance, and correlation coefficient in the original data were reproduced in the synthetic microdata created. Here, synthetic microdata is used to refer to microdata that can be accessed without an application and used without restrictions (called 'trial synthetic microdata' hereinafter). Trial synthetic microdata can be downloaded free, and have been used in higher education and training. However, the previous trial did not include sufficient information about statistical tables such as kurtosis and skewness and therefore did not allow the creation of completely synthetic microdata based on statistical tables. Thus the distribution type of the synthetic microdata was not taken into account in reproducing the distribution type of the original data, and not sufficient for advanced statistical analysis.

In this study, we tested a method for generating random numbers with a distribution close to that of the original data. Random numbers were generated completely from the values posted in the result tables. For this work, Anscombe's quartet was taken into account. Also, based on the numerical values in the result tables, we aimed to establish the closest approach to the distribution type of the original data.

Section 2 discusses the methods used to create existing synthetic microdata by applying microaggregation, problems with the existing microdata, and a correction for those problems. Section 3 describes some new synthetic microdata using the correction introduced in Sect. 2. Section 4 presents an assessment of disclosure risk for new synthetic microdata. Section 5 compares the existing and new synthetic microdata with the original data and evaluates their utility. Section 6 presents our conclusions and a future outlook.

2   Problems with and Improvements to the Trial Synthetic Microdata

2.1   Applicability of Microaggregation to Synthetic Microdata

Microaggregation is one of the disclosure limitation methods adopted for official microdata. Microaggregation entails dividing the individual records into groups larger than a threshold k and replacing the records with common values as measures of the central tendency (e.g., the mean) within each group. The method of microaggregation was developed based on research by Defays and Anwar (1998), Domingo-Ferrer and Mateo-Sanz (2002), and others.

Ito et al. (2008) and Ito (2009) applied the methodology of microaggregation to Japanese official microdata, identified the applicability of microaggregation to synthetic microdata, and evaluated the effectiveness of microaggregation for individual data from the National Survey of Family Income and Expenditure. These studies were the first in Japan to advocate methods using multi-dimensional cross-tabulation to create microaggregated data that closely resembles individual data. The proposed method of microaggregation is as follows. In the first step, records with common values for all types of qualitative attributes based on multi-dimensional cross-tabulation were created. In the second step, records with common values for qualitative attributes were sorted and divided into groups larger than a specific threshold, and the value of each quantitative attribute for records was replaced with an average value within each group.

Microaggregation is generally applied to the quantitative attributes contained in microdata. For such attributes, if the records containing a common attribute value are grouped for every target qualitative attribute and these attribute values are viewed as being replaced with representative values for the group, then grouping of records related to qualitative attribute values can also be positioned as a form of microaggregation. In this case, the microaggregated data are considered to be the set of the same qualitative attribute values within a particular group and the corresponding set of records that contain the mean values of the quantitative attributes. Although this kind of microaggregated data can be viewed as data that conform to individual data consisting of a set of qualitative attribute values and a set of mean values of quantitative attributes, the set of attribute values of each of the records can be positioned as only aggregate values.

Although cross-tabulation tables can be created by the grouped target qualitative attributes, the frequency of the designated cells within a cross-tabulation table matches the number of records within the corresponding group in the microaggregated data. This means that the number of qualitative attributes used for the grouping of records increases as the dimensionality of the cross-tabulation table increases. By expanding on this methodology, we can define "hyper-multidimensional cross-tabulation tables" which are "n-dimensional cross-tabulation tables created by tabulating the set of all attributes of the individual data" (Ito 2008), and we can logically construct a set of microaggregated data that characterize the set of records having a correspondence with the cells contained in the cross-tabulation table. Note that hyper-multidimensional cross-tabulation tables include all dimensions of cross-tabulation tables from 1 to n dimensions (Bethlehem et al. 1990, Höhne 2003). This means that various dimensions of cross-tabulation tables can be created for setting the hyper-multidimensional tabulation tables and can serve as the basis for creating synthetic microdata in the framework of hyper-multidimensional cross-tabulation tables.

As mentioned above, the characteristics of microaggregation are that the records contained in the individual data are grouped into a set of records with a threshold value k, and the individual attribute values in the records of the group are replaced with a representative value such as the mean value. This indicates that the number of records that exist within the set of records with common values for qualitative attributes has a correspondence relationship with the frequency of cells in the hyper-multidimensional cross-tabulation tables created with the same set of attributes. Therefore, once the lower limit on the number of records contained in the set of records with common values for qualitative attributes has been set, this determines the threshold value for the frequency of cells contained in the hyper-multidimensional cross-tabulation table. When the threshold k is set, a cross-tabulation table can be created by appropriately selecting the combination of attributes from the set of attributes that form the aggregation items in the hyper-multidimensional cross-tabulation table in such a way that no cells contained in the hyper-multidimensional cross-tabulation table are zero and all the cells have a frequency of at least k. Furthermore, if cells with a frequency less than the threshold k exist in the hyper-multidimensional cross-tabulation table, then it is possible to perform grouping into the set of records with common values for qualitative attributes with k or higher by performing processing based on "unknowns" in the group of attributes of records corresponding to those cells.

By doing this the creation of data that conforms to individual data based on hyper-multidimensional cross-tabulation data can be methodologically positioned within the microaggregation framework. This demonstrates that microaggregation forms a logical foundation in the method of creating synthetic microdata for education.

2.2 Creating the Trial Synthetic Microdata

Synthetic microdata for public Japanese microdata were created based on the methodology of microaggregation. This section describes how the synthetic microdata were created using multi-dimensional tabulation, in reference to Sect. 3 of Makita et al. (2013). The detailed process for creating synthetic microdata is as follows.

First, quantitative and qualitative attributes to be contained in the synthetic microdata were selected. Second, records with common values for qualitative attributes were sorted into groups with a minimum size of 3. Third, tables were created in order to generate multivariate lognormal random numbers and records for which the values for some quantitative attributes were 0. This process allowed the creation of synthetic microdata with characteristics similar to those of the original microdata (Makita et al. 2013, p. 2).

Figures 1 and 2 present the detailed process of creating the synthetic microdata, as described below.

1. Qualitative attributes were selected from the multi-dimensional statistical tables compiled based on the original microdata. Specifically, 14 qualitative attributes were selected based on the survey items used most frequently by researchers, including gender, age, and employment status. In addition, 184 quantitative attributes were selected, including yearly household income and monthly household expenditures.
2. Records with common values for qualitative attributes were sorted into groups with a minimum size of 3. For records that have common values for some qualitative attributes and that refer to groups with a size of 1 or 2, values for the other qualitative attributes were transformed to 'unknown' (V) in order to create groups with a minimum size of 3.
3. Two types of tables were created in order to generate (1) multivariate lognormal random numbers and (2) records with negative values for some quantitative attributes. Tables of 'Type 1' contain frequency, mean, variance, and covariance of quantitative attributes not including 0. The records on which these tables are based were classified by qualitative attribute in order to generate multivariate lognormal random numbers. Tables of 'Type 2' are tables created by sorting records based on whether values for quantitative attributes are 0 or not 0, and on this basis, the values for some quantitative attributes in the records were transformed to 0 (Makita et al. 2013, p. 3).

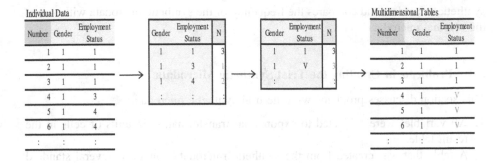

Fig. 1. Sorting records with common values for qualitative attributes into groups with a minimum size of 3. Note: "V" stands for "unknown". Source: Makita et al. (2013).

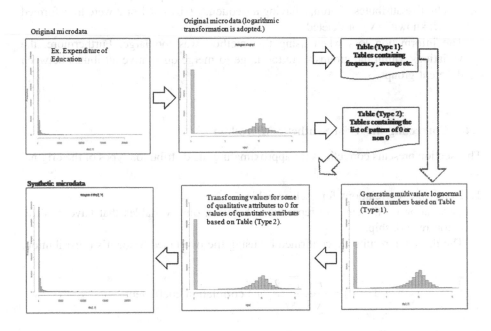

Fig. 2. Creation of the synthetic microdata and comparison between the frequency of the synthetic microdata and that of the original microdata. Source: Makita et al. (2013).

To create the synthetic microdata, logarithmic transformation was applied to the original microdata items. Then, multivariate lognormal random numbers were created based on the above two types of tables, and the values for some quantitative attributes were transformed to 0. As a final step, exponential transformation was conducted. Figure 1 illustrates how to process records with common values for qualitative attributes into groups with a minimum size of 3. Figure 2 shows the creation of the synthetic microdata and compares the frequency of the synthetic microdata with that of the original microdata.

2.3   Problems in Creating the Trial Synthetic Microdata

This section discusses problems with the trial synthetic microdata.

1. All variables were subjected to exponential transformation in units of cells in the result table.
   A table that was created from the synthetic microdata contained several standard deviations that were too large.
2. Correlation coefficients (numerical) between all variables were reproduced.
   Several correlation coefficients were too small. This was because correlation coefficients between uncorrelated variables were also reproduced.
3. Qualitative attributes of groups having a frequency (size) of 1 or 2 were transformed to "Unknown" (V) or deleted.
   The information loss when using this method was too large. Furthermore, the variations within the groups were too large to merge qualitative attributes between different groups.

2.4   Correcting the Trial Synthetic Microdata

This section presents corrections for approximating the distribution types of the original data.

1. Detect non-correlations for each variable.
   Correlation coefficients are reproduced between only variables that have a correlation relationship:
   The detection results are confirmed by using the two-tailed Student's t distribution.

$$T(r,0) = \frac{|r|\sqrt{n-2}}{\sqrt{1-r^2}} \quad r : \text{correlation coefficient}$$

2. Qualitative attributes in groups with a size of 1 or 2 are merged into a group that has a minimum size of 3 in the upper hierarchical level.

Note that Anscombe's quartet shows four groups that have the same frequency, mean, standard deviation, and regression model parameters. However, the distribution types of these groups are different (Fig. 3).

This indicates that second moments can be reproduced based on the mean and standard deviation. However, it also indicates that third and fourth moments (skewness and kurtosis) cannot be reproduced. More specifically, we can see that the numerical values of the kurtosis and skewness differ from those of the original microdata.

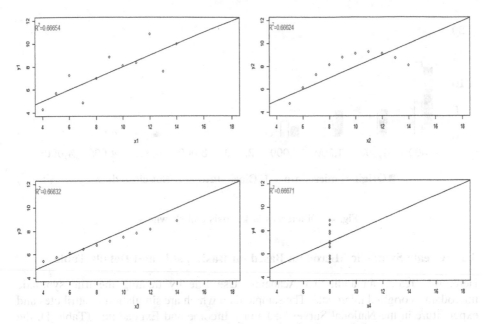

Fig. 3. Scatter plots of numerical examples for Anscombe's quartet.

To resolve these problems, it is necessary for the numerical values of the third and fourth moments to approximate those of the original microdata. The specific indicators are frequency, mean, standard deviation, kurtosis, and skewness. Furthermore, to create the synthetic microdata (here, pseudo-microdata created by microaggregation) based on multivariate normal random numbers is required in order to change the distribution type of the original data into a standard distribution. Note that these indicators are the minimum indicators for reproducing the original microdata, and are not absolute indicators.

3 Creating New Synthetic Microdata

3.1 Create the Synthetic Microdata Based on Kurtosis and Skewness

After creating several multivariate normal random numbers, a random number that approximates the kurtosis and skewness of the original microdata was selected.

From Fig. 4, the synthetic microdata have approximately the same kurtosis and skewness as the original microdata. This figure shows that the contributions to kurtosis and skewness are clear. In this study, a set of new synthetic microdata created by the frequency, mean, standard deviation, kurtosis and skewness are called an 'Academic Use File'.

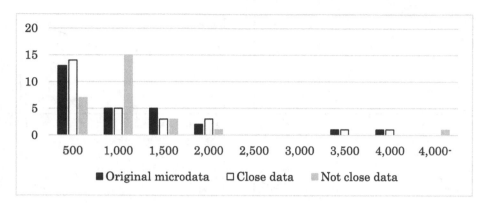

Fig. 4. Differences in kurtosis and skewness.

3.2   Create Synthetic Microdata Based on Basic Table and Details Table

Here, we created an additional Academic Use File by treating the trial synthetic microdata as original microdata. The sample data which are simulation of attributes and expenditure in the National Survey of Family Income and Expenditure (Table 1), the basic table (Table 2) and the details table (Table 3) are shown in order to explain the method for creating the Academic Use File. Note that the method for creating the Academic Use File is the same as for creating the trial synthetic microdata.

In this study, we created an Academic Use File based on the correction described in Sect. 2.4 above.

We employed the following two tables:
Basic table: Frequency, mean, standard deviation, kurtosis and skewness
Details table: Frequency, mean, and standard deviation

Several multivariate normal random numbers were generated based on the mean and standard deviation from the basic table. Next, we selected random numbers that are near the kurtosis and skewness of the original microdata. From this, we performed transformation based on non-correlation detection. Finally, we replaced the random numbers we have been working with up to now with the mean and standard deviation within each group in the details table. By doing this, the numerical values of each of the variables in the synthetic microdata matched the numerical values of the variables in the details table, and we obtained multivariate microdata. Furthermore, the mean and standard deviation were the same and the kurtosis and skewness were approximately the same at the same level of dimensionality as the basic table (number of multivariate cross fields).

Note that if there were groups of size 1 or 2 in the details table, those qualitative attributes were not transformed to unknown (V). Furthermore, those records were also not deleted. This was because groups of size 1 or 2 were merged into groups at the same level as the basic table (upper hierarchy level).

Table 1. Sample data (individual data).

Group No.	A	B	C	D	E	F	Living expenditure	Food
1	2	1	1	2	5	1	125,503.5	29,496.1
	2	1	1	2	5	1	255,675.9	25,806.2
	2	1	1	2	5	1	175,320.4	38,278.2
2	2	1	1	3	6	1	181,085.6	74,122.1
	2	1	1	3	6	1	124,471.0	33,256.8
	2	1	1	3	6	1	145,717.7	46,992.8
3	2	1	1	3	7	1	319,114.3	113,177.1
	2	1	1	3	7	1	253,685.2	67,253.6
	2	1	1	3	7	1	236,447.6	61,129.8
4	3	1	1	1	5	1	137,315.3	27,050.1
	3	1	1	1	5	1	253,393.7	47,205.6
	3	1	1	1	5	1	232,141.8	52,259.6
	3	1	1	1	5	1	214,540.4	54,920.9
5	3	1	1	1	6	1	234,151.4	74,993.0
	3	1	1	1	6	1	278,431.0	78,916.1
	3	1	1	1	6	1	197,180.8	72,909.6
6	3	1	1	2	5	1	118,895.1	48,821.6
	3	1	1	2	5	1	130,482.8	47,798.5
	3	1	1	2	5	1	147,969.1	50,277.9
	3	1	1	2	5	1	150,973.7	48,291.0

A: 5-year age groups; B: employment/unemployed; C: company classification; D: company size; E: industry code; F: occupation code

Table 2. Basic table (matches with original mean and standard deviation, approximate correlation coefficients for each variable).

	Living expenditure	Food
Mean	195,624.8	54,647.8
SD	59,892.6	21,218.1
Kurtosis	-1.004164	1.628974
Skewness	0.346305	0.992579
Frequency	20	20
Correlation coefficients	Living expenditure	Food
Living expenditure	1	
Food	0.643	1

SD: Standard Deviation. The same hereinafter.

Table 3. Details table (means and standard deviations for creating synthetic microdata for multidimensional cross fields).

Groups	Living expenditure			Food		
	Frequency	Mean	SD	Frequency	Mean	SD
1	3	185,499.9	65,680.5	3	31,193.5	6,406.9
2	3	150,424.8	28,599.3	3	51,457.2	20,795.2
3	3	269,749.0	43,611.7	3	80,520.1	28,447.0
4	4	209,347.8	50,580.8	4	45,359.0	12,618.4
5	3	236,587.8	40,679.9	3	75,606.2	3,049.8
6	4	137,080.2	15,119.7	4	48,797.2	1,071.9

3.3 Create Synthetic Microdata Based on Multivariate Normal Random Numbers

This is a method for creating the trial synthetic microdata. Refer to Sect. 2.2 above for details. Since kurtosis and skewness are used as stated in Sect. 3.1, we do not use exponential transformation. Furthermore, we also tested other methods and specifically looked at microaggregation with a threshold of 3. This is a method of sorting the values of variables in ascending order, dividing them into groups of minimum size 3, and creating synthetic microdata based on the means and standard deviations in these groups. This method is very simple and useful, but it was not suitable for creating synthetic microdata based on public statistics result tables because the multivariate variables cannot be sorted in ascending order for each variable. As a result, this method was excluded from this research.

4 Disclosure Risk of Academic Use File

The Academic Use File reproduces the original distribution type, so the standard deviation, kurtosis and skewness can be estimated. However, the combination of data values has a risk of disclosure if the standard deviation, skewness, and kurtosis are known (see Shirakawa et al., 2016).

5 Comparison Between Various Sets of Synthetic Microdata

To compare various sets of synthetic microdata, we selected synthetic microdata that most closely approximated the original microdata. Furthermore, we selected indicators for creating the optimal synthetic microdata. We compared the characteristics with the original data in order to establish how easy the synthetic microdata are to use. Table 4 shows various indicators for the original microdata and three sets of synthetic microdata.

The number of observation values was 20 in all of the microdata, and the means and standard deviations were also the same. Furthermore, the correlation coefficients

Table 4. Comparison of original microdata and each set of synthetic microdata

No.	1 Original microdata		2 Hierarchization, kurtosis, and skewness		3 Kurtosis and skewness		4 Multivariate lognormal random numbers	
	Living expenditure	Food	Living expenditure	Food	Living expenditure	Food	Living expenditure	Food
1	125,503.5	29,496.1	110,487.8	25,143.0	107,684.0	23,459.9	133,549.9	38,559.9
2	255,675.9	25,806.2	232,691.8	37,905.5	281,880.8	56,520.4	123,716.6	42,930.1
3	175,320.4	38,278.2	213,320.2	30,531.9	254,267.3	37,419.4	152,784.8	67,263.8
4	181,085.6	74,122.1	183,430.4	75,469.1	294,589.9	112,843.9	195,764.8	8,286.1
5	124,471.0	33,256.8	134,867.6	39,568.9	193,191.6	54,363.3	202,865.8	75,558.0
6	145,717.7	46,992.8	132,976.4	39,333.7	189,242.7	53,980.3	193,003.4	70,994.2
7	319,114.3	113,177.1	242,622.5	68,472.2	151,183.6	55,303.2	191,620.1	52,311.7
8	253,685.2	67,253.6	320,055.9	113,008.5	271,338.1	79,991.4	72,773.7	13,621.6
9	236,447.6	61,129.8	246,568.6	60,079.7	157,306.9	50,650.9	201,114.6	74,899.0
10	137,315.3	27,050.1	144,192.6	32,572.9	167,431.0	36,116.3	217,530.7	60,736.0
11	253,393.7	47,205.6	267,708.8	60,344.8	270,301.8	78,246.4	297,608.7	77,464.3
12	232,141.8	52,259.6	212,050.7	37,656.3	223,946.8	43,827.9	175,993.6	71,416.6
13	214,540.4	54,920.9	213,439.1	50,862.2	225,103.2	63,861.2	297,653.0	86,400.5
14	234,151.4	74,993.0	205,595.0	73,919.1	165,972.3	49,350.6	123,197.1	31,645.5
15	278,431.0	78,916.1	282,652.7	79,126.9	249,749.1	73,474.1	277,501.6	69,910.5
16	197,180.8	72,909.6	221,515.6	73,772.7	183,281.1	48,672.3	235,221.1	58,700.6
17	118,895.1	48,821.6	127,964.3	50,240.7	115,639.3	71,059.5	182,363.2	49,433.2
18	130,482.8	47,798.5	159,328.0	48,533.5	170,231.1	38,723.5	158,939.4	45,131.8
19	147,969.1	50,277.9	133,795.5	47,660.6	125,789.2	22,188.5	212,194.2	37,995.6
20	150,973.7	48,291.0	127,232.9	48,754.2	114,366.4	42,903.1	267,100.1	59,697.3
Mean	195,624.8	54,647.8	195,624.8	54,647.8	195,624.8	54,647.8	195,624.8	54,647.8
SD	59,892.6	21,218.1	59,892.6	21,218.1	59,892.6	21,218.1	59,892.6	21,218.1
Kurtosis	-1.004164	1.628974	-0.810215	1.473853	-1.220185	1.721354	-0.212358	-0.052164
Skewness	0.346305	0.992579	0.310913	1.050568	0.160612	0.949106	0.035785	-0.709361
Correlation coefficients	0.642511		_0.689447_		0.642511		0.642511	
Maximum	319,114.3	113,177.1	320,055.9	113,008.5	294,589.9	112,843.9	297,653.0	86,400.5
Minimum	118,895.1	25,806.2	110,487.8	25,143.0	107,684.0	22,188.5	72,773.7	8,286.1

were either the same (column numbers 3 and 4) or approximately the same (column number 2) as those of the original microdata.

Note that the correlation coefficients for all of the synthetic microdata were the same as those for the original microdata. However, because the synthetic microdata for column number 2 was transformed from the means and standard deviations in the six groups in the details table and not from the means and standard deviations in the basic table after creating the random numbers, they do not match due to variations in the values between when the random numbers were created and after transformation. In addition, the indicators for the skewness, kurtosis, maximum value, and minimum value differ between the different microdata.

The most useful microdata from the indicators in Table 4 are in column number 2. Next are those in column number 3, and finally column number 4. Note that for reference, column number 4 is the same as trial synthetic microdata method.

From Fig. 5, column number 2 approximates the original microdata, and column numbers 3 and 4 contain several outliers. This result shows that kurtosis and skewness are useful indicators for synthetic microdata, and furthermore that transformation using the mean and standard deviation from the details table (lower hierarchical level) is required after creating the random numbers. Note that Table 5 shows an example of the result table for creating the optimal synthetic microdata.

Table 5. Example of the result table for creating the Academic Use File.

No.	A	B	C	D	E	F	Frequency	Mean	SD	Frequency	Mean	SD
			Items					Living expenditure			Food	
1	2	1	1	2	5	1	3	185,499.9	65,680.5	3	31,193.5	6,406.9
	2	1	1	3			6	210,086.9	73,208	6	65,988.7	27,387.3
2	2	1	1	3	6	1	3	150,424.8	28,599.3	3	51,457.2	20,795.2
	2	1	1	3	7	1	3	269,749.0	43,611.7	3	80,520.1	28,447.0
	3	1	1	1			7	221,022.1	45,197.7	7	58,322.1	18,550.2
3	3	1	1	1	5	1	4	209,347.8	50,580.8	4	45,359.0	12,618.4
	3	1	1	1	6	1	3	236,587.8	40,679.9	3	75,606.2	3,049.8
4	3	1	1	2	5	1	4	137,080.2	15,119.7	4	48,797.2	1,071.9
Mean								195,624.8			54647.8	
Standard deviation								59,892.6			21218.1	
Kurtosis								-1.004			1.629	
Skewness								0.346			0.993	
Correlation coefficients								0.643				

A: 5-year age groups; B: employment/unemployed; C: company classification; D: company size; E: industry code; F: occupation code

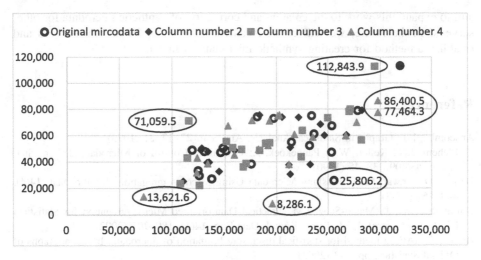

Fig. 5. Scatter plots of living expenditure and food for each set of microdata.

6 Conclusions and Future Outlook

In this paper, we focused on improvements to synthetic microdata created by the National Statistics Center for statistics education and training. The synthetic microdata created by National Statistics Center are not a duplicate of the original microdata, but rather a substitute suitable for statistics education and training. More specifically, these synthetic microdata were created by using microaggregation, which is a disclosure limitation method for public statistical microdata. However, the synthetic microdata do not always reproduce the distribution type of the original data.

We attempted to create an Academic Use File using several methods that adhere to this disclosure limitation method. The results show that kurtosis and skewness are useful in addition to the frequency, mean, standard deviation, and correlation coefficient that have previously been used as indicators. In addition, the results show that adjustment using the mean and standard deviation from strata in the details table is required for utility of the Academic Use File. However, there are no examples containing the indicators we examined in this work (kurtosis and skewness) in result tables of Japanese public statistics. The conclusion of this paper is to take the tabulation table with the kurtosis and skewness added to the conventional indicators as a basic table. Furthermore, for correlation relationships, the correlation coefficients (numerical values) between variables are reproduced based on detection of non-correlations. Transformations to the frequency, mean, and standard deviation in each group are based on a details table (multi-dimensional cross fields). By doing this, it is possible to create an Academic Use File that approximates the original microdata.

Problems for the future are deciding number of cross fields (dimensionality) of the basic table and details table and the style (indicators to tabulate) of the result table according to the statistical fields in the public survey. The reason is that new trials will be necessary if there is a lack of indicators based on this conclusion. Furthermore, we aim to expand this work to the creation and correction of synthetic microdata for other surveys. In the future, we will create Academic Use Files for several surveys and establish a method for creating synthetic microdata in Japan.

References

Anscombe, F.J.: Graphs in Statistical Analysis. Am. Stat. **27**, 17–21 (1973)

Bethlehem, J.G., Keller, W.J., Pannekoek, J.: Disclosure Control of Microdata. J. Am. Stat. Assoc. **85**(409), 38–45 (1990)

Defays, D., Anwar, M.N.: Masking Microdata Using Micro-Aggregation. J. Offic. Stat. **14**(4), 449–461 (1998)

Domingo-Ferrer, J., Mateo-Sanz, J.M.: Practical Data-oriented Microaggregation for Statistical Disclosure Control. IEEE Trans. Knowl. Data Eng. **14**(1), 189–201 (2002)

Höhne, J.: SAFE-a method for statistical disclosure limitation of microdata. In: Monographs of Official Statistics, pp. 1–3 (2003)

Hundepool, A., van de Wetering, A., Ramaswamy, R., Franconi L., Polenttini, S., Capobianchi, A., de Walf, P. P., Domingo-Ferrer, J., Torra, V., Brand, R., Giessing, S.: μ-ARGUS Version 4.2 Software and User's Manual. Statistics Netherlands, Vooburg NL (2008). http://neon.vb.cbs.nl/casc/Software/MuManual4.2.pdf

Ito, S., Isobe, S., Akiyama, H.: A Study on Effectiveness of microaggregation as disclosure avoidance methods: based on national survey of family income and expenditure. NSTAC Working Paper 10, pp. 33–66 (2008). (in Japanese)

Ito, S.: On Microaggregation as Disclosure Avoidance Methods. J. Econ. Kumamoto Gakuen Univ. **15**(3/4), 197–232 (2009) (in Japanese)

Makita, N., Ito, S., Horikawa, A., Goto, T., Yamaguchi, K.: Development of synthetic microdata for educational use in Japan. Paper Presented at 2013 Joint IASE/IAOS Satellite Conference, Macau Tower, Macau, China, pp. 1–9 (2013)

Shirakawa, K., Abe, Y., Ito, S.: Empirical analysis of sensitivity rules: cells with frequency exceeding 10 that should be suppressed based on descriptive statistics. In: Privacy in Statistical Databases, Dubrovnik, Croatia, September 2016

COCOA: A Synthetic Data Generator for Testing Anonymization Techniques

Vanessa Ayala-Rivera[✉], A. Omar Portillo-Dominguez, Liam Murphy,
and Christina Thorpe

Lero@UCD, School of Computer Science, University College Dublin, Dublin, Ireland
vanessa.ayala-rivera@ucdconnect.ie,
{andres.portillodominguez,liam.murphy,christina.thorpe}@ucd.ie

Abstract. Conducting extensive testing of anonymization techniques is critical to assess their robustness and identify the scenarios where they are most suitable. However, the access to real microdata is highly restricted and the one that is publicly-available is usually anonymized or aggregated; hence, reducing its value for testing purposes. In this paper, we present a framework (COCOA) for the generation of realistic synthetic microdata that allows to define multi-attribute relationships in order to preserve the functional dependencies of the data. We prove how COCOA is useful to strengthen the testing of anonymization techniques by broadening the number and diversity of the test scenarios. Results also show how COCOA is practical to generate large datasets.

1 Introduction

The increasing availability of microdata has attracted the interest of organizations to collect and share this data for mining purposes. However, current legislation requires personal data to be protected from inappropriate use of disclosure. Anonymization techniques help to disseminate data in a safe manner while preserving enough utility for its reuse. For this reason, a plethora of methods for anonymizing microdata exists in the literature. Each of these techniques claims a particular superiority over the others (e.g., improving data utility or computational resources). However, their performance can vary when they are tested with different datasets [13]. This fact makes difficult for data controllers to generalize the conclusions of the performance evaluation to decide which algorithm is best suited for their requirements.

All performance evaluations are limited in one way or the other (due to time/effort/cost constraints). A common limitation is the number and the diversity of the datasets used, as the process to obtain good quality datasets can be burden and time consuming. For example, the access to real microdata is highly restricted to protect the privacy of individuals. Agencies grant access only for certain period and once this expires, the provided files must be destroyed [2,4,10] which does not always allow for reproducibility of experimental results.

Consequently, researchers often use real datasets that are publicly-available or synthetic data generated in an ad-hoc manner [20]. Both solutions are to

© Springer International Publishing Switzerland 2016
J. Domingo-Ferrer and M. Pejić-Bach (Eds.): PSD 2016, LNCS 9867, pp. 163–177, 2016.
DOI: 10.1007/978-3-319-45381-1_13

some extend biased as they are constrained by an specific data distribution: either a real one which consists of a single scenario (e.g., demographics of a single country) or an ideal synthetic data which may never be found in the real world. Moreover, the applicability of most real datasets is often limited for testing anonymization techniques as datasets available in research are usually aggregated and pre-anonymized to protect the privacy of people (which is exactly the use case of privacy-preserving methods). Thus this data lacks of enough diversity to simulate attacking scenarios or to evaluate the robustness of the proposed techniques (e.g., data sparsity and outliers). For these reasons, synthetic data is a valuable resource for conducting testing in multiple areas, as it allows to manipulate the data to meet specific characteristics that are not found in the real data but still need to be considered for testing (hypothetical future scenarios). To be adequate substitutes for real data, the functional dependencies of the data need to be preserved.

To tackle these issues, our research work has centered on developing techniques to create realistic synthetic datasets applicable to the data privacy area. The aim has been to help researchers and practitioners (hereinafter referred as users) to improve the testing of anonymization techniques by decreasing the effort and expertise needed to create useful datasets in order to be able to perform more robust testing. As a first step in that direction, in our previous work [12] we described the process followed to mimic the distribution of the aggregated statistics from the 2011 Irish Census. The work proposed in this paper builds on top of that work by presenting a framework (COCOA) that facilitates the creation of realistic datasets in the privacy area. Internally, COCOA leverages on a set of supported domains to automatically generate datasets of different characteristics. The contributions of this paper are the following:

1. A framework (COCOA) to generate realistic synthetic datasets (at record-level) that allows to define multi-attribute relationships in order to preserve the functional dependencies of the data.
2. A comprehensive practical evaluation of COCOA, consisting of a prototype and a set of experiments to assess it in terms of the benefits it brings to the testing of anonymization techniques as well as its costs.
3. Three sets of datasets (publicly available [1]), each one composed of 72 different datasets (based on real data and offering diverse data distributions and sizes) which can be useful for the research community to perform more comprehensive validations in the data privacy area.

2   Related Work

Numerous approaches have been proposed to generate synthetic data. Some of those works are general-purpose synthetic data generators [17,24]. That is, they do not target a specific application area. However, this generality may reduce the accuracy in the results when specific features are required. Depending on the domain, it is important to preserve certain characteristics in the generated datasets to make it realistic and thus, suitable for their intended use. In the data privacy community, synthetic data generation is used as a strategy for disseminating data while ensuring confidentiality. The work in this research area focuses on generating synthetic populations that reproduce certain statistical properties of the original data. This kind of approaches was firstly introduced in [26]. Afterward, many techniques and tools have been proposed for generating synthetic datasets and populations [7,8]. Similarly, other research efforts have focused on evaluating the privacy of synthetic methods and avoid the risk of disclosure [18,22]. In our work, we focus on generating synthetic microdata, not as a technique of data protection, but as a mechanism to improve the testing of anonymization techniques. Furthermore, our solution focuses on categorical data due to its relevance (e.g., a valuable asset of data mining) and because methods for numerical data have been well studied in the literature [23,27].

3   COCOA: A Synthetic Data Generator

Here we provide the context of our solution, discuss its internal workings, and describe the generated datasets.

3.1   Overview

The goal of this research work has been to develop a framework for synthetically generating datasets (COCOA) that can be used to diversify the set of characteristics available in microdata. This strategy would help researchers to improve the testing of anonymization techniques. In Fig. 1, we depict the conceptual view of our solution. It can be seen how COCOA follows an iterative process (see Sect. 3.2) to build a set of datasets based on the *information base* provided by the user. The information base is composed of all the input parameters required by the chosen dataset domain (e.g., dataset size).

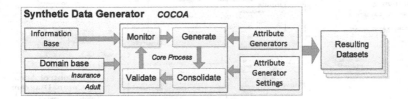

Fig. 1. COCOA - conceptual view.

The key element of COCOA is its *domain base*, which encapsulates the expert knowledge of the supported business domains (e.g., census, healthcare, finance). This element allows COCOA to be easily extensible and capable of incorporating multiple business cases (even for the same domain e.g., Irish census, USA census), which might be suitable to different test scenarios. In this context, a *domain* defines the rules and constraints required to generate a dataset that preserves the functional dependencies of a business case. Each domain is characterized by a name, a set of attributes and their corresponding attribute generators.

To generate data, the domains make use of the available set of *attribute generators*. These elements are supporting logic which offer miscellaneous strategies to generate the values of attributes. For example, a generator might focus on diversifying the data in an attribute by fitting it into a data distribution (e.g., normal distribution). In this example, several generators can be combined to offer different data distributions (as the appropriate distribution might vary depending on the usage scenario). In case an attribute generator requires any particular settings to work properly (e.g., the default values for its applicable parameters), this information (e.g., mean and variance for a normal distribution) can also be captured by the framework (as an *attribute generator setting*).

3.2 Core Process

From a workflow perspective, COCOA has a core process (depicted in Fig. 2a). It requires three user inputs related to desired characteristics of the resulting dataset: the domain (selected among the ones available in the framework), the size, and the dataset name. An additional optional user input is the set of specific parameters that a domain might need to initialize a dataset. As an initial step, the process sets all the configured input parameters and generates an empty dataset for the chosen domain. Next the loop specified in the monitor, generate, consolidate and validate phases is performed: First, the size of the new dataset (number of tuples) is monitored. This check is done in order to determine when the dataset has reached the target size and finish the process. Next, the applicable generators to create the attributes' data are retrieved from the knowledge database (as they will depend on the domain). Then, their relationships are identified in order to sort the generators and execute them in the correct order,

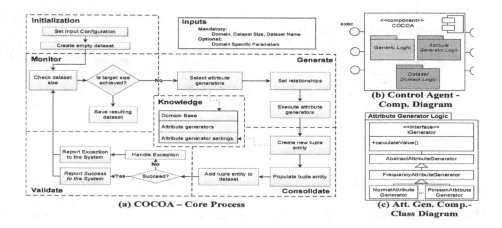

(a) COCOA – Core Process

(b) Control Agent - Comp. Diagram

(c) Att. Gen. Comp.- Class Diagram

Fig. 2. COCOA architecture

so that the proper chain of functional dependencies is executed (see Sect. 3.4). Once there are new values for all the attributes, a new entity object is created to represent the new tuple. Next, its attributes are populated with the new values and the tuple entity is added to the dataset. Moreover, any exceptions are internally handled and reported. This core process continues iteratively until the new dataset has been fully generated. As a final step, the dataset is saved to disk.

3.3 Architecture

COCOA is complemented by the architecture presented in the component diagram [9] of Fig. 2b. COCOA is composed of three main components: The generic component contains the control logic and all supporting functionality which is independent of the supported domains and attribute generators (e.g., the monitor and validate phases of the core process). Regarding the logic that interfaces with the domains, it needs to be customized per domain (by defining a tuple entity and a dataset generator). Similar case with the supported attribute generators. Therefore, these two logics are encapsulated in their respective components to minimize the required code changes. To complement this design strategy, the components are only accessed through interfaces. This is exemplified in Fig. 2c, which presents the high-level structure of the attribute generator component. It contains a main interface *IGenerator* to expose all required actions and an abstract class for all the common functionality. This hierarchy can then be extended to support specific attribute generators.

3.4 Attribute Generators

As discussed in Sect. 3.1, attribute generators are supporting logic which offer miscellaneous strategies to generate the values of attributes. Among the alternative choices to develop attribute generators for COCOA, we have initially concentrated on implementing the following three:

Distribution-based generator. This type of generator produces independent data, so it is applicable for attributes whose values do not depend on others. In this case, the strategy used to diversify the data is to mimic a probability distribution. This generator can also be a good fit for data that comes from a previously consolidated real data source (like most of the datasets currently available). This is because, the objective in those cases is commonly to diversify the frequencies of the existing values without modifying the actual values or the cardinality of the attribute. A distribution-based generator requires two parameters: a distribution (which will be used to re-distribute the frequencies of the values) and a sorting strategy (which will be used to order the candidate values before applying the distribution). For the initial version of COCOA, 11 commonly-used data distributions are supported: normal, beta, chi, chi square, exponential (exp), gamma, geometric, logarithmic (log), poisson, t-student (Tstu), and uniform (uni) [30]. An additional supported distribution is the "original" one. When it is used, the generator only mirrors the given input distribution. This is useful for generating new datasets (of different sizes) for existing domains (e.g., adult and german credit) without modifying the original distribution present in the real data. Regarding the sorting strategies, COCOA currently supports three: alpha-numeric, reverse alpha-numeric, and no-sort. The usage of alpha-numeric and reverse alpha-numeric allows a user not only to logically sort the categorical values, but also helps to further diversify the tested behaviors by mixing the supported sorts and distributions. For instance, the usage of an alpha-numeric sorting and an exponential distribution can generate a J-Shaped distribution, while a reverse J-Shaped distribution can be easily generated by switching to the reverse alpha-numeric sorting. Finally, the no-sort strategy respects the original order of the data (following the same line of thought as the original distribution previously discussed).

Attribute-based generator. This type of generator produces dependent data. It is applicable to cases when there are functional dependencies that need to be preserved between attributes (so that the generated data can be realistic) as well as cases when an attribute needs to be derived from others. For instance, consider a dataset with information about individuals owned by an insurance company. The aim is to derive a class attribute that identifies groups with a higher risk of having an accident. One way to derive this class (e.g., low, medium, high) would be to use the values from attributes such as age, occupation, and hobby. The parameters required by this type of generator might vary, but normally they will take most of their required input information from other attributes. As discussed in Sect. 3.2, COCOA executes the generators in such an order that an attribute-based generator has available its required information (i.e., the attributes it depends on) before its execution.

Distribution and attribute-based generator. This type of generator also produces dependent data. This is because it is a hybrid of the previous two generators. It is useful to capture more complex relationships where the generation of a value is influenced not only by a frequency distribution, but also by the value of one or more attributes. For instance, the place where a hobby is practiced does not only depend on the performed activity but also on a certain distribution. For example, based on historical information, soccer is mostly practiced on outside fields, and in less degree, in other places such as an indoor facility or beach. Another example is the salary of a person, which is influenced by multiple factors such as her occupation and years of work experience. This kind of relationships can be easily captured in COCOA by this type of generator.

3.5   Supported Domains

The following sections describe the domains currently supported by COCOA. The datasets generated for all the domains are publicly available [1].

Irish census. This domain, initially presented on [12], is composed of 9 attributes belonging to the Irish Census 2011. Appendix B lists the attributes and their generators. It is worth noticing how different types of generators were used to capture the relationships among the data.

Insurance domain. To further exemplify the capabilities of COCOA and its different attribute generators, we have designed the insurance domain. It represents information of interest to an insurance company carrying out a risk assessment on potential clients. This domain is based on real information obtained from the Payscale USA website [6]. The list of attributes, and the type of generator used for generating their data, are shown in Appendix B. It is worth highlighting the extensive usage of different attribute generators to retain the realistic characteristics of the generated data. For example, the gender assigned to a tuple depends on a person's occupation (e.g., nursing is overwhelmingly female).

Adult domain. This domain is based on the Adult census dataset from the UCI Machine Learning Repository [21], which has become one of the most widely-used benchmarks in the privacy area. This domain is composed of 9 socio-economic demographic attributes (e.g., occupation, education, salary class). To mimic the values of this dataset, this domain exclusively leverages on distribution-based data generators.

German credit domain. This domain is based on the German credit dataset from the UCI Machine Learning Repository [21]. This domain is composed of 20 credit-related attributes (e.g., employment, purpose of credit, credit class). To mimic the values of this dataset, this domain exclusively leverages on distribution-based data generators.

4 Experimental Evaluation

Here we present the experiments performed to assess the benefits and costs of using COCOA. Firstly, we evaluated how well COCOA achieved its objective of generating diverse datasets within a domain. Secondly, we assessed how useful the resulting datasets are to strengthen the testing of anonymization algorithms. Thirdly, we evaluated COCOA's costs.

4.1 Experimental Setup

Here we present the developed prototype, the test environment and the parameters that defined the evaluated experimental configurations. We also describe the evaluation criteria used in our experiments.

Prototype: Our current prototype supports the domains discussed in Sect. 3.5, and the three types of attribute generators discussed in Sect. 3.4. To simplify the configuration of the distribution-based attributes, as well as to exemplify the benefits of using attribute generator settings, representative default values were configured for the parameters applicable to each distribution (e.g., mean and standard deviation for the normal distribution). From a technical perspective, we built our prototype on top of the Benerator tool [3]. This solution was chosen because it is open source and developed in Java, characteristics which facilitated its integration with other used libraries (e.g., the OpenForecast library [5], which is used for all statistical calculations). Developing the prototype in Java also makes our solution highly portable, as there are Java Virtual Machines (JVM) available for most contemporary operating systems.

Environment: All experiments were performed in an isolated test environment using a machine with an Intel Core i7-4702HQ CPU at 2.20 Ghz, 8 GB of RAM and 450 GB of HD; running 64-bit Windows 8.1 Professional Edition, and Hotspot JVM 1.7.0_67 with a 1 GB heap.

Configurations: As evaluation data, we used the insurance, adult and german credit domains. This was done with the aim of diversifying the evaluated domains and test the generality of the benefits and costs of using COCOA. For the adult and german datasets, the richest attribute (in terms of diversity of categorical values) was used as quasi-identifier (QID) for the anonymization and for the data generation: occupation for adult, and purpose for german credit. In contrast, as the insurance dataset offers a broader set of interesting categorical attributes, four of them were selected: occupation (occ), workplace, activity (act) and place of activity. For each domain, 72 different datasets were created. This was achieved by varying the frequencies of the QIDs (by using the 12 distributions discussed in Sect. 3.4) and generating 6 different dataset sizes (5 K, 10 K, 20 K, 30 K, 50 K and 100 K). As anonymization settings, we selected a popular greedy multidimensional algorithm called Mondrian [20] (from the UTD Anonymization Toolbox [11]). It partitions the domain space recursively into regions that contain at least k records (k-anonymity [28,29]). We tested different levels of privacy, varying the k-values $\in [2..100]$.

Evaluation Criteria: To evaluate the diversity among the generated datasets, we used the Principal Components Analysis (PCA), which is a technique commonly used in the literature [14–16] to assess the (dis)similarity among benchmarks (such as our datasets). PCA is a multivariate statistical procedure that decreases a X dimensional space into a lower dimensional uncorrelated space (hence simplifying the analysis). Moreover, it generates a positive or negative weight associated with each metric. These weights transform the original higher dimension space into Y principal components. In our case, the chosen constituent metrics were the average and standard deviations of the frequencies of all the attributes in a domain. This strategy is similar to the one used in [14]. To measure the utility remaining in the data after anonymization, we used Generalized Information Loss (GenILoss) [19], a widely-used general-purpose metric that captures the penalty incurred when generalizing a specific attribute. In terms of costs, our main metrics were execution time, CPU (%) and memory (MB) utilizations. Garbage collection (GC) was also monitored as it is an important performance concern in Java [25].

4.2   Experimental Results

Here we discuss our experimental results in terms of the diversity of the generated datasets, the benefits of using diverse datasets in the testing of anonymization algorithms, and the costs of using COCOA. Due to space constraint, we only present the most relevant results.

Dataset diversity. To assess the diversity of the datasets generated by COCOA, we calculated the PCA for all the datasets (grouped by domain and size). PCA computes the principal components (i.e., PC1, PC2, etc.) and identifies them in order of significance (i.e., PC1 is the most determinative component). Typically, most of the total variance is explained by the first few principal components. In our case, PC1 and PC2 accounted for at least 80% of the total variance in all the cases, so our analysis centered on them. Our hypothesis was that the usage of the different types of attribute generators (see Sect. 3.4) should lead to variances in the data regardless the domain and size. This was confirmed by the results of this analysis. Although there were some (expected) differences in the degree of variance that COCOA achieved across the tested domains, in all cases a fair diversity of characteristics was achieved. Similar behavior was exhibited with all the sizes of the generated datasets. The main difference when comparing them was that the "scales" of the differences varied (precisely due to the size differences). This behavior is visually illustrated in Fig. 3a and b, which shows how the insurance datasets differ in a two-dimensional space (PC1 vs. PC2) for the 5 K and 100 K sizes. Intuitively, the farther the distance that separates the datasets within a domain size, the more different they are with respect to the metrics. Thus, as the datasets are well dispersed in both figures, the datasets differ independently of the size. For further reference, Appendix A presents the constituent metrics used for PCA.

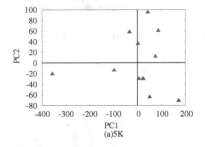 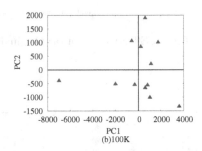

Fig. 3. PC1 vs. PC2 for insurance dataset with sizes (a) 5 K and (b) 10 K

Anonymization. We performed two sets of anonymizations. Firstly, to exemplify the usefulness of testing with multiple different datasets, five datasets of size 5 K for each domain were anonymized. Secondly, to exemplify the benefits of testing with different sizes (i.e., scalability), we used a single variant of the insurance dataset (ActPoissonOccNormal) with 6 different sizes. It is important to remark that our intention was not to exhaustively investigate the data utility of the Mondrian algorithm (nor the effectiveness of k-anonymity), but to document how the resulting datasets are useful to broaden the testing of anonymization algorithms.

Our first analysis focused on assessing the data utility. Figure 4a and b offer a high-level view of the GenILoss obtained, per domain, at each tested

k-value. Figure 4a shows the results obtained by using the original versions of the domains, while Fig. 4b shows the results achieved by using all the different generated datasets per domain. It can be clearly noticed how diverse values can be obtained by using a broad range of datasets (depicted by the standard deviations in Fig. 4b). This test strategy can then help to derive more generic conclusions from the results.

To further exemplify the risks of evaluating with single datasets, Fig. 5a, b, and c presents the GenILoss obtained for each anonymized dataset version, per domain, using a k-value = 5. It can be seen how the GenIloss values considerably fluctuate among the dataset variants. This shows the variability of results that can be obtained when different datasets are considered. This further motivates the usage of benchmarks that provide an adequate coverage of test scenarios.

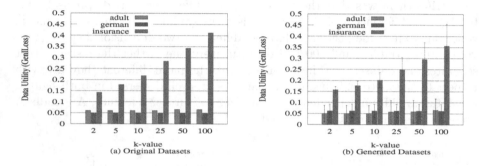

Fig. 4. Effectiveness of (a) Original and (b) Generated datasets of 5 K size

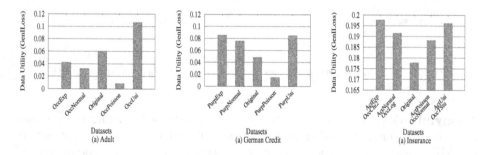

Fig. 5. Effectiveness of anonymized datasets (a) Adult, (b) German and (c) Insurance

Normally, the most important aspect of a scalability testing is to assess the costs of using an anonymization technique. These results are shown in Fig. 6a, b, and c which depict the execution time, average CPU, and memory utilization, respectively. It can be noticed how Mondrian experiences a relatively exponential growth in terms of execution time, while requiring a low amount of resources (as both CPU and memory do not considerable grow with respect to the dataset size). Finally, the analysis of GC behavior showed that its performance cost was only significant for the 100 K versions of the datasets. In those cases, the time spent performing MajorGC (MaGC) was 5 % of the execution time (meaning that the processes can benefit from additional memory).

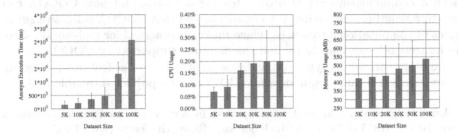

Fig. 6. Efficiency of anonymization w.r.t. (a) Execution time, (b) CPU and (c) Memory

Costs. Fig. 7a, b, and c depict the execution time, average CPU, and memory utilization of the data generation process (per dataset size). It can be seen how COCOA experienced a relatively linear growth in all metrics. Furthermore, COCOA proved to be lightweight in terms of CPU and execution time. For instance, the generation of the largest datasets (i.e., 100 K) took an average of 4 s (with a standard deviation of 1.5 s). Similarly, the CPU never exceeded 10 % (meaning that there was a considerable amount of idle resources to support larger dataset sizes). In terms of memory, COCOA only used approximately 35 % of the available memory. It also never triggered a MaGC, which was another indicator that the memory settings were always appropriate for COCOA. Costs were also analyzed per domain. Although comparable across domains, the biggest costs were experienced by the german credit. This was because it contains the largest number of attributes. A second factor influencing the execution time was the complexity of the attribute generators. In this sense, the distribution-based ones tend to be less expensive (in terms of resources) than the other two types.

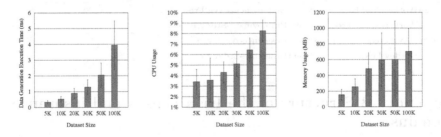

Fig. 7. Efficiency of COCOA w.r.t. (a) Execution time, (b) CPU and (c) Memory

5 Conclusions and Future Work

This paper presented COCOA, a framework for the generation of realistic synthetic microdata that can facilitate the testing of anonymization techniques. Given the characteristics of the desired data, COCOA can effectively create multi-dimensional datasets. Our experiments demonstrated the importance of using a comprehensive set of diverse testing datasets, and how COCOA can help to strengthen the testing. Finally, we showed that COCOA is lightweight in terms of computational resources, which makes it practical for real-world usage. As future work, we plan to expand the domains supported by COCOA, investigate other correlation types in the dataset, integrate other features for testing anonymization techniques, and release COCOA as a publicly-available tool.

Acknowledgments. This work was supported, in part, by Science Foundation Ireland grant 10/CE/I1855 to Lero - the Irish Software Research Centre (www.lero.ie)

Appendix A PCA Constituent Metrics

Tables 1, 2, and 3 list the constituent metrics used to perform the PCA analysis of the datasets generated by COCOA (discussed in Sect. 4.2).

Table 1. Constituent metrics for PC analysis of German credit domain.

German credit domain metrics		
age-Avg.Freq.	installmentRate-Avg.Freq.	property-Avg.Freq.
ageRange-Avg.Freq.	installmentRate-Stdev.Freq.	property-Stdev.Freq.
ageRange-Stdev.Freq.	job-Avg.Freq.	purpose-Avg.Freq.
age-Stdev.Freq.	job-Stdev.Freq.	purpose-Stdev.Freq.
costMatrix-Avg.Freq.	numberExistingCredits-Avg.Freq.	savings-Avg.Freq.
costMatrix-Stdev.Freq.	numberExistingCredits-Stdev.Freq.	savings-Stdev.Freq.
creditAmount-Avg.Freq.	numberLiablePeople-Avg.Freq.	statusAndSex-Avg.Freq.
creditAmount-Stdev.Freq.	numberLiablePeople-Stdev.Freq.	statusAndSex-Stdev.Freq.
creditHistory-Avg.Freq.	otherDebtors-Avg.Freq.	statusCheckAccount-Avg.Freq.
creditHistory-Stdev.Freq.	otherDebtors-Stdev.Freq.	statusCheckAccount-Stdev.Freq.
duration-Avg.Freq.	otherInstallmentPlans-Avg.Freq.	telephone-Avg.Freq.
duration-Stdev.Freq.	otherInstallmentPlans-Stdev.Freq.	telephone-Stdev.Freq.
foreignWorker-Avg.Freq.	presentEmployment-Avg.Freq.	
foreignWorker-Stdev.Freq.	presentEmployment-Stdev.Freq.	
housing-Avg.Freq.	presentResidence-Avg.Freq.	
housing-Stdev.Freq.	presentResidence-Stdev.Freq.	

Appendix B Structure of the Irish census and insurance domains

Tables 4 and 5 list the attributes and the type of generators used for producing data for the Irish census and insurance domains, respectively (discussed in Sect. 3.5).

Table 2. Constituent metrics for PC analysis of adult domain.

Adult census domain metrics		
age-Avg.Freq.	income-Avg.Freq.	race-Avg.Freq.
age-Stdev.Freq.	income-Stdev.Freq.	race-Stdev.Freq.
country-Avg.Freq.	marital-Avg.Freq.	sex-Avg.Freq.
country-Stdev.Freq.	marital-Stdev.Freq.	sex-Stdev.Freq.
education-Avg.Freq.	occupation-Avg.Freq.	typeEmployer-Avg.Freq.
education-Stdev.Freq.	occupation-Stdev.Freq.	typeEmployer-Stdev.Freq.

Table 3. Constituent metrics for PC analysis of insurance domain.

Insurance domain metrics		
activity-Avg.Freq.	placeOfActivity-Avg.Freq.	riskOfAccidentClass-Avg.Freq.
activity-Stdev.Freq.	placeOfActivity-Stdev.Freq.	riskOfAccidentClass-Stdev.Freq.
age-Avg.Freq.	Workplace-Avg.Freq.	semAge-Avg.Freq.
age-Stdev.Freq.	Workplace-Stdev.Freq.	semAge-Stdev.Freq.
gender-Avg.Freq.	salaryCatClass-Avg.Freq.	
gender-Stdev.Freq.	salaryCatClass-Stdev.Freq.	
occupation-Avg.Freq.	salaryClass-Avg.Freq.	
occupation-Stdev.Freq.	salaryClass-Stdev.Freq.	

Table 4. Irish census domain structure.

Attribute name	Distr.-based	Distr./Att.-based	Att.-based
given name		x (age,native country)	
family name		x (native country)	
age	x		
gender	x		
marital status		x (age)	
education level		x (age)	
county		x(age)	
economic status		x (age)	
industral group		x (gender)	
field of study area		x (gender)	
field of study		x (field of study area)	
native country	x		

Table 5. Insurance domain structure.

Attribute name	Freq.-based	Freq./Att.-based	Att.-based
gender		x (occupation)	
age	x		
age range			x (age)
occupation	x		
workplace		x (occupation)	
activity	x		
place of activity		x(activity)	
years of experience		x (age)	
salary		x (occupation,YoE)	
risk of accident			x (age, occ, workplace, activity, placeOfAct)
salary class			x (salary)
risk of accident class			x (riskOfAccident)

References

1. COCOA Datasets. https://github.com/ucd-pel/COCOA/
2. CSO. Access to Microdata. http://www.cso.ie/en/aboutus/dissemination/access-tomicrodatarulespoliciesandprocedures/accesstomicrodata/
3. Data Benerator Tool. http://databene.org/databene-benerator
4. Eurostat. Access to Microdata. http://ec.europa.eu/eurostat/web/microdata
5. OpenForecast Library. http://www.stevengould.org/software/openforecast/
6. Payscale USA. http://www.payscale.com/research/US/
7. simPop: Simulation of Synthetic Populations for Survey Data Considering Auxiliary Information. https://cran.r-project.org/web/packages/simPop
8. synthpop: Generating Synthetic Versions of Sensitive Microdata for Statistical Disclosure Control. https://cran.r-project.org/web/packages/synthpop
9. UML basics: The component diagram. https://www.ibm.com/developerworks/rational/library/dec04/bell/
10. US Census. Restricted-Use Microdata. http://www.census.gov/research/data/restricted_use_microdata.html
11. UTD Anonym ToolBox. http://cs.utdallas.edu/dspl/cgi-bin/toolbox/
12. Ayala-Rivera, V., McDonagh, P., Cerqueus, T., Murphy, L.: Synthetic data generation using benerator tool. arXiv preprint arXiv:1311.3312 (2013)
13. Ayala-Rivera, V., McDonagh, P., Cerqueus, T., Murphy, L.: A systematic comparison and evaluation of k-anonymization algorithms for practitioners. Trans. Data Priv. **7**(3), 337–370 (2014)
14. Blackburn, S.M., Garner, R., Hoffmann, C., Khang, A.M., McKinley, K.S., Bentzur, R., Diwan, A., Feinberg, D., Frampton, D., Guyer, S.Z., et al.: The dacapo benchmarks: Java benchmarking development and analysis. ACM Sigplan Not. **41**, 169–190 (2006). ACM
15. Chow, K., Wright, A., Lai, K.: Characterization of java workloads by principal components analysis and indirect branches. In: Workshop on Workload Characterization, pp. 11–19 (1998)

16. Eeckhout, L., Georges, A., De Bosschere, K.: How java programs interact with virtual machines at the microarchitectural level. ACM SIGPLAN Not. **38**, 169–186 (2003)

17. Hoag, J.E., Thompson, C.W.: A parallel general-purpose synthetic data generator. ACM SIGMOD Rec. **36**(1), 19–24 (2007)

18. Hu, J., Reiter, J.P., Wang, Q.: Disclosure risk evaluation for fully synthetic categorical data. In: Domingo-Ferrer, J. (ed.) PSD 2014. LNCS, vol. 8744, pp. 185–199. Springer, Heidelberg (2014)

19. Iyengar, V.S.: Transforming data to satisfy privacy constraints. In: International Conference on Knowledge Discovery and Data Mining, pp. 279–288 (2002)

20. LeFevre, K., DeWitt, D., Ramakrishnan, R.: Mondrian multidimensional K-Anonymity. In: International Conference Data Engineering, p. 25 (2006)

21. Lichman, M.: UCI Machine Learning Repository (2013)

22. Machanavajjhala, A., Kifer, D., Abowd, J., Gehrke, J., Vilhuber, L.: Privacy: theory meets practice on the Map. In: International Conference on Data Engineering, pp. 277–286 (2008)

23. Mateo-Sanz, J.M., Martínez-Ballesté, A., Domingo-Ferrer, J.: Fast generation of accurate synthetic microdata. In: Domingo-Ferrer, J., Torra, V. (eds.) PSD 2004. LNCS, vol. 3050, pp. 298–306. Springer, Heidelberg (2004)

24. Pedersen, K.H., Torp, K., Wind, R.: Simple and realistic data generation. In: Proceedings of the 32nd International Conference on Very Large Data Bases. Association for Computing Machinery (2006)

25. Portillo-Dominguez, A.O., Perry, P., Magoni, D., Wang, M., Murphy, J.: Trini: an adaptive load balancing strategy based on garbage collection for clustered java systems. Softw. Pract. Exp. (2016)

26. Rubin, D.B.: Discussion of statistical disclosure limitation. J. Off. Stat. **9**(2), 461–468 (1993)

27. Sakshaug, J.W., Raghunathan, T.E.: Nonparametric generation of synthetic data for small geographic areas. In: Domingo-Ferrer, J. (ed.) PSD 2014. LNCS, vol. 8744, pp. 213–231. Springer, Heidelberg (2014)

28. Samarati, P.: Protecting respondents identities in microdata release. Trans. Knowl. Data Eng. **13**(6), 1010–1027 (2001)

29. Sweeney, L.: k-Anonymity: a model for protecting privacy. Int. J. Uncertain. Fuzziness Knowl.-Based Syst. **10**(05), 557–570 (2002)

30. Walck, C.: Handbook on statistical distributions for experimentalists (2007)

Remote and Cloud Access

photographs, videos, audio, etc. The use of multiple geosocial media sources for information extraction and knowledge generation in various application domains is a challenging task, both in terms of data management and analysis and in terms of knowledge production. Croitoru et al. [23], states there is a growing interest to apply these technologies to social and politics.

Form the analysis of the reviewed literature we can conclude there is growing interest for geospatial media applications which are developed for various contexts, purposes and scenarios. In some cases they are meant for specific purposes and in other they are designed for general use, like Waze and Foursquare. These applications use georeferenced information over maps and make frequently use of the Global Positioning System (GPS) of mobile devices. The information that should be conveyed (for whatever purpose) is stored in order to be used in a place (the same one or another) describing the place where something interesting is located or something happened. We consider that geo-localization information adds value to the shared information which can be used for an effective communication in the scenario of intercultural adjustment of foreign students having to adapt themselves to a new physical and cultural environment.

3 AUM: Anxiety/Uncertainly Management Theory

AUM theory has been applied to improve communication quality of people adjusting themselves to life in new cultures, [10, 11]. It has also been applied to improve cross-cultural interpersonal and intergroup relationships, to characterize cultural-dependent communication styles [11], and for studying cases of exchange students at schools [24].

According to [10], strangers have uncertainty about host's attitudes, feelings, beliefs, values and behaviors. Strangers need to be able to predict which of several alternative behavior patterns hosts' will employ. Strangers also need to be able to explain hosts, attitudes, feelings and behaviors. Whenever strangers try to figure out why hosts behaved the way they did, strangers are engaging in explanatory uncertainty reduction. When strangers communicate with hosts, strangers not only experience uncertainty, they also experience anxiety (usually based on negative expectations, is the affective (emotional) equivalent of uncertainty). Anxiety is the tension, feelings of being uneasy, or apprehension strangers have about what will happen when they communicate with hosts. To adjust to another culture, strangers do not want to try to totally reduce their anxiety and uncertainty. At the same time, strangers cannot communicate effectively with hosts if their uncertainty and anxiety are too high. If uncertainty is too high, strangers cannot accurately interpret hosts' messages or make accurate predictions about hosts' behaviors. When anxiety is too high, strangers communicate on automatic pilot and interpret hosts' behaviors using their own cultural frames of reference. Also, when anxiety is too high, the way foreigner people process information is very simple, thereby limiting their abilities to predict hosts' behavior. When uncertainty is too low, strangers become over-confident that they understand hosts' behaviors and do not question whether their predictions are accurate. When anxiety is too low, strangers are not motivated to communicate with hosts. If strangers'

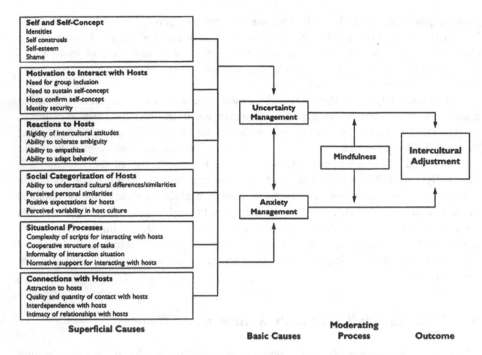

Fig. 1. A schema of the basic AUM theory proposed by [10].

anxiety is high, they must cognitively manage their anxiety to communicate effectively and adjust to the host cultures. Managing anxiety requires that strangers become mindful (i.e., think about our communication and continually work at changing what we do in order to become more effective). Mindfulness is the way that in-group members and strangers can reduce their anxiety and uncertainty to optimum levels. When strangers have managed their anxiety, they need to try to develop accurate predictions and explanations for hosts' behaviors. When strangers communicate on automatic pilot, they predict and interpret hosts' behaviors using their own frames of reference. When strangers are mindful, in contrast, they are open to new information and aware of alternative perspectives, and they can make accurate predictions.

In AUM theory, managing uncertainty and anxiety are the "basic causes" of strangers' effective communication with hosts and intercultural adjustment, while "superficial causes" influence the amount of uncertainty and anxiety foreign people experience. The amount of uncertainty and anxiety strangers experience in their interactions with hosts is a function of many superficial causes; e.g. self-concepts, motivation, reactions to hosts, social categorization, situational processes, and connections with hosts. Figure 1 provides a summary of the basic theory.

AUM theory presents 96 axioms explaining the "superficial causes" and how they affect or relate to the "basic causes" (i.e. anxiety and uncertainty), [10]. The next section explains how "basic causes" and some axioms of the "superficial causes" have been used as fundamental design principles of EMHC.

Towards a National Remote Access System for Register-Based Research

Annu Cabrera[✉]

Department of Standards and Methods,
Statistics Finland, Helsinki, Finland
annu.cabrera@stat.fi

Abstract. In Finland, a lot of register-based data is available for research purposes, e.g. through Statistics Finland's remote access system. When a huge quantity of highly detailed micro data is available, maintaining the confidentiality must be carefully ensured. One common national remote access system would improve the safety of data and simplify data access for researchers. The infrastructure project Finnish Microdata Access Services (FMAS) has been launched for this purpose. The project has been slowed down by the lack of funding but the work and co-operation between different register keeping agencies continues.

Keywords: Remote access system · Register-based research · Data confidentiality

1 Introduction

Statistics Finland offers more and more data for research via a remote access system. The majority of these data come from national registers maintained by different governmental agencies but utilised by Statistics Finland for statistical purposes. Remote access is a safer option to offer register-based data for research than releasing the data outside of Statistics Finland's servers. Maintaining the confidentiality of the data is of course of highest priority.

Today not all register keepers' data are available through Statistics Finland's remote access system. Register keepers also have their own ways of releasing data for research purposes. When a researcher wants to use data from different register keepers, he or she must sometimes go through several data application processes at different agencies and data with identification numbers must be sent from one agency to another. Lately steps have been taken towards a national remote access system for register-based research that would simplify the data application and release process both for researchers and register keepers.

The aim of this paper is to describe the current state of the remote access system at Statistics Finland and the infrastructure project "Finnish Microdata Access Services (FMAS)" on improving the possibilities to use the microdata gathered in different registers for research purposes. First the register-based data and their use for research are discussed. In the third section, the current remote access system at Statistics Finland with its latest developments and challenges is described. The fourth section is about the

© Springer International Publishing Switzerland 2016
J. Domingo-Ferrer and M. Pejić-Bach (Eds.): PSD 2016, LNCS 9867, pp. 181–189, 2016.
DOI: 10.1007/978-3-319-45381-1_14

objectives of and steps taken towards a national remote access system. Finally, the paper concludes with a short summary.

2   Register-Based Data and Research

In Finland, there are many national registers on persons, enterprises, buildings, etc. that cover the whole population. Even though these registers are maintained by different governmental agencies, e.g. Tax Administration and Population Register Centre, the information from different registers can be combined with the help of commonly used identification numbers. All residents and businesses in Finland have a unique identification number, Personal Identification Number (PIN) or Business Identification Number (BIN), that is used by all governmental agencies and also by many commercial service providers.

Being primarily meant for administrative purposes the national registers are also a rich source of research data. In Finland, it is not unusual to carry out research only with register data. Of course, research use of register data requires that researchers have access to the data. At the moment, lots of register-based data can be accessed through Statistics Finland. The Finnish Personal Data Act [5], Section 8 allows Statistics Finland to process personal data since processing is necessary for compliance with a task provided by law (Statistics Finland Act, Section 2, [8]). Furthermore, the Statistics Act [6], Section 4 stipulates that Statistics Finland, first of all, has the right to process data with identification numbers and, secondly, is allowed to collect data from administrative registers. Finally, the Statistics Act, Section 13 states that statistical authorities may release confidential data collected for statistical purposes for scientific research without having to ask for consent from the data subjects.

Of course, data confidentiality sets some limits on how the data can be released for research purposes. Survey and register data can be released and combined with other data but this requires that survey respondents have been informed that their data might be combined with register data and the data might be used for research purposes (Statistics Act, Section 9). Data may not be released in such form that statistical units can be directly identified. However, Statistics Finland may give permission to use (via remote access) confidential data from which the statistical unit could be indirectly identified (Statistics Act, Section 13).

Statistics Finland as a statistical authority is obliged to extend the use of the data collected for statistical purposes in scientific studies and statistical surveys on social conditions (Statistics Act, Section 3). Amongst other things, this means that Finland should promote a simpler access and delivery system for the data.

3   Current Remote Access System at Statistics Finland

Statistics Finland may give permission to use its data for scientific research via a remote access system. The current remote access system was established in 2010 at the same time that research services were centralised inside the agency and became the Researcher Services unit. Providing good services for researchers was also set as one of the agency's strategic goals.

Through the remote access system Statistics Finland provides both register and survey data for research purposes. The data available are data collected for statistical purposes. It is also common that researchers want to include their own data or data from other register keepers (which they have already received a permit for) in research data. In such cases, the register keepers and/or researcher have sent their data to Statistics Finland where all data are combined using common ID-numbers. After combining, all data it are made available to the researcher via Statistics Finland's remote access system. Before making the data available, all identifiers will be changed into matching pseudo codes.

3.1 Statistics Act Amendment in 2013

Before 2013, all research data released by Statistics Finland or used via a remote access system had to be completely anonymised. The Amendment of the Statistics Act in 2013 made it possible for researchers to get permission to use data where statistical units can be indirectly identified. In practice, this means that, via a remote access system researchers may use data that include the whole population and only direct identifiers (PIN/BIN, name, address, etc.) have been removed. No other statistical disclosure control (SDC) methods, such as reducing the level of detail in classifications, are required. This amendment made Statistics Finland's remote access system more popular among researchers since the quality of research datasets no longer had to be "ruined" with SDC methods. This doesn't mean that confidentiality and privacy of people and businesses would no longer be of any importance, quite the contrary. Since more and more data are available through the remote access system, the emphasis in ensuring confidentiality has moved from implementing traditional SDC to making sure the research environment is safe.

Regardless of well-functioning remote access systems researchers may still get data released on e.g. CD-ROM. Compared to data obtained via remote access, the released data cannot be as detailed or large since total anonymisation is still required. Reasons for researchers to choose released data over remote access might be costs of the remote access system or that they are simply used to, and more comfortable with, released data.

3.2 Confidentiality Measures in the Remote Access System

There are three types of confidentiality measures related to the remote access system: (1) data security provided by the system, (2) terms and conditions ensuring confidentiality defined in the licence to use Statistics Finland's data, and (3) statistical disclosure control methods applied to output produced by researchers based on datasets in the remote access systems.

Data security provided by the remote access system. In the remote access system, researchers are able to handle confidential data through a secure connection to a server at Statistics Finland. This server is separate from servers used in statistics production and contains only data for research purposes. Statistics Finland opens its firewall to connections from IP-addresses defined in a contract between Statistics Finland and the research organisation. Connection is available only from these IP-addresses.

Statistics Finland grants a user licence only to researchers named in the permit application for research data. These researchers obtain a user ID, and for each logon, a new disposable password in sent to the researcher's mobile phone.

The researchers can see the micro data on the remote desktop and perform analyses with the statistical software available in the system. Researchers cannot copy the data out from the system. Only keyboard and mouse signals coupled with a screen image are transferred between the system and the researcher's computer.

All data transfers into or out from the remote access system (e.g. programme codes, statistical output) are done through the personnel of Researcher Services or Statistics Finland's IT department.

Terms and conditions in the user licence. In the Licence to use Statistics Finland's data files there are terms and conditions to guarantee that researchers will handle confidential data with necessary care. These terms and conditions state that

- The data may only be used for the purpose stated in the decision on granting the licence;
- The researcher can access the research project's files only for the duration of the permit;
- The researchers must give a written pledge of secrecy;
- No attempt must be made to identify the data subjects from the material;
- Material to be published based on the data files must be protected so that no individual persons or enterprises can be identified;
- All data in remote access must be handled according to the rules of Research Services, including guidelines on data disclosure control.

All researchers using the data must follow the above mentioned terms. The user licence also states that a sanction is prescribed for violating the secrecy, non-disclosure and prohibition of use of the data. The punishment for researcher for these offences can be a fine or imprisonment for at most one year. Use violating the licence or breaching against the rules has an effect on the rights of the organisation and/or researcher to obtain a licence to Statistics Finland's data in the future [7].

Output checking. Researchers must ensure that the research results and statistical outputs they want to publish or just transfer out from the remote access system contain no unit-level data or possibility of their disclosure. Researcher services apply an output checking process of research results, which ensures the implementation of data protection in the outputs produced by the researcher from the data. All output is screened, and after screening, sent to the researcher by e-mail.

3.3   Recent Development

Statistics Finland is currently producing ready-made data modules. Each module consist of basic and/or most popular variables describing a certain subject, e.g. families, working life, etc. The idea behind the modules is to make it easier for researchers to define what kind of dataset he or she needs for the research when he/she can define the research data using modules instead of defining each variable individually. Ready-made modules also speed up the process of establishing research datasets at Researcher Services and helps the researchers to gain access to the data faster.

The ready-made modules that have been prepared in 2016 include variables belonging to the following subject areas: Income, Degrees, Families and Employment. One module includes basic socio-economic and demographic information for the years 1987 to 2015 while another one includes basic census data for the years 1970 to 1985. Each module includes 25 to 50 variables. More modules are planned to be available to researchers via the remote access system in coming years.

3.4 Challenges in Ensuring Confidentiality

There are a few challenges already met by Statistics Finland on how to maintain confidentiality when increasing the amount of data that are made available through the remote access system. These challenges are strongly related to the 'need-to-know' principle. According to the need-to-know principle, Statistics Finland provides, for each particular research project, only the data that are really necessary for that research. By following this principle no unnecessary confidential data will be seen or processed by researchers. Problems can arise when defining what is necessary and what is not.

Even if in some cases Statistics Finland was allowed to provide datasets covering the whole population it shouldn't be done if the research could be conducted using only sample data. Sometimes it's not clear whether or not total population data are necessary and if not, what sample size would be enough. Normally in these cases, the researcher is asked to give valid reasons to use total population data. Without these reasons, Statistics Finland may provide only sample data.

Statistics Finland always provides data for particular, well-defined, and limited research projects. Another issue relating to the 'need-to-know' principle arises from defining a research project. Some researchers may apply for data for a single but huge project including lots of data and many researchers. This huge project is later planned to be divided into smaller sub-projects and it's not sure that all researchers will participate in all sub-projects. However, to avoid excessive bureaucracy, researchers may want to make only one application for the huge project instead of applying for data for each sub-project individually. This can lead to a situation where some members of the research group have access to data that they don't actually need. Extra bureaucracy is not in Statistics Finland's interests but a good balance must be found between avoiding a too heavy application process and providing data only to researchers that actually need it.

4 Finnish Microdata Access Services

In 2013, Statistics Finland together with the National Archives proposed a new unified national service for register research, the Finnish Microdata Access Services (FMAS, [3]). The idea of FMAS is to facilitate the use of register and statistical data maintained by different authorities. FMAS would be a single channel for the entire research process, from planning to data analysis. FMAS would consist of four parts:

- A metadata catalogue that can be used to find information on available data repositories;
- An electronic permit application service for obtaining permissions from various public authorities;
- A remote access system for the combination and analysis of various licensed data;
- An information and support service providing assistance and advice on all issues related to register research.

The FMAS project obtained infrastructure funding for development of the services for the year 2014 (Fig. 1).

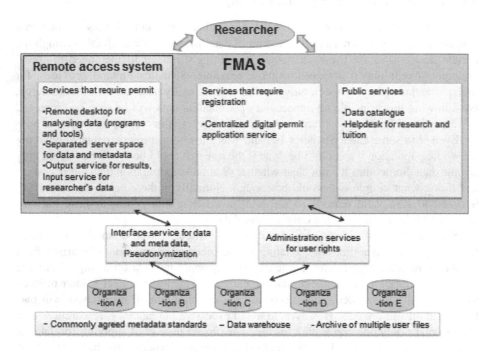

Fig. 1. The Finnish Microdata Access Services, [4]

In addition to making data access a simpler process for the researcher, FMAS would also improve the protection of register-based personal data since data containing PINs were no longer needed to be transferred from one agency to another. In the FMAS system, some major register keepers (e.g. Statistics Finland, the National Institute for Health and Welfare and the Social Insurance Institution) would manage their own users in the remote access system and provide their data straight to the system without third party processing. This is one of the biggest differences between FMAS and the current situation, where Statistics Finland as the only administrator of the remote access system, maintains all data from other register keepers that is desired to be used together with Statistics Finland's data. These data must first be transferred (with PINs) to Statistics Finland and then Statistics Finland will provide the data to the researcher through the remote access system.

4.1 Towards FMAS

Even though FMAS is still far from ready, Statistics Finland and the National Archives have already taken several steps towards it.

Outsourcing the remote access system. The remote access system was developed and maintained by the IT department of Statistics Finland. However, Statistics Finland in not the most suitable organisation for the technical management of the system since it is not the core competence area of Statistics Finland's IT staff. Sometimes problems have occurred in the system and solving them have been quite burdensome and taken resources away from Statistics Finland's core competence: management of statistics production systems.

In 2015, Statistics Finland decided to outsource the technical maintenance of the remote access system to the IT Center for Science (CSC). CSC is a non-profit, state-owned company administered by the Ministry of Education and Culture. It maintains and develops the state-owned centralised IT infrastructure and uses it to provide nationwide IT services for research, libraries, archives, museums and culture, as well as information, education and research management. CSC also has the task of promoting the operational framework of Finnish research, education, culture and administration [1].

When changing over to CSC, the remote access system was given a new name FIONA (FInnish ONline Access).

Together with CSC, Statistics Finland continues developing FIONA to become a national system. Even though the microdata provided for researchers will be maintained at CSC, Statistics Finland still remains the administrator of the system.

Other development. A metadata catalogue and an electronic permit application service were also part of the original FMAS plan. Statistics Finland has published a beta version of a metadata catalogue in March 2016. The National Archives has been developing the permit application system. A demo version of the system already exists but it hasn't been tested yet. It's planned that later the metadata catalogue will work together with the application system so that the researcher can choose the required datasets and variables from the catalogue and they will automatically be printed in the permit application.

Recently, other register keepers have shown an interest to make their data available through FIONA. Finnish Customs and the Ministry of Employment and the Economy have prepared some datasets that are ready to be used in FIONA. Also the Finnish Defence Forces and the National Institute for Health and Welfare have similar plans.

4.2   Challenges Met by FMAS

Funding. After 2014, the FMAS project hasn't received any funding. This naturally has slowed things down. Now, Statistics Finland is looking for funding from new sources. One option could be the Finnish innovation fund Sitra operating directly under the Finnish Parliament. Sitra has already launched a project to prepare the establishment of a service provider that will focus on gathering and co-ordinating national well-being data [2]. This project has some similar features with FMAS and co-operation with Sitra could benefit also FMAS.

Careful planning of co-operation. The plan for FMAS is that instead of only Statistics Finland several register keepers would use the remote access system together to make their data available to researchers. Good co-operation between agencies must naturally be planned carefully. For example, there is not yet a clear plan how responsibilities, profits and costs should be divided between the different agencies. At the moment, Statistics Finland covers all costs of FIONA such as payments for CSC, for system maintenance and salaries of personnel preparing datasets and taking care of output checking. On the other hand, Statistics Finland also receives all payments from the researchers using FIONA.

The legal basis for full co-operation must also be ensured. At the moment, slightly different laws regulate how different register keepers can release data for research purposes. Differences can occur in the required level of anonymisation or the type of research (scientific, commercial, etc.) data can be obtained for. Some amendments in legislation might be required.

5   Summary

Even though in Finland, a lot of register-based data is already available and utilised for research purposes, e.g. through Statistics Finland's remote access system FIONA, some actions could be done to improve the process of applying and obtaining access to the data on a national level. When a huge quantity of highly detailed micro data is available, maintaining the confidentiality must be carefully ensured. One common national remote access system would improve the safety of data and simplify data access for researchers. The infrastructure project FMAS has been launched for this purpose. The project has been slowed down by the lack of funding but the work and co-operation between different register keeping agencies continues.

While Statistics Finland seeks new funding for the project, other register keepers have shown an interest to make their data available through FIONA. Also, a metadata catalogue and electronic application service are under development. These are the next steps for FIONA to become a national remote access system.

References

1. CSC - IT Center For Science. https://www.csc.fi/home
2. Finnish Innovation Fund Sitra. http://www.sitra.fi/en
3. FMAS forum. http://fmas-foorumi.fi/. (in Finnish)
4. Johnson, M.: Creating a national remote access system for register-based research. In: Joint UNECE/Eurostat Work Session on Statistical Data Confidentiality, Helsinki (2015). http://www1.unece.org/stat/platform/display/SDCWS15/Statistical+Data+Confidentiality+Work+Session+Oct+2015+Home;jsessionid=0E3AC537CB04D410AD34AAB7DA641313
5. Personal Data Act (523/1999). http://www.finlex.fi/fi/laki/kaannokset/1999/en19990523.pdf
6. Statistics Act (280/2004). http://tilastokeskus.fi/meta/lait/2013-09-02_tilastolaki_en.pdf
7. Statistics Finland. http://www.tilastokeskus.fi/index_en.html
8. Statistics Finland Act (48/1992). http://www.finlex.fi/fi/laki/ajantasa/1992/19920048. (in Finnish)

Accurate Estimation of Structural Equation Models with Remote Partitioned Data

Joshua Snoke[1($\boxtimes$)], Timothy Brick[2], and Aleksandra Slavković[1]

[1] Department of Statistics, Pennsylvania State University, State College, USA
{snoke,sesa}@psu.edu
[2] Department of Human Development and Family Studies,
Pennsylvania State University, State College, USA
tbrick@psu.edu

Abstract. This paper focuses on a privacy paradigm centered around providing access to researchers to remotely carry out analyses on sensitive data stored behind firewalls. We develop and demonstrate a method for accurate estimation of structural equation models (SEMs) for arbitrarily partitioned data. We show that under a certain set of assumptions our method for estimation across these partitions achieves identical results as estimation with the full data. We consider two situations: (i) a standard setting with a trusted central server and (ii) a round-robin setting in which none of the parties are fully trusted, and extend them in two specific ways. First, we formulate our methods specifically for SEMs, which have become increasingly common models in psychology, human development, and the behavioral sciences. Secondly, our methods work for horizontal, vertical, and complex partitions without needing different routines. In application, this method will serve to increase opportunities for research by allowing SEM estimation without transfer or combination of data. We demonstrate our methods with both simulated and real data examples.

Keywords: Statistical disclosure control · Partitioned data · Structural Equation Models · Distributed Maximum Likelihood Estimation

1 Introduction

Due to privacy constraints researchers often cannot access data directly because they carry sensitive information about individuals or organizations. Current access to such data either comes through the use of publicly available versions that have undergone statistical disclosure control, or private versions available after complicated and restrictive user agreements. To address this issue, statistical privacy methods have arisen. These methods are aimed at increasing research opportunities and encouraging reproducible research while maintaining protection for individuals' sensitive information, and they have taken various forms, most commonly associated with methods reported in the literature of Statistical Disclosure Control (SDC) and Privacy Preserving Data Mining (PPDM). Some

© Springer International Publishing Switzerland 2016
J. Domingo-Ferrer and M. Pejić-Bach (Eds.): PSD 2016, LNCS 9867, pp. 190–209, 2016.
DOI: 10.1007/978-3-319-45381-1_15

standard methods include, perturbation, microaggregation, swapping, suppression, or synthetic data (see [16] or [26] for more detailed overviews).

Privacy restrictions are typically intensified when researchers plan to combine multiple data sources, a common practice in the behavioral and social sciences. Public data for behavioral research from many sources, such as the U.S. Census Bureau, the National Center for Health Statistics, the National Center for Educational Statistics, or the Bureau of Labor Statistics, can only be undertaken with serious privacy guarantees, as noted by the National Human Research Protections Advisory Committee (NHRPAC)[1]. These privacy guarantees often require that data not leave certain specific locations (e.g. Census Research Data Centers[2]), making it difficult or impossible to perform a centralized join of participant data. In other cases, data may not be publicly available at all in raw form, and it is unclear how a given set of analyses will be affected by using data that have undergone different disclosure control procedures.

We propose a method for accurately estimating models while leaving the data behind their firewalls. We develop our methods for Structural Equation Models (SEMs), an increasingly popular choice of modeling framework in the behavioral and social sciences. Some researchers in the behavioral sciences have even begun to explicitly argue for analysis in distributed settings both for ensuring confidentiality and for furthering research possibilities; see [3] on distributed estimation with primary applications to structural equation models.

Examples of previous work on partitioned data models include [8,17,18,20, 25]. We formulate our methods in a similar paradigm as these works, but the nature of our models leads us to unique risk situations and possible solutions. In contrast to the previous work, our methods can be used for accurate estimation of any type of remote partitioned data, whether it be horizontal, vertical, or complex. These terms are commonly used to denote the different ways in which data can be partitioned: horizontal implies a partitioning of rows or individuals among databases where the same variables have been measured in the partitions, vertical implies separate variables are stored in different databases on the same set of individuals, and complex implies some combination of both horizontal and vertical, which is the most likely in a real world situation; see Fig. 1. In this paper we assume records in different partitions are linked by a unique ID.

The rest of this paper is organized as follows: Sect. 2 describes the two possible partitioned database routines we consider. Section 3 describes the methodology behind accurate estimation of structural equation models with horizontally, vertically, or complexly partitioned data. Section 4 describes two algorithms for estimation under the settings described in Sects. 2.1 and 2.2. Section 6 gives simulation results for computational complexity and shows accurate reproduction of published results with real data. Section 7 discusses further improvements, privacy implications, and future directions.

---

[1] See http://www.hhs.gov/ohrp/archive/nhrpac/documents/dataltr.pdf.

[2] See https://www.census.gov/about/adrm/fsrdc/about/secure_rdc.html.

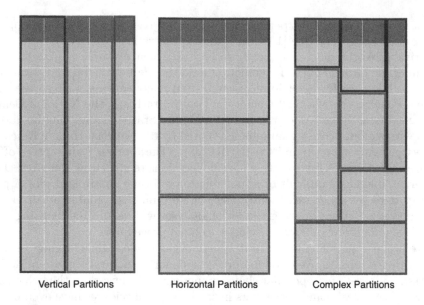

Fig. 1. Data partition types, with rows for individual entries and columns for attributes.

2 Partitioned Database Routines

A number of papers in the statistical literature have presented methods and algorithms for estimation over multiple partitioned databases. Methods for secure log-linear and logistic regression were presented by [8], methods for vertically partitioned data using secure matrix computations by [9,18,25] exhibited secure logistic regression, and [17] used distributed maximum likelihood estimation to estimate general linear models across horizontal and vertical partitions. Some have even produced functioning implementations of these concepts and are moving towards making it possible for researchers to utilize these methods, see [10,23].

Much of the previous work relied on two paradigms under which these methods could be implemented. One assumes a trusted central server which can receive summary statistics from separate parties and use these to estimate the model parameters, such as [10,23]. In this situation, the primary risk comes from the need for encryption and to ensure that the model parameters provided by the separate parties are not disclosive. Work such as [9,17], for example, assume instead that the models are being estimated without a trusted central party. In this case, the distributed databases will pass intermediate statistics to each other in a round robin fashion relying on the Secure Multi Party Computation (SMPC) methods until parameters can be estimated. SMPC originated in cryptography, see [13,27], and has since been used in both statistical disclosure limitation and privacy-preserving data-mining, e.g. [20]. It involves a group of parties estimating an aggregate outcome based on their collective information without any party accessing or learning another party's particular information, and it requires each party to be semi-honest. Semi-honesty means that parties will not collude, and they will participate honestly in the routine and not pass false information.

We build on the ideas from the aforementioned works, and extend them by designing algorithms for estimation of structural equation models (SEMs) for both the central-server setting and the round-robin information sharing setting. While traditionally formulated for continuous data, the SEM framework can be easily extended to include binary outcome measures, see [19], and therefore can be thought of as a generalization of logistic regression. Our work also relates particularly to [17], since we use distributed Maximum Likelihood Estimation (MLE) to estimate our models as they did for regression models. In the following sections we describe in more detail the two settings, and in Sect. 4 we present algorithms for SEMs estimation in these settings and show that accurate estimation can be achieved for horizontal, vertical, and complex partitions.

2.1 Trusted Central Party Setting

We first consider the setting where a trusted centralized server distributes models to the locations housing each partition and receives back intermediate statistics and likelihoods in order to estimate the model parameters. An example would be a controlled research center where researchers could submit analyses to be run on the controlled servers or a node run by a trusted source such as the NIH or Census Bureau. The central server would send out the model specifications to the different data warehouses, and receive back the appropriate statistics to estimate the parameters and goodness-of-fit statistics for each model. The results could then be distributed back to the parties of interest. This setting has been discussed in more details, for example, by [5,10] who are already moving towards implementation. As the amount of data collected grows along with researchers desire to link datasets collected by different agencies, this set up eases logistics to create networks of data warehouses with trusted central servers to perform analyses. On the other hand, new technological advances and proposals for data collection (and possible ownerships) by individual participants themselves make it more likely that individuals will not have a common trusted party. See [6,22], or [4] for more on smart technology and users' opinions on privacy and research. Thus, although the centralized-server scenario poses fewer risk barriers, we do not expect to rely on this scenario in practice.

In this setting, the data holders could be passive participants who do not wish to release their data but are willing to participate by providing intermediate statistics for the research models provided by the central server. Alternatively, one or more of the data holders may have an active interest in the results, and might be willing to participate in exchange for final estimates from the central server. These situations would carry different risks due to the different information being shared with each party. We will discuss this further in Sect. 7.1. Figure 2 gives a visual representation of the trusted central party routine. We provide an algorithm for accurately estimating SEMs with a trusted central server in Sect. 4.1.

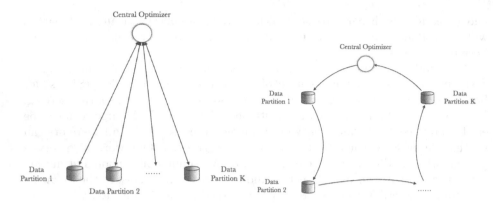

Fig. 2. Trusted central server routine   **Fig. 3.** Round robin routine

Risk is minimized in this setting primarily because the central party is trusted to see intermediate statistics of the data, while the researcher or data holders will only receive final estimates. While these estimates can still be disclosive under certain situations, particularly if multiple models are queried, this form of disclosure is frequently considered within the range of reasonable disclosure for scientific purposes. Our distributed method for this situation allows each separate data warehouse to maintain their data separately either for privacy or computational reasons.

2.2   Round Robin Setting

We also consider the setting where there is no trusted central server and the separate data warehouses do not fully trust either each other or the central server. Rather than receiving model parameters from a trusted central server and returning intermediate statistics, the separate data holders will pass parameters and statistics in a round robin fashion to each other. In our setting, each round constitutes one step in an optimization routine, with successive rounds until optimization is complete and estimates are obtained. We still make the assumption that each data server is semi-honest or complicit in the process. Classically this problem has been addressed under the framework of Secure Multi-Party Computation (SMPC), which similarly assumes semi-honesty. In this paper we do not use SMPC methods in our estimation algorithms, but for the non-central trusted server settings these methods are likely to be necessary for minimizing disclosure risk and will be implemented in future work; see Sect. 7.1 for more on this.

An example of this setting would be an analysis using data across different hospitals or proprietary data from several companies. These entities may collect the same information on different groups of individuals or different information about the same individuals, and either an outside researcher or the data holders themselves may wish to perform some analyses using the data from multiple sources. This situation is more likely, since most data holders will not have a readily available third party whom they all trust to act as the central server described in the previous section.

As with the previous routine, horizontal, vertical, or complex data partitions are all possible, such as compiling individuals measured at different hospitals (horizontal) or a tracking the treatment of individuals as they move from a hospital to maintained therapy at a separate institution (vertical). Research opportunities could greatly expand if data custodians can adequately assure privacy such that analyses can be carried out without the actual data crossing the firewalls of respective institutions, and without any outsider being able to reconstruct the data from shared statistics. We give an algorithm for accurately estimating SEMs in a round robin setting (see Fig. 3 for a visual representation) in Sect. 4.2.

3   Methodology

3.1   Structural Equation Models

Structural Equation Models (SEMs) have become increasingly popular in social science fields, particularly Psychology and Human Development, and they are flexible to include a number of different model types, such as common factor models, latent growth curve models, or ACE (additive genetics, common estimation, and unique environment) models. SEMs also allow for estimation of regression models, therefore much of the work previously derived in the literature for partitioned estimation can easily be incorporated into our framework. The SEM framework has gained prominence in recent years because it represents a generalized form of linear covariance modeling that subsumes traditional regression models and latent variable models such as factor analyses. Of particular importance in the behavioral sciences, SEM specifications can be represented visually as directed box-and-arrow graphs known as *Path Diagrams*. The isomorphism between path diagram representations and distributional expectations for the covariance and mean structure for the data is defined using one of several processes. For the purposes of this paper, we use the Reticular Action Model (RAM) specification developed by [21] (see the Technical Appendix for more details).

SEMs allow researchers to model a relationship structure between latent (unmeasured) and manifest (measured) variables. There are various ways to estimate these model parameters, but in this paper we use maximum likelihood estimation. The model is parameterized using the three matrices of the RAM algebra, representing the means, (co)variances, and regression weights among the parameters respectively. We transform this parameterization into model-implied mean and covariance matrices using RAM algebra, and then use an optimization procedure such as gradient descent to find parameter values that optimize the likelihood of the data given the model-implied distribution. We assume the data fit a multivariate normal distribution, a standard assumption for SEMs. These models hold for non-normal data in some situations, such as ordinal data or data that are approximately normal under a transformation. The simplifying

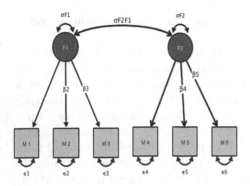

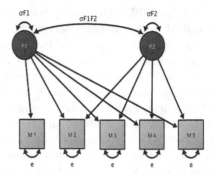

Fig. 4. Two factor model - path diagram

Fig. 5. Latent growth curve model - path diagram

assumption of multivariate normality and maximum likelihood estimation over alternative choice enables our method for estimation across partitioned databases, as we discuss in the next section.

3.2 Distributed Likelihood Estimation

Distributed Likelihood Estimation (DLE) is an extension of classic Maximum Likelihood Estimation (MLE). It assumes the same model as MLE with the exception that the data are separated, so the likelihoods for the data partitions are first calculated separately and then combined for the maximization step. For horizontally partitioned data (separated observations) the method requires only a simple combination of likelihoods from each partition. The method for vertically partitioned data (separated variables) uses combinations of marginal and conditional likelihoods across the partitions. Complex partitions can also be formulated as a simple combination of horizontal and vertical DLE. Applying this to the routines discussed in Sect. 2, a researcher would initialize model parameters and send them to the partitions, who would return separate or compiled (see Sects. 2.1 and 2.2) likelihood values which are then used to optimize over the parameter space and estimate the model using classic MLE.

Full Information Maximum Likelihood. In order to calculate these separate likelihoods we use Full Information Maximum Likelihood (FIML), originally named so by [1]. FIML is a simple extension of the tradition concept of likelihood that reformulates the traditional log-likelihood equation as a sum across each observation for a data set allowing the mean and covariance parameters to vary in dimension for each data row. As the names suggests, this method was originally proposed as a full data method for missing data. The variable mean and covariance matrices allow for the log-likelihood to be calculated on rows with different subsets of variables, as would be expected with missing data, so the mean and covariance parameters need to have different dimension for each row.

The log-likelihood function for multivariate normal data is shown in Eq. (1) for a data matrix $X \in R^{n \times p}$ with mean parameter $\mu \in R^{1 \times p}$ and covariance parameter $\Sigma \in R^{p \times p}$. It is important to note here that these parameters are model-implied parameters throughout our methods and not sample means or covariances. These model-implied parameters are defined using the RAM algebra transformation, which takes a SEM defined with paths for both manifest and latent variables and transforms these paths to mean and covariance matrices for the manifest variables only; see the Appendix for more on this transformation.

The FIML reformulation is shown in Eqs. (2) and (3) for a given data row $X_i \in R^{1 \times q}$ with parameters $\mu_i \in R^{1 \times q}$ and $\Sigma_i \in R^{q \times q}$ where q is the number of variables in the row, $1 \leq q \leq p$. If the data are complete, there is no difference between Eqs. (1) and (3).

$$logL(\mu, \Sigma | X) = -\frac{n}{2} log(2 * \pi) - \frac{n}{2} log(|\Sigma|) - \frac{1}{2}(X - \mu)^T \Sigma^{-1}(X - \mu) \qquad (1)$$

$$logL_i(\mu_i, \Sigma_i | x_i) = -\frac{p}{2} log(2 * \pi) - \frac{1}{2} log(|\Sigma_i|) - \frac{1}{2}(x_i - \mu_i)^T \Sigma_i^{-1}(x_i - \mu_i) \qquad (2)$$

$$logL^*(\mu, \Sigma | X) = \sum_{i=1}^{n} logL_i(\mu_i, \Sigma_i | x_i) \qquad (3)$$

FIML is specifically needed for estimating models with vertically or complexly partitioned data. We can estimate the total log-likelihood in separate parts across data partitions, using only the subset of the model-implied mean and covariance parameters which belong to the variables in the partition. If the data are simply horizontally partitioned, we can use the complete data log-likelihood at each partition and simply sum the different parts as we would in the normal case. Due to these qualities of the log-likelihood function we can estimate the exact same parameters of a model with DLE across partitions as would be obtained using MLE on non-partitioned data.

Marginal and Conditional Likelihoods for Vertical Partitions. When we cannot use the complete data log-likelihood, as is the case for vertically or complexly partitioned data, we need to partition the joint model-implied mean and covariance matrices. We cannot simply split them according to the variables in the corresponding partitions, since this would be unable to account for the implied relationship between variables in the different partitions. To solve this we rely on the relationship between the joint distribution of a set of variables to their marginal and conditional distribution as noted in Eq. (4).

$$f_{X,Y}(x,y) = f_X(x)f_{Y|X}(y|x) \qquad (4)$$

Suppose then we have a theoretical complete database such that there are two partitions containing different variables on the same individuals, $Z = (X, Y)$.

We can rewrite Eqs. (1) and (3) as Eqs. (5) and (6), which we use to calculate the FIML of the model-implied parameters given Z.

$$logL(\mu, \Sigma | Z) = logL(\mu_x, \Sigma_x | X) + logL(\mu_{y|x}, \Sigma_{y|x} | Y) \tag{5}$$

$$logL^*(\mu, \Sigma | Z) = \sum_{i=1}^{n} logL_i(\mu_x, \Sigma_x | x_i) + \sum_{i=1}^{n} logL_i(\mu_{y|x}, \Sigma_{y|x} | y_i) \tag{6}$$

μ_x and $\mu_{y|x}$ are the model-implied mean vectors for the variables in X and the variables in Y conditioned on the variables in X respectively, and Σ_x and $\Sigma_{y|x}$ are the model-implied covariance matrices for the variables in X and the variables in Y conditioned on the variable in X respectively. As with the horizontal partition, maximizing over this sum of the separate log-likelihood functions is equivalent to standard MLE with a non-partitioned data matrix.

Separating the parameters in this way relies only on standard conditional distribution properties. Recall that under the SEM framework we assume a multivariate normal data distribution. Subsetting the parameters as shown in Eq. (7) for the combined dataset over X and Y, we can write the conditional parameters for the variables in Y given the variables in X as shown in Eqs. (8) and (9). These relationships come from the properties of the multivariate normal distribution and are also known as the Schur Complement in matrix theory, see [15, 24].

$$\Sigma = \begin{pmatrix} \Sigma_{xx} & \Sigma_{xy} \\ \Sigma_{yx} & \Sigma_{yy} \end{pmatrix}, \mu = \begin{pmatrix} \mu_x & \mu_y \end{pmatrix}, \tag{7}$$

Σ_{xx}, Σ_{yy}, and Σ_{xy} are the model-implied marginal covariance elements for the variables in X, the variables in Y, and covariances between the variables in the two subsets respectively. μ_x and μ_y are the mean parameters for the variables in X and the variables in Y, respectively.

It is important to note again here that these mean and covariance parameters are not the observed sample estimates from the data, since it would be impossible to calculate Σ_{xy} across partitioned data without sharing data. They are model-implied based on parameters chosen to specify the SEM and then optimized over the FIML. The exception, though, are the conditional mean parameters that are estimated using real data, see Eq. (9), and thus are denoted with the hat notation. $\hat{\mu}_{y|x}$ is not directly reversible to obtain x, but there is potential risk here that a nefarious party might be able to approximate x particularly with an iterative sharing of $\hat{\mu}_{y|x}$ as is necessary for optimization. We discuss this issue further in Sect. 7.1. Based on these methods shown here, the concept of distributed likelihood estimation can be used for estimating structural equation models on horizontal, vertical, or complex partitions of data.

$$\Sigma_{y|x} = \Sigma_{yy} - \Sigma_{xy} \Sigma_{xx}^{-1} \Sigma_{yx} \tag{8}$$

$$\hat{\mu}_{y|x} = \mu_y + \Sigma_{xy} \Sigma_{xx}^{-1} (x - \mu_x) \tag{9}$$

4   Implementation Workflows

We combine each of these theoretical elements to produce two practical implementations for distributed estimation. In addition to the previously discussed theoretical assumptions, we assume in these implementations that the partitioned data is linked by observation, so the different partitions know which individuals would belong together in a theoretical combined database. In future work this assumption may be relaxed, since methods exist for linking records in a privacy preserving way without prior knowledge, see [14]. We give implementations for two algorithms following the settings described in Sects. 2.1 and 2.2. For computation purposes, calculations will need to happen both at a central location and locally at the data partitions. We refer to these locations as the central node and external nodes in the following paragraphs and assume computation is possible at both nodes.

4.1   Trusted Central Server

In this setting a central node exists which either holds researchers' results or produces results for external researchers, and we assume that each data warehouse trusts this server with statistics of their data but for scalability or other reasons do not wish to combine their data with other warehouses.

The researcher at the central node defines the model along with all parameters and submits this to the central node. Using RAM matrix algebra, an expected mean vector and covariance matrix are generated given the chosen model parameters (a unique transformation). Recall these matrices correspond only to the observed variables, namely those stored in the external nodes. The central node then partitions these means and covariances based on the variables housed in each data partition, and it computes the conditional means and covariances for each partition based on a predetermined order using Schur complementation. A complex partition can be treated as a set of separate vertical partitions for each horiztonal break. In this case the central server would compute conditional means and covariances for each group of vertical partitions.

These matrices are sent to each corresponding node holding the data. The external nodes then compute the full information likelihood using the received model instances and their data, and they each return a single value to the central node. The central node then sums these values and repeats the same process iteratively until achieving a maximum. This process is summarized in Algorithm 1.

4.2   Round Robin

Now consider a central node either controlled by the researcher or available for input, but without trust from the external servers to view sufficient statistics. The process is similar to the one outlined above except the central node sends the expected means and covariance matrices to the first external node housing a data partition. From here, the node partitions the received matrices into two

---

Algorithm 1. Trusted Central Server

1: $Node_{central}$ chooses initial parameter estimates for given model
2: $Node_{central}$ calculates Σ and μ, uniquely defined by the initial parameters
3: $Node_{central}$ sends Σ and μ to $Node_1$ who splits them into marginal para-
   meters Σ_{x_1}, μ_{x_1} and conditional parameters $\Sigma_{x_{-1}|x_1}$, $\hat{\mu}_{x_{-1}|x_1}$
4: $Node_1$ computes the full information likelihood, $-2LL_1$, for its data using
   Σ_{x_1} and μ_{x_1}
5: $Node_1$ sends $-2LL_1$, $\Sigma_{x_{-1}|x_1}$, and $\hat{\mu}_{x_{-1}|x_1}$ $Node_{central}$
6: $Node_{central}$ sends $\Sigma_{x_{-1}|x_1}$ and $\hat{\mu}_{x_{-1}|1}$ to $Node_2$
7: **for** i in $2, ..., k$ partitions **do**
8:     $Node_i$ repeats steps 3 - 5
9:     $Node_{central}$ repeats step 6
10: **end for**
11: $Node_{central}$ calculates $\Sigma_{i=1}^{k} - 2LL_i$.
12: New parameter estimates are chosen and steps 2 - 10 are repeated until
    $\Sigma_{i=1}^{k} - 2LL_i$ minimized

---

parts: one containing the expected marginal mean and covariance for the variables it holds, and the second containing everything else. Using these matrix partitions, the node estimates by Schur complementation the conditional mean and covariance matrices for all the rest of the variables dependent on the variables in its partition. It also estimates the marginal log-likelihood values for all observations using the marginal mean and covariance matrices and the data it holds. Both of these are sent on to the next external node.

The next external node, and each following one, performs the same steps as the first node. The received conditional model matrices are partitioned between the data the external node holds and those in the proceeding nodes. Thus conditioning all following nodes on all previous ones. This also implies the log-likelihood values calculated at each node are conditional on all previous data. Each node adds its total log-likelihood value from all observations to the total received from the previous node. In this way, the joint distribution is attained and the total log-likelihood for the whole data set is estimated exactly the same as if all the data was combined. We summarize the process in Algorithm 2.

5 Cannabis Study Example

We present a simple example using real data from The Brisbane Longitudinal Twin Study (BLTS), [12]. This study collects data on twins in Australia with a focus on cannabis use, abuse, and dependence, as well as mental disorders. While these data have been collected in actuality by only one study, it is easy to imagine a situation where such data were collected by multiple sources who do not wish to combine their data. For example, consider a study looking at the relationship between cannabis use initiation, cannabis use disorder, and psychosis. Different measurements about cannabis use and problems could be collected by hospitals,

---

Algorithm 2. Round Robin

1: Researcher defines the SEM and chooses initial parameter estimates
2: $Node_{central}$ calculates Σ and μ, uniquely defined by the initial parameters
3: $Node_{central}$ sends Σ and μ to $Node_1$ who splits them into marginal parameters Σ_{x_1}, μ_{x_1} and conditional parameters $\Sigma_{x_{-1}|x_1}$, $\hat{\mu}_{x_{-1}|x_1}$
4: $Node_1$ computes the full information likelihood, $-2LL_1$, for its data using Σ_{x_1} and μ_{x_1}
5: $Node_1$ sends $-2LL_1$, $\Sigma_{x_{-1}|x_1}$, and $\hat{\mu}_{x_{-1}|x_1}$ $Node_2$
6: **for** i in 2, ..., (k-1) partitions **do**
7:     $Node_i$ repeats steps 3 - 5 with $\Sigma_{j=1}^{i} - 2LL_j$ in place of $-2LL_1$
8: **end for**
9: $Node_k$ repeats steps 3 - 4
10: $Node_k$ sends $\Sigma_{i=1}^{k} - 2LL_i$ to $Node_{central}$
11: New parameter estimates are chosen and steps 2 - 10 are repeated until $\Sigma_{i=1}^{k} - 2LL_i$ minimized

---

general practitioners, researchers, and medical specialists. Indeed, it is possible that such different sources would have information about the same individuals.

Figure 6 shows path diagrams for two possible models with the manifest variables in possible partitions. These models are adapted from a presentation, [11], given on causation models using the Brisbane data set. The original models were twin models with genetic factors included, and for simplicity we estimate simple factor models without the twin genetic elements. The first model assumes a hybrid causation between cannabis use and psychosis symptoms, while the second model only assumes causation from cannabis use to psychosis symptoms.

We estimate the parameters using code we developed as a proof of concept for the round robin procedure. We estimate parameters once using three partitions as shown, once with all the data combined, and once with the data combined and the *OpenMx* software, a standard SEM package available in R, [2]. These estimates are shown in Table 1; we obtain identical parameters up to four decimal precision for the partitioned, non-partitioned, and *OpenMx* results. We test the accuracy of our code further with simulations in the next section.

These results occurred as expected because mathematically the estimates should be the same whether the likelihood estimation is done jointly or separately (as described in Sect. 3.2). In this implementation we do not perturb any model parameters or likelihoods, so results will be identical. Recall we assume here that the partitioned data are linked by an ID column across the different datasets. In other words, the different data holders would have a way of matching the conditional mean statistics with the correct individuals. This of course opens the door for further privacy questions and adds an extra burden to implementations. Possible solutions to link data in a private way do exist such as [14], and it bears consideration for future work what effect imperfectly linked data has on parameter estimates (utility).

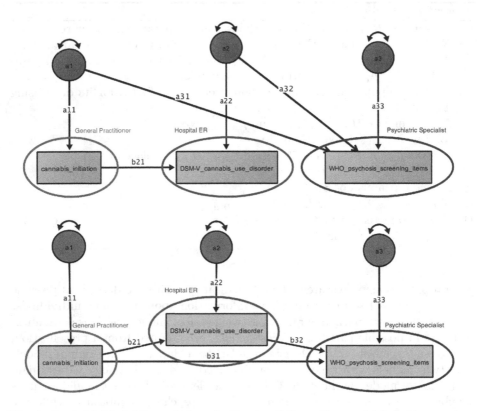

Fig. 6. Cannabis Model Formulations for Partioned Data. Top model: hybrid causation between cannabis use and psychosis symptoms, and bottom model: causation from cannabis use to psychosis symptoms.

Table 1. Cannabis Models Parameter Estimates

Parameter	Model 1	Model 2
a11	1.2061	1.2061
a22	0.6872	0.6872
a33	0.4270	0.4270
b21	0.3957	0.3956
a31	0.1235	
a32	0.0530	
b31		0.0719
b32		0.0771

6   Simulations for Accuracy and Computational Complexity

We developed code as a proof of concept for estimating SEMs with horizontal, vertical, or complexly partitioned data. Hopefully it may also serve as a very basic first step towards a future implantation of this framework. The code is freely available at github.com/jsnoke/Firewall.

To assess both the accuracy of parameter estimates and the computational complexity of our functions, we ran simulations using Latent Growth Curve Models, see Fig. 5. The type of model is only for consistency, and we ran the simulations only for vertical partitions. Horizontal partitions simply divide the joint likelihood into sums without the extra step of conditioning, and complex partitions are combinations of vertical and horizontal. From a computational standpoint, the speed and accuracy of the vertical partitions is the important piece.

For accuracy we compared parameter estimates from our code with estimates using the standard SEM R package *OpenMx*. This comparison is for the sake of ascertaining the performance of our code against standard software, rather than ascertaining the accuracy of the partitioned estimation, since we obtain identical results with our functions using non-partitioned versus partitioned data sets. This held true (within numerical precision) regardless of the number of partitions, k, manifest variables, p, or observations, n.

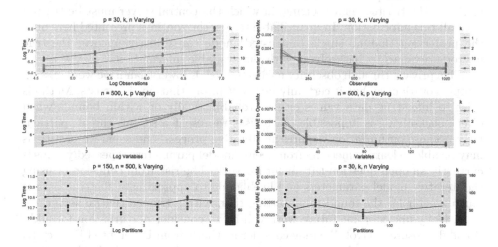

Fig. 7. Simulations for run time and parameter accuracy, with varying sample size, n, number of manifest variables p, and number of partitions, k.

The simulations for accuracy, right column of Fig. 7, show that our estimates converge to those obtained with state-of-the-art software as the number of observations or variables increases. We also see the accuracy is not affected by the number of partitions. For run time, our code is much slower than the *OpenMx* software, but it scales as expected, scaling $O(n)$ with observations and $O(n^3)$

with variables. These match expectations under usual matrix inversion and multiplication steps. Importantly, increasing partitions does not increase run time, particularly as the number of observations grows large. The bottom left graph shows no increase from $k = 1$ to $k = 150$ when $p = 150$. This makes sense, since we are trading extra computations for smaller covariance matrix inversions and multiplies. While the current implementation is a proof of concept and would not serve as an actual implementation, the simulations show our estimation routines can function without the loss of accuracy or speed given partitions.

7   Discussion

7.1   Privacy Question

This paper puts forth methods for extending the partitioned modeling framework to structural equation models and offers a proof of concept implementation, but many privacy questions remain open before these routines could become reality. While we currently do not implement perturbation or secure sharing methods, our methods do not fit the usual privacy situations. Due to the nature of SEMs, we pass model parameters and some conditional statistics, but no one node has all the pieces with which to produce the original data. The model parameters and the likelihoods together can lead back to the data, but no single node has access to both of those pieces except for their own partition of data. The exception to this is the trusted-server setting, in which the central server must be trusted with all this information. In quantifiable terms, the routines presented here offer the protection that no single (untrusted) node can directly uncover the true microdata values by backsolving with the available information.

This privacy through obscurity means the risk is lower than if we were passing sample covariances, but it certainly *does not* imply that no risk exists. Methods exist to infer true data values either probabilistically or within bounds when direct backsolving is impossible. Such risks represent a serious threat, particularly because optimization requires an iterative process, and separate partitions may be able to learn something from seeing model parameters repeatedly. This is currently an open question left for future work. The risk from collusion or using the iterative model parameters and likelihoods to approximate the true data is much greater for the round-robin setting than with a trusted central server.

A likely first step for future work will be to consider perturbation for additional disclosure control. We may consider methods from Differential Privacy [7] to ensure the privacy of shared model parameters and likelihoods. In this setting we could perturb either the data or the resulting likelihoods, so the shared information has a controllable level of associated risk. Alternatively we could implement SMPC as much of the previous work with partitioned models has done in order to safely share likelihoods or model matrices. These protection cases should be evaluated based on the risk-utility tradeoff. In the routines presented here, utility is maximized given that model parameter estimates are identical between partitioned and non-partitioned optimizations. If perturbation or secure sharing are implemented, risk will decrease but utility will likely also decrease. As with other statistical disclosure control implementations, this tradeoff should be optimized based on situational constraints such as the severity of disclosure consequences or the desirable utility.

There are additional questions, as we alluded to earlier, concerning the risk due to linking the data across partitions. Risk may exist simply from one group knowing the existence of individuals or variables in another database. This problem expands with the possibility of complex or overlapping partitions that share information on individuals. In general the risk is somewhat mediated in a distributed data scenario because the lack of a single point of failure means that the payoff for a successful disclosure attack is reduced. In contrast to the public data release format, an attacker will likely not be able to reveal all the information pertaining to individuals in the data set in a single attack. Instead, an attacker can reveal at most a subset–only the data available at a given node. This is not to say that the situation has no risk but simply that the stakes are lower with partitioned data. Substantial further work is needed to diagnose the risks and possible solutions for these situations.

7.2   Future Work

We present a potential solution here to facilitate research when data reside behind two or more firewalls, such that the data cannot be combined for privacy or proprietary reasons. The two settings provided here, either using a trusted central server or a round robin, are both plausible for a real implementation, but other formulations are also possible. Specifically for estimating SEMs using distributed MLE, these two settings provide the most direct routines. In the previous section we discussed the many existing privacy questions that must be addressed in future work, but we hope to lay the groundwork with this paper by first showing accurate estimation is possible, and then the privacy questions can be further pursued and the implementation altered accordingly.

While our current software serves as a proof of concept, it would not serve for any actual implementation of this method. We run the process on one machine with separate memory slots that are persevered to function as partitions, but in a real instance the implementation would use piping to pass information from multiple different machines. We hope our code may provide a basis for understanding the proposed algorithms and can aid future work towards creating such an infrastructure. Such future work should also consider the computational speed of the algorithm, which relates to the workflow of the implementation, since certain methods of passing elements may lead to computational gains. A vast field of literature exists on methods for partitioned model estimation apart from privacy concerns, and our work here should be able to blend with existing work.

Acknowledgements. This work was supported in part by NSF grants Big Data Social Sciences IGERT DGE-1144860 to Pennsylvania State University, and BCS-0941553 and SES-1534433 to the Department of Statistics, Pennsylvania State University. The work was also in part supported by the National Center for Advancing Translational

Sciences, National Institutes of Health, through Grant UL1 TR000127. The content is solely the responsibility of the authors and does not necessarily represent the official views of the NIH.

A   Appendix

A.1   RAM Algebra

We briefly exhibit here the method we use for defining SEMs and transforming the model parameters to model implied means and covariance matrices. These model implied matrices are then used to calculate log likelihoods iteratively given the data. Optimizing over these matrices is equivalent to optimizing over the model parameters, giving us our estimates.

The SEM path diagram has a one-to-one relationship with the Multivariate normal mean and covariance matrices for the manifest variables. We construct this relationship through the use of RAM matrix algebra. For this we define five matrices denoted A, S, F, M, and I. These matrices contain both fixed and free model parameters. The free parameters are to be estimated and will be changed during optimization, while the fixed parameters do not change. In these matrices, free parameters are denoted with a greek symbol and the fixed parameters are designated by a constant number.

Recall the path diagram shown in Fig. 4. For this example model, the RAM algebra proceeds as follows. The A ("asymmetric") matrix defines all regression parameters or one-headed arrows in the path diagram. It has number of rows and columns equal to the number of combined latent and manifest variables, with the column designating the path origin and the row designation the destination.

$$
A = \begin{array}{c}
 \\
F_1 \\
F_2 \\
M_1 \\
M_2 \\
M_3 \\
M_4 \\
M_5 \\
M_6
\end{array}
\begin{array}{cccccccc}
F_1 & F_2 & M_1 & M_2 & M_3 & M_4 & M_5 & M_6 \\
\left(\begin{array}{cccccccc}
0 & 0 & 0 & 0 & 0 & 0 & 0 & 0 \\
0 & 0 & 0 & 0 & 0 & 0 & 0 & 0 \\
1 & 0 & 0 & 0 & 0 & 0 & 0 & 0 \\
\beta_2 & 0 & 0 & 0 & 0 & 0 & 0 & 0 \\
\beta_3 & 0 & 0 & 0 & 0 & 0 & 0 & 0 \\
0 & 1 & 0 & 0 & 0 & 0 & 0 & 0 \\
0 & \beta_4 & 0 & 0 & 0 & 0 & 0 & 0 \\
0 & \beta_5 & 0 & 0 & 0 & 0 & 0 & 0
\end{array}\right)
\end{array}
$$

The S ("symmetric") matrix defines are variance parameters or two-headed arrows in the path diagram in the same way as the A matrix.

$$
S = \begin{array}{c}
\begin{array}{cccccccc} F_1 & F_2 & M_1 & M_2 & M_3 & M_4 & M_5 & M_6 \end{array} \\
\begin{array}{c} F_1 \\ F_2 \\ M_1 \\ M_2 \\ M_3 \\ M_4 \\ M_5 \\ M_6 \end{array}
\left(\begin{array}{cccccccc}
\sigma_{F1} & \sigma_{F12} & 0 & 0 & 0 & 0 & 0 & 0 \\
\sigma_{F12} & \sigma_{F2} & 0 & 0 & 0 & 0 & 0 & 0 \\
0 & 0 & \epsilon_1 & 0 & 0 & 0 & 0 & 0 \\
0 & 0 & 0 & \epsilon_2 & 0 & 0 & 0 & 0 \\
0 & 0 & 0 & 0 & \epsilon_3 & 0 & 0 & 0 \\
0 & 0 & 0 & 0 & 0 & \epsilon_4 & 0 & 0 \\
0 & 0 & 0 & 0 & 0 & 0 & \epsilon_5 & 0 \\
0 & 0 & 0 & 0 & 0 & 0 & 0 & \epsilon_6
\end{array}\right)
\end{array}
$$

The F ("filter") matrix acts a filter for the manifest variables. It has columns equal to the combined number of latent and manifest variables but rows equal only the number of manifest variables. For each manifest variable it has a one on the diagonal.

$$
F = \begin{array}{c}
\begin{array}{cccccccc} F_1 & F_2 & M_1 & M_2 & M_3 & M_4 & M_5 & M_6 \end{array} \\
\begin{array}{c} M_1 \\ M_2 \\ M_3 \\ M_4 \\ M_5 \\ M_6 \end{array}
\left(\begin{array}{cccccccc}
0 & 0 & 1 & 0 & 0 & 0 & 0 & 0 \\
0 & 0 & 0 & 1 & 0 & 0 & 0 & 0 \\
0 & 0 & 0 & 0 & 1 & 0 & 0 & 0 \\
0 & 0 & 0 & 0 & 0 & 1 & 0 & 0 \\
0 & 0 & 0 & 0 & 0 & 0 & 1 & 0 \\
0 & 0 & 0 & 0 & 0 & 0 & 0 & 1
\end{array}\right)
\end{array}
$$

The M ("mean") matrix defines the mean parameters if any for the latent and manifest variables. These are not always included in the path diagrams.

$$
M = \begin{array}{c}
\begin{array}{cccccccc} F_1 & F_2 & M_1 & M_2 & M_3 & M_4 & M_5 & M_6 \end{array} \\
\left(\begin{array}{cccccccc}
\mu_{F1} & \mu_{F2} & 1 & 0 & 0 & 0 & 0 & 0
\end{array}\right)
\end{array}
$$

Finally an I ("identity") matrix is included, with columns and rows equal to the number of combined latent and manifest variables.

$$
I = \begin{array}{c}
\begin{array}{cccccccc} F_1 & F_2 & M_1 & M_2 & M_3 & M_4 & M_5 & M_6 \end{array} \\
\begin{array}{c} F_1 \\ F_2 \\ M_1 \\ M_2 \\ M_3 \\ M_4 \\ M_5 \\ M_6 \end{array}
\left(\begin{array}{cccccccc}
1 & 0 & 0 & 0 & 0 & 0 & 0 & 0 \\
0 & 1 & 0 & 0 & 0 & 0 & 0 & 0 \\
0 & 0 & 1 & 0 & 0 & 0 & 0 & 0 \\
0 & 0 & 0 & 1 & 0 & 0 & 0 & 0 \\
0 & 0 & 0 & 0 & 1 & 0 & 0 & 0 \\
0 & 0 & 0 & 0 & 0 & 1 & 0 & 0 \\
0 & 0 & 0 & 0 & 0 & 0 & 1 & 0 \\
0 & 0 & 0 & 0 & 0 & 0 & 0 & 1
\end{array}\right)
\end{array}
$$

Using these matrices, we obtain the corresponding model implied mean (μ) and covariance matrices (Σ) of the manifest variables based on the chosen parameters. The following equations give this crucial relationship.

$$\Sigma = F * (I - A)^{-1} * S * ((I - A)^{-1})^T * F^T \tag{10}$$

$$\mu = F * (I - A)^{-1} * M \tag{11}$$

References

1. Arbuckle, J.L., Marcoulides, G.A., Schumacker, R.E.: Full information estimation in the presence of incomplete data. Adv. Struct. Equ. Model. Issues Tech. **243**, 277 (1996)
2. Boker, S., Neale, M., Maes, H., Wilde, M., Spiegel, M., Brick, T., Spies, J., Estabrook, R., Kenny, S., Bates, T., et al.: Openmx: an open source extended structural equation modeling framework. Psychometrika **76**(2), 306–317 (2011)
3. Boker, S.M., Brick, T.R., Pritikin, J.N., Wang, Y., von Oertzen, T., Brown, D., Lach, J., Estabrook, R., Hunter, M.D., Maes, H.H., et al.: Maintained individual data distributed likelihood estimation (middle). Multivar. Behav. Res. **50**(6), 706–720 (2015)
4. CALIT. Personal data for the public good. Technical report, California Institute for Telecommunications and Information Technology (2014)
5. de Montjoye, Y.-A., Shmueli, E., Wang, S.S., Pentland, A.S.: OpenPDS: protecting the privacy of metadata through safeanswers. PloS one **9**(7), e98790 (2014)
6. Dufau, S., Duñabeitia, J.A., Moret-Tatay, C., McGonigal, A., Peeters, D., Alario, F.-X., Balota, D.A., Brysbaert, M., Carreiras, M., Ferrand, L., et al.: Smart phone, smart science: how the use of smartphones can revolutionize research in cognitive science. PloS one **6**(9), e24974 (2011)
7. Dwork, C.: Differential privacy: a survey of results. In: Agrawal, M., Du, D.-Z., Duan, Z., Li, A. (eds.) TAMC 2008. LNCS, vol. 4978, pp. 1–19. Springer, Heidelberg (2008)
8. Fienberg, S.E., Fulp, W.J., Slavkovic, A.B., Wrobel, T.A.: "Secure" log-linear and logistic regression analysis of distributed databases. In: Domingo-Ferrer, J., Franconi, L. (eds.) Privacy in Statistical Databases. LNCS, vol. 4302, pp. 277–290. Springer, Heidelberg (2006)
9. Fienberg, S.E., Nardi, Y., Slavković, A.B.: Valid statistical analysis for logistic regression with multiple sources. In: Gal, C.S., Kantor, P.B., Lesk, M.E. (eds.) ISIPS 2008. LNCS, vol. 5661, pp. 82–94. Springer, Heidelberg (2009)
10. Gaye, A., Marcon, Y., Isaeva, J., LaFlamme, P., Turner, A., Jones, E.M., Minion, J., Boyd, A.W., Newby, C.J., Nuotio, M.-L., et al.: DataSHIELD: taking the analysis to the data, not the data to the analysis. Int. J. Epidemiol. **43**(6), 1929–1944 (2014)
11. Gillespie, N.: Direction of causation and comorbidity models mutualism, sibling / spousal interaction. Presentation at Advanced Genetic Epidemiology Statistical Workshop 2015, Richmond, VA (2015)
12. Gillespie, N.A., Henders, A.K., Davenport, T.A., Hermens, D.F., Wright, M.J., Martin, N.G., Hickie, I.B.: The brisbane longitudinal twin study: pathways to cannabis use, abuse, and dependence project–current status, preliminary results, and future directions. Twin Res. Hum. Genet. **16**(01), 21–33 (2013)
13. Goldwasser, S.: Multi party computations: past and present. In: Proceedings of the Sixteenth Annual ACM Symposium on Principles of Distributed Computing, pp. 1–6. ACM (1997)
14. Hall, R., Fienberg, S.E.: Privacy-preserving record linkage. In: Domingo-Ferrer, J., Magkos, E. (eds.) PSD 2010. LNCS, vol. 6344, pp. 269–283. Springer, Heidelberg (2010)
15. Haynsworth, E.V.: On the schur complement. Technical report, DTIC Document (1968)

16. Hundepool, A., Domingo-Ferrer, J., Franconi, L., Giessing, S., Nordholt, E.S., Spicer, K., De Wolf, P.-P.: Statistical Disclosure Control. John Wiley & Sons, Hoboken (2012)
17. Karr, A.F., Fulp, W.J., Vera, F., Young, S.S., Lin, X., Reiter, J.P.: Secure, privacy-preserving analysis of distributed databases. Technometrics **49**(3), 335–345 (2007)
18. Karr, A.F., Lin, X., Sanil, A.P., Reiter, J.P.: Privacy-preserving analysis of vertically partitioned data using secure matrix products. J. Official Stat. **25**(1), 125 (2009)
19. Kupek, E.: Beyond logistic regression: structural equations modelling for binary variables and its application to investigating unobserved confounders. BMC Med. Res. Methodol. **6**(1), 1 (2006)
20. Lindell, Y., Pinkas, B.: Secure multiparty computation for privacy-preserving data mining. J. Priv. Confidentiality **1**(1), 5 (2009)
21. McArdle, J.J., McDonald, R.P.: Some algebraic properties of the reticular action model for moment structures. Br. J. Math. Stat. Psychol. **37**(2), 234–251 (1984)
22. Miller, G.: The smartphone psychology manifesto. Perspect. Psychol. Sci. **7**(3), 221–237 (2012)
23. Raab, G.M., Dibben, C., Burton, P.: Running an analysis of combined data when the individual records cannot be combined: practical issues in secure computation. In: Statistical Data Confidentiality Work Session, UNECE, October 2015
24. Schur, I.: Neue begründung der theorie der gruppencharaktere (1905)
25. Slavkovic, A.B., Nardi, Y., Tibbits, M.M.: "Secure" logistic regression of horizontally and vertically partitioned distributed databases. In: Seventh IEEE International Conference on Data Mining Workshops (ICDM Workshops 2007), pp. 723–728. IEEE (2007)
26. Willenborg, L., De Waal, T.: Statistical Disclosure Control in Practice, vol. 111. Springer, New York (1996)
27. Yao, A.C-C.: Protocols for secure computations. In: FOCS 82, pp. 160–164 (1982)

A New Algorithm for Protecting Aggregate Business Microdata via a Remote System

Yue Ma[1]([✉]), Yan-Xia Lin[1], James Chipperfield[1,2], John Newman[2], and Victoria Leaver[2]

[1] National Institute for Applied Statistics Research Australia,
School of Mathematics and Applied Statistics, University of Wollongong,
Wollongong, Australia
ym894@uowmail.edu.au
[2] Australian Bureau of Statistics, Canberra, Australia

Abstract. Releasing business microdata is a challenging problem for many statistical agencies. Businesses with distinct continuous characteristics such as extremely high income could easily be identified while these businesses are normally included in surveys representing the population. In order to provide data users with useful statistics while maintaining confidentiality, some statistical agencies have developed online based tools to allow users to specify and request tables created from microdata. These tools only release perturbed cell values generated from automatic output perturbation algorithms in order to protect each underlying observation against various attacks, such as differencing attacks. An example of the perturbation algorithms has been proposed by Thompson et al. (2013). The algorithm focuses largely on reducing disclosure risks without addressing much on data utility. As a result, the algorithm has limitations, including a limited scope of applicable cells and uncontrolled utility loss. In this paper we introduce a new algorithm for generating perturbed cell values. As a comparison, The new algorithm allows more control over utility loss, while it could also achieve better utility-disclosure tradeoffs in many cases, and is conjectured to be applicable to a wider scope of cells.

Keywords: Business data · Output perturbation · Remote access · Continuous tabular data · Statistical disclosure control

1 Introduction

Disseminating data containing confidential information is a challenging issue for many statistical agencies. On the one hand, the released data should not reveal confidential information to the public; on the other hand, the released data

---

Disclaimer: Views expressed in this paper are those of the author(s) and do not necessarily represent those of the Australian Bureau of Statistics. Where quoted or used, they should be attributed clearly to the author.

© Springer International Publishing Switzerland 2016
J. Domingo-Ferrer and M. Pejić-Bach (Eds.): PSD 2016, LNCS 9867, pp. 210–221, 2016.
DOI: 10.1007/978-3-319-45381-1_16

should carry enough statistical information to reflect behaviours of the population. To achieve these two conflicting objectives, statistical agencies release confidentialised data to data users. The confidentialised data conceals sensitive information from the public at the expense of some data utility.

However, it is not easy to confidentialise business data. Typically, some industries will be dominated by large businesses whose information is difficult to conceal by existing data masking methods. Non-perturbative data masking, such as top coding (Klein et al. 2014), suppression (Salazar-González 2005) and micro-aggregation (Defays and Nanopoulos 1993), significantly reduce information of continuous data items such as turnover or profit, which are of key interest to data users. Perturbation methods, such as data swapping (Moore 1996), synthetic data (Rubin 1993) and noise addition (Kim and Winkler 1995), cannot efficiently protect businesses with distinct continuous-valued characteristics. As a result, most statistical agencies have taken a cautious approach to releasing business data, and the majority of business data is still released in the form of broad-level tables.

The emergence of remote access may provide a solution to releasing business microdata. Remote access (Blakemore 2001; Reiter 2004) is a virtual system that provides a data analyst with access to a remote system built by a data agency. The data agency stores microdata in the remote system, and the data analyst communicates with the remote system through a query system. The analyst is restricted from viewing the underlying microdata. Instead, the analyst could only obtain statistical outputs of underlying microdata through the following model (see O'Keefe and Chipperfield 2013; Chipperfield and O'Keefe 2014): (1) an analyst submits a query (i.e. request for a table) to the remote system; (2) the remote system modifies or restricts estimates using an automatic algorithm; (3) the system sends the modified output to the analyst. An example of remote access system is American FactFinder (Hawala et al. 2004), which releases confidentialised tabulations of census data to data users.

The reason for a remote system to release confidentialised statistics is to prevent disclosure of confidential values via various methods of attack, the most significant of which is a differencing attack (Lucero et al. 2009, Sect. 4.1). A differencing attack reveals a confidential value by taking the difference of two cell totals whose contributing values differ by one. A differencing attack could be very effective on a remote system as the attacker is able to obtain statistical outputs of different underlying microdata with a high degree of freedom.

In this paper we base our discussion on releasing confidentialised totals from business data through a remote system. We assume a remote system could allow users to specify and request tables created from business microdata. Each cell of a table contains a perturbed survey estimate of total computed from a set of surveyed business values as specified by a data user. The algorithm in Thompson et al. (2013) has been shown to perform particularly well for confidentialising totals from business-typed microdata, and hence we have investigated the prospect of implementing it on remote systems to perturb business totals. The algorithm works by adding a perturbation amount to the unperturbed cell estimate to produce a perturbed cell estimate. The perturbation amount follows a parametric distribution which could be adjusted according to the distribution of underlying business values to produce the best result.

It has been proposed that an algorithm of generating perturbed statistical estimates should satisfy the definition of ϵ-differential privacy (Dwork et al. 2006). Such algorithms include adding Laplace distributed noises (Dwork et al. 2006) and other similar variations (Soria-Comas and Domingo-Ferrer 2013; Nissim et al. 2007) to perturb statistical estimates. These algorithms generally sacrifice a large degree of data utility for data confidentiality (see, for example, Sarathy and Muralidhar 2011). The algorithm in Thompson et al. is not designed to achieve ϵ-differential privacy; however, it achieves good utility-disclosure trade-offs for many cells. The details of the algorithm are introduced in Sect. 2.

A distinct feature of the algorithm in Thompson et al. is that, for a given cell of a table, the algorithm achieves its best performance if the optimal set of parameters for perturbing the cell is used. As developing a program of searching the optimal set of parameters to be used to perturb each cell value is non-trivial, a recent study investigated the outcomes of using one set of parameters to perturb all cells. The set of parameters was selected upon satisfying the requirement of disclosure risks for a few benchmark cells, and the study examined its impact on utility losses of different cells through empirical studies (see Chipperfield et al. 2016).

However, there are issues with this configuration. The issues are: 1. The algorithm cannot always generate legitimate cell estimates which fulfill the requirements of both utility loss and disclosure risk. 2. The algorithm could still produce a very perturbed cell value even though the requirement of utility loss is satisfied. 3. The way it trades data utility for data confidentiality may not be the most efficient one. As a result, we are looking for alternative algorithms which could help to solve these issues.

It needs to be mentioned that the first issue is not easy to solve completely. The reason is that cells contributed to by a small number of business values, some of which strongly dominate the cell value, are very difficult to perturb in a reasonable manner. Methods for confidentialising such cells require further studies. In this paper, we do not consider this kind of cells. Instead, we focus on perturbing common cells which do not contain strong dominant contributors.

In this paper, a new algorithm for perturbing business totals is proposed. The main point of the new algorithm is to limit the loss of utility, and simulation results show that the new algorithm addresses the issues mentioned above more effectively than the algorithm in Thompson et al. The new algorithm allows a better control over cell utility losses, achieves better utility-disclosure tradeoffs for many cells, and is conjectured to be able to legitimately perturb a wider range of cells.

This paper is organized as follow. Section 2 describes the algorithm in Thompson et al. of generating perturbed cell values. Section 3 introduces measures of disclosure risk and utility loss. Sections 4 describes the new algorithm.

Section 5 discusses advantages of the new algorithm compared with the algorithm in Thompson et al. through simulations. Section 6 concludes the paper.

2 The Algorithm in Thompson et al.

Consider any particular cell in a table and let there be n sample units contributing to the cell, where the units are indexed by $i = 1, 2, \cdots n$. Define a continuous valued characteristic (e.g. income or turnover) for the ith unit (e.g. business) by y_i, the estimation weight of y_i by w_i, and the survey estimate of the total is $\hat{s} = \sum_{i=1}^{n} w_i y_i$. We assume $y_1 w_1 \geq y_2 w_2 \geq \cdots \geq y_n w_n$. We call the weighted business values $(y_1 w_1, \cdots y_n w_n)$ **contributor values** to the cell value $\hat{s}$. The algorithm in Thompson et al. (2013) generates a perturbed cell value in the following way:

1. The algorithm identifies the parameters (K, m) to be used.
2. The algorithm generates a perturbation amount p^* from a random variable P^*, and add p^* to the total $\hat{s}$ to generate a perturbed cell value $\hat{s}^*$.

The random variable P^* has the expression $P^* = \sum_{i=1}^{K} (m_i D_i^* H_i^*) y_i w_i$, where K is the number of top contributors in the cell that are used in calculation of P^*; $m = (m_1, \cdots, m_K)$ is a magnitude vector; D_i^* is a random variable taking the value -1 and 1 with equal probability; H_i^* is a random variable centred on 1 and for the purpose of this paper we set H_i to have a symmetric triangular probability density function centered at 1 with width 0.6.

The optimal set of parameters to be used for each cell depends on the distribution of contributing values to the cell estimate. The optimal set of parameters guarantees that the perturbed estimates have the lowest average utility loss subject to having an acceptable disclosure risk. Examples of the optimal choices of magnitude vector when $K = 3$ for different contributor values are given in Table 1.

As mentioned in the Introduction, developing a program of searching the optimal set of parameters to perturb each cell value is non-trivial. One possible remedy is to use a fixed set of parameters to perturb all cells. However, this configuration certainly limits the efficacy of the algorithm as the choice of parameters is not always the optimal one for perturbing many cells.

To evaluate the validity of perturbed estimates generated by an algorithm, we need to define measures of utility loss and disclosure risk. It is important that the perturbed estimates should satisfy both an acceptable level of utility loss and an acceptable level of disclosure risk. In next section, we introduce these measures.

3 Measuring Disclosure Risk and Utility Loss

3.1 Differencing Attack

We measure the disclosure risk with respect to a 'Differencing Attack'. Throughout the paper, without loss of generality, we assume the attacker's target is the largest contributor value $y_1 w_1$, and the attacker also knows the weight of y_1 is equal to one ($w_1 = 1$). The reason for these assumptions is that, normally speaking, the largest contributor value $y_1 w_1$ has the highest disclosure risk against differencing attack than any other contributor value.

Differencing Attack: The attacker uses the difference between two perturbed cell estimates, $\hat{y}_1 = \hat{s}^* - \hat{s}^*_{-1}$ as an estimate of y_1, where $\hat{s}^*_{-1}$ is defined as the same as $\hat{s}^*$ except that the attacker's target, y_1, is dropped from the cell.

To define disclosure risk, we conservatively assume that: 1. the target is in the sample. and 2. the attacker could uniquely identify the target in terms of a set of quasi-identifiers. So the only protection available in a remote system is perturbation. Consequently, perturbation is the focus of how disclosure risk is measured.

3.2 Defining Disclosure

We first describe the process of conducting a differencing attack on a remote system.

Consider the following scenario: suppose a continuous valued characteristic of the ith sample unit is y_i and there are n sample units, and the estimation weight of y_i is w_i. Define $y = (y_1, y_2, \cdots, y_n)$ and $y_1 w_1 \geq y_2 w_2 \geq \cdots \geq y_n w_n > 0$ with $w_1 = 1$. The attacker estimates y_1 by taking the difference of two perturbed estimates. The estimate of y_1 is a realization of $\hat{Y}_1 = \hat{S}^* - \hat{S}^*_{-1} = y_1 + P^* - P^*_{-1}$, where $\hat{S}^*$ is the underlying random variable for the cell consisting of $(y_1 w_1, \cdots, y_n w_n)$ from which a perturbed cell value is drawn; P^* is the perturbation random variable for perturbing the cell; $\hat{S}^*_{-1}$ is the underlying random variable for the cell consisting of $(y_2 w_2, \cdots, y_n w_n)$; and P^*_{-1} is the perturbation random variable for perturbing the cell.

Disclosure occurs if the realization of $\hat{Y}_1$ reveals the value of y_1. It is not necessary for the realization of $\hat{Y}_1$ to be exactly equal to y_1- the degree of accuracy required for disclosure must be determined by the statistical agency. The following definition of disclosure risk we adopt is similar to that used by Lin and Wise (2012) and Klein et al. (2014).

Disclosure Risk: We say that the disclosure risk against a differencing attack is the probability that a realization of $\hat{Y}_1$ is within $100\alpha\%$ of the true value y_1. If we define disclosure risk of attacking target value y_1 as $D(y_1)$, then

$$D(y_1) = P(|P^* - P^*_{-1}| < \alpha y_1)$$

We say that α is the definition of disclosure and R is the acceptable disclosure risk. Different values of (R, α) could be justified on the basis of whether the attack is likely to occur. We say that perturbed cell estimates have an acceptable disclosure risk if $D(y_1)$ is less than R.

3.3 Defining Utility Loss

We define the **utility loss** of perturbing a cell as the relative distance between the perturbed cell value and unperturbed cell value. Measuring utility loss by percentage difference between the perturbed estimate and the unperturbed estimate has been widely used in many applications. It is formally introduced by Domingo-Ferrer and Torra (2001) and widely used by other authors in their studies (see Kim and Winkler 1995; Yancey et al. 2002).

As an algorithm produces a perturbed cell value randomly, in order to assess the general performance of the algorithm to perturb a cell in terms of utility loss, we look at the **average utility loss**, which is the expected utility loss of perturbing the cell using the algorithm. It is preferable that the average utility loss to be as low as possible given that $D(y_1) < R$.

4 A New Algorithm to Generate Perturbed Cell Estimates

Now we introduce a new algorithm to generate perturbed cell estimates. Suppose an ordinary cell has contributor values $(y_1 w_1, y_2 w_2, \cdots, y_n w_n)$, $w_1 = 1$. The new algorithm perturbs the cell value as follows:

1. The statistical agency sets a parameter value of β.
2. Define $\lambda = \hat{s}\beta$, where $\hat{s}$ is the cell value. If n is an even number, the new algorithm generates a perturbed amount $p^* = d_1 z_1$ and adds it to $\hat{s}$ to produce a perturbed cell estimate, where d_1 and z_1 are random samples drawn from random variables D_1 and Z_1, respectively. The random variable D_1 takes values -1 or 1 with equal probabilities and random variable Z_1 is distributed as U_1 or U_2 with equal probabilities, where $U_1 \sim U(0, 0.5\lambda)$ and $U_2 \sim U(1.5\lambda, 2\lambda)$. If n is an odd number, then the new algorithm generates a perturbed amount $p^* = z_2 d_2$ and adds it to $\hat{s}$ to produce a perturbed cell estimate, where z_2 and d_2 are samples drawn from random variables Z_2 and D_2, respectively. The random variable Z_2 has distribution $U(0.5\lambda, 1.5\lambda)$ and D_2 has the same distribution as D_1.

The value of the parameter β is actually the average utility loss of using the algorithm to perturb a cell. Therefore, the value of β could be set according to the requirements of the statistical agency. The disclosure risk of y_1 could also be mathematically determined and the mathematical expressions are given in Tables 2 and 3.

The advantage of splitting up odd and even cases is addressing the differencing attack by guaranteeing that the counts of contributors will go from odd to even or even to odd. It makes it much harder for the perturbation under the set of n contributors and the set of $n-1$ contributors to cancel out if the largest contributor is not strongly dominating the cell.

In next section, we discuss the advantages of using the new algorithm compared to the algorithm in Thompson et al.

5   Discussion of the New Algorithm Against the Algorithm in Thompson et al.

5.1   Controlled Utility Loss of Perturbing a Cell

For a given cell with cell value $\hat{s}$, it can be easily seen that the perturbation amount p^* generated by the new algorithm is bounded in $(-2\beta\hat{s}, 2\beta\hat{s})$. As a result, the utility loss of perturbing the cell through the new algorithm is bounded in $(0, 2\beta)$. As a result, both the average utility loss and the maximum utility loss of perturbing the cell could be controlled.

To illustrate this advantage, suppose the contributor values to a cell are $(y_1 w_1, \cdots y_8 w_8) = (25, 25, 25, 25, 25, 25, 25, 25)$. From Table 1, the optimal magnitude values are $(m_1, m_2, m_3) = (0.5, 0.4, 0.3)$. If the cell is perturbed by the algorithm in Thompson et al. with the optimal magnitude values, the average utility loss is 7.54 %. However, simulation results show that it is possible that the algorithm generates a extremely high perturbation amount which leads to a 22.3 % utility loss to the cell. A perturbed result with such a high utility loss may mislead some data users. In contrast, perturbing the cell by the new algorithm with $\beta = 0.0754$ also gives an average utility loss of 7.54 % while the maximum utility loss of perturbing the cell is only 15.08 %.

5.2   A Conjectured Wider Applicability

It is conjectured that the new algorithm could legitimately perturb a wider range of cells. A legitimately perturbed cell value should satisfy the requirements of both disclosure risk and utility loss. Recall that we assume the attacker's target is the largest contributor value y_1, and $w_1 = 1$. In the following, we say that a cell could be legitimately perturbed by an algorithm if the average utility loss is less than T, and the disclosure risk of y_1 against differencing attack is less than R given a specified definition of disclosure α. We illustrate this conjectured wider applicability by comparing the performances of the two algorithms on different cells.

Suppose the statistical agency set (T, R, α) to be $(10\%, 15\%, 11\%)$ for a perturbed cell estimate to be legitimate. Recall that in a recent study, we use one set of parameters for the algorithm in Thompson et al. for perturbing all cells as it is more practical. We stick to this way in this section and the parameters were set to be $K = 3$, $m = (0.4, 0.3, 0.2)$. The parameter of the new algorithm was set to be $\beta = 0.1$. Recall that $\beta = 0.1$ guarantees a 10 % utility loss for each cell.

Cell 1: The contributor values of cell 1 consist of $(y_1 w_1, y_2 w_2, \cdots, y_6 w_6) = (30, 30, 30, 10, 5, 5)$. Using the algorithm in Thompson et al., the average utility loss and disclosure risk of releasing perturbed cell estimates are 12.4 % and 9.4 %, respectively. It means that, even though the perturbed cell estimates satisfy a required level of disclosure risk, they do not carry enough data utility as required by the statistical agency. Using the new algorithm, the average utility loss and disclosure risk of releasing perturbed cell estimates are 10 % and 6.5 %. It means that, both the requirements of utility loss and disclosure risk are satisfied, and it is legitimate to release a perturbed cell value generated by the new algorithm.

Cell 2: The contributor values of cell 2 consist of $(y_1 w_1, y_2 w_2, \cdots, y_7 w_7) = (25, 25, 25, 25, 1, 1, 1)$. Using the algorithm in Thompson et al., the average utility loss and disclosure risk of releasing perturbed cell estimates are 10.9 % and 12.0 %, respectively. That means perturbed cell estimates do not carry enough data utility as required by the statistical agency. Using the new algorithm, the average utility loss and disclosure risk of releasing perturbed cell estimates are 10 % and 11.4 %. Both the requirements of utility loss and disclosure risk are satisfied, and it is legitimate to release a perturbed cell value generated by the new algorithm.

The above two cells are used to show that the new algorithm could help to generate legitimate cell estimates that are not achievable by the algorithm in Thompson et al. However, we next show that it is possible that, when a cell contains a dominant contributor value, the algorithm in Thompson et al. is better. For illustration, we set (T, R, α) to be $(15 \%, 12 \%, 11 \%)$, and $\beta = 0.15$ for the next cell.

Cell 3: The contributor values of cell 3 consist of $(y_1 w_1, y_2 w_2, \cdots y_9 w_9) = (60, 20, 20, 15, 15, 10, 10, 10, 10)$. The average utility loss and disclosure risk of releasing perturbed cell estimates generated by the algorithm in Thompson et al. are 14.1 % and 9.5 %, while the counterparts generated by the new algorithm are 15 % and 13.1 %. In this case the algorithm in Thompson et al. is the better algorithm perturbing the cell.

The reason the new algorithm is not favourable for **Cell 3** is that, when the ratio $y_1 w_1 / \hat{s}$ gets large, the disclosure risk of using the new algorithm goes up dramatically. To see this, without loss of generality, we assume n is even, $\beta \geq 2\alpha$, $y_1 w_1 < min(\frac{\lambda}{1.5\beta + \alpha}, \frac{\lambda}{2\alpha})$. From Table 3, the disclosure risk is $P_{C11} = \frac{1}{2\lambda\lambda_1} \alpha y_1^2 w_1^2 \beta$. It is evident that the ratio $y_1 w_1 / \hat{s}$ largely impact the value of disclosure risk. Possible future research would be to use either the algorithm in Thompson et al. or the new algorithm to perturb a cell estimate subject to a condition involving the value of $y_1 w_1 / \hat{s}$.

5.3 Better Utility-Disclosure Trade-Offs

We compare the utility-disclosure tradeoffs of the two algorithms on different cells through simulations. In order to obtain utility-disclosure plots, we gradually changed the values in the magnitude vector m used by the algorithm in Thompson et al. and the parameter β used by the new algorithm. We recorded the average utility losses and disclosure risks given different parameter values. Moreover, we provide utility-disclosure plots for $\alpha = 0.11$ and $\alpha = 0.18$, respectively.

Simulation 1: The contributor values of a cell are $(y_1 w_1, y_2 w_2, \cdots, y_8 w_8) = (25, 25, 25, 25, 25, 25, 25, 25)$. We set the magnitude vector to be $m = (0.3 + 0.01i, 0.2 + 0.01i, 0.1 + 0.01i)$, where $i = 1, 2 \cdots 40$. We recorded the utility loss and disclosure risk of releasing perturbed cell estimates generated by the algorithm in Thompson et al. for each value of i for generating the utility-disclosure plot. When $i = 20$, $m = (0.5, 0.4, 0.3)$, which is the optimal magnitude vector as shown in Table 1. Similarly we set $\beta = 0.04 + i/400, i = 1, 2 \cdots 40$; and we obtained the utility-disclosure plot for the new algorithm. We use a box-plot to represent the utility-disclosure plot of the algorithm in Thompson et al. and a dotted plot to represent utility-disclosure plot of the new algorithm and these symbols also apply to Figs. 2 and 3 discussed in Simulations 2 and 3. The utility-disclosure plots for $\alpha = 0.11$ and 0.18 are provided in Fig. 1.

Simulation 2: The contributor values of a cell are $(y_1 w_1, y_2 w_2, \cdots, y_9 w_9) = (40, 20, 20, 15, 15, 10, 10, 10, 10)$. We set the magnitude vector to be $m = (0.15 + 0.01i, 0.1 + 0.01i, 0.05 + 0.01i)$, where $i = 1, 2 \cdots 40$. The parameter of new algorithm is set to be $\beta = 0.03 + i/300, i = 1, 2 \cdots 40$. We follow the same procedure as in Simulation 1 to obtain the utility-disclosure plots of the two algorithm for $\alpha = 0.11$ and 0.18. The plots are given in Fig. 2.

Simulation 3: The contributor values are $(y_1 w_1, y_2 w_2, \cdots, y_{49} w_{49}) = (60, 20, 10, 10, \cdots, 10)$. We set the magnitude vector to be $m = (0.05 + 0.05i, 0.05i, 0.05i)$, where $i = 1, 2 \cdots 40$. The parameter of the new algorithm is set to be $\beta = 0.01 + i/150, i = 1, 2 \cdots 40$. The utility-disclosure plots for $\alpha = 0.11$ and 0.18 are given in Fig. 3.

From Fig. 1, we see that the new algorithm leads to a better utility-disclosure trade-off when the contributor values to a cell estimate are uniformly distributed. From Fig. 2, we see that this advantage is reduced when the largest contributor value dominates the cell estimate. From Fig. 3, we see that the new algorithm again offers a better utility-disclosure trade-off even though the largest contributor value is significantly larger than all other contributor values, as in this case the largest contributor value does not dominate the cell estimate.

6   Conclusion

In this paper we introduced a new algorithm to generate perturbed cell estimates. The advantages of the new algorithm are discussed compared with the algorithm in Thompson et al. It is conjectured that the new algorithm could be widely used in many remote systems for creating tables from business microdata. Possible future research would be on combining the new algorithm with the algorithm in Thompson et al. to perturb survey estimate of population totals from business microdata.

Acknowledgements. We sincerely thank Sybille McKeown and all the others in the Australian Bureau of Statistics for providing so many good feedbacks for the paper.

Appendix

Table 1. Magnitude values that guarantee 15 % disclosure risk given $\alpha = 0.11$ and minimise the average utility loss for different distributions of top contributor values.

Distribution	Relative size of top contributors				Optimal magnitude values		
	1st	2nd	3rd	4th	m1	m2	m3
1	90	5	5		0.15	0.1	0.1
2	80	10	5	5	0.15	0.1	0.1
3	70	20	10		0.15	0.1	0.1
4	60	20	10	10	0.2	0.1	0.1
5	60	40			0.25	0.15	0.1
6	50	20	20	10	0.25	0.15	0.1
7	40	30	30		0.3	0.2	0.1
8	30	30	30	10	0.4	0.3	0.2
9	25	25	25	25	0.5	0.4	0.3

Table 2. Probability expressions for Table 3. $m_1 = y_1 w_1$ and $\lambda_1 = \hat{s}_{-1}\beta$.

P_{C11}	$\frac{1}{2\lambda\lambda_1}\alpha m_1^2\beta$
P_{C12}	$\frac{1}{2\lambda\lambda_1}(0.125\lambda^2 + 0.5\lambda\alpha m_1 + 0.5\alpha^2 m_1^2 - 0.25\lambda\lambda_1 + 0.125\lambda_1^2 - 0.5\alpha m_1\lambda_1)$
P_{C13}	$\frac{1}{2\lambda_1\lambda}(1.125\lambda_1^2 + 1.5\lambda_1\alpha m_1 + 0.5\alpha^2 m_1^2 - 2.25\lambda\lambda_1 + 1.125\lambda^2 - 1.5\alpha m_1\lambda)$
P_{C21}	$\frac{\alpha m_1}{2\lambda}$
P_{C22}	$\frac{1}{2\lambda\lambda_1}(-0.5\alpha^2 m_1^2 - \lambda_1 m_1\alpha + 3\lambda\lambda_1 - 2\lambda_1^2 - 1.125\lambda^2 + 1.5\lambda\alpha m_1)$
P_{C23}	$\frac{3\alpha m_1^2\beta}{2\lambda\lambda_1}$
P_{C24}	$\frac{1}{2\lambda\lambda_1}(0.125\lambda_1^2 + 0.5\alpha^2 m_1^2 + 0.5\alpha m_1\lambda_1 - 0.25\lambda\lambda_1 + 0.125\lambda^2 - 0.5\alpha m_1\lambda)$
P_{C25}	$\frac{1}{2\lambda\lambda_1}(0.75\lambda\lambda_1 + 0.5\alpha m_1\lambda_1 - 0.875\lambda_1^2)$
P_{C26}	$\frac{1}{2\lambda\lambda_1}(1.125\lambda^2 + 0.5\alpha^2 m_1^2 + 1.5\alpha m_1\lambda - 2.25\lambda\lambda_1 + 1.125\lambda_1^2 - 1.5\lambda_1\alpha m_1)$

Table 3. Disclosure risk of perturbed estimates generated by the new algorithm.

	$\beta \geq 2\alpha$	$\frac{\alpha}{1.5} < \beta < 2\alpha$	$\beta \leq \frac{\alpha}{1.5}$
n is odd, $\frac{\lambda}{4\beta-2\alpha} < y_1 w_1 < min(\frac{\lambda}{1.5\beta+\alpha}, \frac{\lambda}{2\alpha})$	P_{C21}	$P_{C24} + P_{C21}$	not possible
n is odd, $\frac{\lambda}{4\beta+2\alpha} < y_1 w_1 < \frac{\lambda}{4\beta-2\alpha}$	P_{C22}	$P_{C24} + P_{C22}$	not possible
n is odd, $\frac{\lambda}{4\beta+2\alpha} < y_1 w_1 < min(\frac{\lambda}{1.5\beta+\alpha}, \frac{\lambda}{2\alpha})$	P_{C22}	$P_{C24} + P_{C22}$	$P_{C24} + P_{C25}$
n is odd, $y_1 w_1 < \frac{\lambda}{4\beta+2\alpha}$	P_{C23}	$P_{C24} + P_{C23}$	$P_{C24} + P_{C26}$
n is even, $y_1 w_1 < min(\frac{\lambda}{1.5\beta+\alpha}, \frac{\lambda}{2\alpha})$	P_{C11}	P_{C12}	$P_{C12} + P_{C13}$

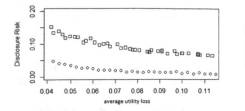
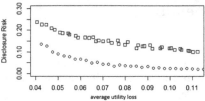

(a) Utility-Disclosure plots for Simulation 1 (b) Utility-Disclosure plots for Simulation 1
with $\alpha = 0.11$ with $\alpha = 0.18$

Fig. 1. Utility-disclosure plots for Simulation 1 with different α values. The box-plot represents results generated by the Thompson et al. algorithm and the dotted plot represents results generated by the new algorithm.

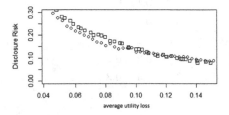
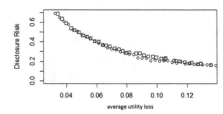

(a) Utility-Disclosure plots for Simulation 2 (b) Utility-Disclosure plots for Simulation 2
with $\alpha = 0.11$ with $\alpha = 0.18$

Fig. 2. Utility-disclosure plots for Simulation 2 with different α values.

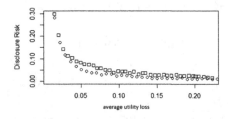
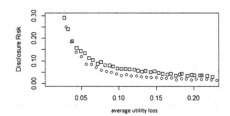

(a) Utility-Disclosure plots for Simulation 3 (b) Utility-Disclosure plots for Simulation 3
with $\alpha = 0.11$ with $\alpha = 0.18$

Fig. 3. Utility-disclosure plots for Simulation 3 with different α values.

References

Blakemore, M.: The potential and perils of remote access. In: Doyle, P., Lane, J., Zayatz, L., Theeuwes, J. (eds.) Confidentiality, Disclosure, and Data Access: Theory and Practical Applications for Statistical Agencies, pp. 315–340. Elsevier Science B. V., Amsterdam (2001)

Chipperfield, J.O., O'Keefe, C.M.: Disclosure-protected inference using generalised linear models. Int. Stat. Rev. **82**(3), 371–391 (2014). doi:10.1111/insr.12054

Chipperfield, J., Newman, J., Thompson, G., Ma, Y., Lin, Y.-X.: Prospects for Protecting Aggregate Business Microdata via a Remote Server (2016). (working paper)

Decays, D., Nanopoulos, P.: Panels of enterprises and confidentiality: the small aggregates method. In: Proceedings of 92 Symposium on Design and Analysis of Longitudinal Surveys, pp. 195–204. Statistics, Canada (1993)

Domingo-Ferrer, J., Torra, V.: Disclosure protection methods and information loss for microdata. In: Doyle, P., Lane, J.I., Theeuwes, J.J.M., Zayatz, L. (eds.) Confidentiality, Disclosure and Data Access: Theory and Practical Applications for Statistical Agencies, pp. 91–110. North-Holland, Amsterdam (2001)

Dwork, C., McSherry, F., Nissim, K., Smith, A.: Calibrating noise to sensitivity in private data analysis. In: Halevi, S., Rabin, T. (eds.) TCC 2006. LNCS, vol. 3876, pp. 265–284. Springer, Heidelberg (2006)

Hawala, S., Zayatz, L., Rowland, S.: American FactFinder: disclosure limitation for the advanced query system. J. Official Stat. 20(1), 115–124 (2004)

Kim, J.J., Winkler, W.E.: Masking microdata files, American statistical association. In: Proceedings of the Section on Survey Research Methods, pp. 114–119 (1995)

Klein, M., Mathew, T., Sinha, B.: Noise multiplication for statistical disclosure control of extreme values in log-normal regression samples. J. Priv. Confidentiality 6, 77–125 (2014)

Lin, Y.X., Wise, P.: Estimation of regression paremeters from noise multiplied data. J. Priv. Confidentiality 4, 61–94 (2012)

Lucero, J., Singh, L., Zayatz, L.: Recent Work on the Microdata Analysis System at the Census Bureau. Research Report Series (Statistics #2009-09) (2009)

Moore, R.: Controlled Data Swapping Techniques for Masking Public Use Microdata Sets. U.S. Bureau of the Census, Washington, DC (1996). http://www.census.gov/srd/papers.pdf.rr96-4.pdf

Nissim, K., Raskhodnikova, S., Smith, A.: Smooth sensitivity and sampling in private data analysis, In: Johnson, D.S., Feige, U. (eds.) 39th ACM Symposium on Theory of Computing-STOC 2007, pp. 75–84. ACM (2007)

O'Keefe, C.M., Chipperfield, J.O.: A summary of attack methods and confidentiality protection measures for fully automated remote analysis systems. Int. Stat. Rev. 81(3), 1–30 (2013). doi:10.1111/insr.12021

Reiter, J.: New approaches to data dissemination: a glimpse into the future (?). Chance 17, 12–16 (2004)

Rubin, D.B.: Discussion: statistical disclosure limitation. J. Official Stat. 9, 461–468 (1993)

Salazar-González, J.-J.: A unified mathematical programming framework for different statistical disclosure limitation methods. Oper. Res. 53(5), 819–829 (2005)

Sarathy, R., Muralidhar, K.: Evaluating laplace noise addition to satisfy differential privacy for numeric data. Trans. Data Priv. 4, 1–17 (2011)

Soria-Comas, J., Domingo-Ferrer, J.: Optimal data-independent noise for differential privacy. Inf. Sci. 250, 200–214 (2013)

Thompson, G., Broadfoot, S., Elazar, D.: Methodology for the automatic confdentialisation of statistical outputs from remote servers at the australian bureau of statistics. In: UNECE Work Session on Statistical Data Confidentiality, Ottawa, pp. 28–30, October 2013

Yancey, W.E., Winkler, W.E., Creecy, R.H.: Disclosure risk assessment in perturbative microdata protection. In: Domingo-Ferrer, J. (ed.) Inference Control in Statistical Databases. LNCS, vol. 2316, pp. 135–152. Springer, Heidelberg (2002)

Disclosure Risk Assessment

Rank-Based Record Linkage
for Re-Identification Risk Assessment

Krishnamurty Muralidhar[1]([✉]) and Josep Domingo-Ferrer[2]

[1] Department of Marketing and Supply Chain Management,
Price College of Business, University of Oklahoma, Norman, OK, USA
krishm@ou.edu
[2] UNESCO Chair in Data Privacy, Department of Computer Engineering
and Mathematics, Universitat Rovira i Virgili, Tarragona, Catalonia, Spain
josep.domingo@urv.cat

Abstract. There is a tradition of data administrators using record linkage to assess the re-identification risk before releasing anonymized microdata sets. In this paper we describe a record linkage procedure based on ranks, and we compare the performance of this rank-based record linkage against the more usual distance-based record linkage to re-identify records masked using several different masking methods. We try to elicit the reasons why RBRL performs better than DBRL for certain methods and worse than DBRL for other methods.

Keywords: Data masking · Rank-based record linkage · Re-identification risk

1 Introduction

In statistical disclosure limitation, record linkage procedures are used to assess the effectiveness of the masking mechanism in preventing re-identification by the adversary [15], that is, preventing the adversary from linking a particular masked record in the anonymized microdata set with the corresponding original record in the original microdata set. Such an assessment makes sense because an adversary may have access to the values of some of the original attributes together *with identification information* from an auxiliary source; the adversary could then attempt to link this identified original record to the masked records using the original attributes whose values he knows (these attributes are called quasi-identifiers). If successful, the adversary would have linked a masked record as belonging to an identified subject: in that case, the adversary could link the confidential attributes in the masked record (e.g. health condition, salary, religion, etc.) to the subject's identity. This is clearly a privacy violation whose risk needs to be evaluated.

Record linkage procedures are often implemented assuming knowledge of the entire original and masked data sets. The procedure then attempts to link records from the masked data set to the original data set. The level of information available to the adversary is one common disagreement when performing record linkage. Data administrators often argue that, to realistically assess disclosure risk, record linkage should assume a level of knowledge by the adversary that is plausible in practice. In addition, when a linked pair is identified, the adversary would be unable to confirm whether the

© Springer International Publishing Switzerland 2016
J. Domingo-Ferrer and M. Pejić-Bach (Eds.): PSD 2016, LNCS 9867, pp. 225–236, 2016.
DOI: 10.1007/978-3-319-45381-1_17

link is correct/incorrect. For these reasons, data administrators dismiss record linkage results as being worst-case and overly pessimistic. While there is considerable merit in these arguments, we view record linkage as being useful for two reasons. First, an assessment of the worst-case disclosure scenario is, in our opinion, important information for the data administrator. Second, even if record linkage is not used for assessing the risk of disclosure, it serves an important purpose by providing the administrator with a means to compare the disclosure protection offered by different masking mechanisms. Hence, for the purposes of this study, we will stick to the tradition and perform record linkage assuming that the entire original and masked data are available.

Our objective in this paper is to evaluate the performance of an alternative approach for record linkage, namely, perform record linkage based on the ranks of the original and masked data. Admittedly, we need to assume that the values of attributes can be ranked, but such an assumption is less problematic than it appears: for numerical or categorical ordinal attributes, ranking is straightforward; for categorical nominal attributes, semantic distance metrics are available that can be used to rank them (for instance, the marginality distance [3, 14]). While rank based record linkage (RBRL) has been used in the literature, there are no comprehensive evaluations of its performance. This is surprising considering that there are rank-based procedures (such as rank swapping [9, 12]) that are frequently used to mask numerical data. When assessing identity disclosure, the commonly used record linkage procedure is distance-based record linkage (DBRL). Our goal is to investigate the application of RBRL for re-identification of masked data and compare its performance to DBRL.

The rest of this paper is structured as follows. Section 2 gives some background on record linkage, including probabilistic and distance-based record linkage. Section 3 describes rank-based record linkage. Section 4 presents empirical results of apply RBRL to re-identify records masked using several different masking methods. Conclusions and future research directions are summarized in Sect. 5.

2 Background on Record Linkage

Record linkage procedures are used to measure the ability of an adversary to link the original and the masked records. The most common methods of record linkage are probabilistic record linkage and distance based record linkage [15].

Distance-based record linkage (DBRL) methods link a masked record with the original record at the shortest distance, for some previously chosen distance. The main difficulty with such methods is to establish appropriate distances for the attributes under consideration: for numerical attributes, the Euclidean distance seems a natural option, but for non-numerical attributes deciding on a distance is less obvious (although, as mentioned above, distances for categorical values are available). An advantage of DBRL is that it allows including subjective information in the re-identification process (e.g. the attributes can be given different weight when computing distances, that is, those attributes that are considered more relevant for re-identification can be given more weight).

Probabilistic record linkage [7, 10, 16], unlike DBRL, does not require standardizing all attributes to a common range or specifying weights for each attribute or choosing a distance. The user (the data administrator in our case) only needs to provide two probabilities as input: the maximum acceptable probability that a pair of records that is not a real match is classified as a linked pair (probability of false positive) and the maximum acceptable probability that a pair of records that is a real match is classified as a non-linked pair (probability of false negative). Probabilistic record linkage works for any type of attribute (numerical, categorical, etc.) without any special adaptation. However, the linkage algorithm itself is less simple and less intuitive than DBRL, which remains the most usual choice if all attributes are numerical.

As mentioned above, for numerical data the most common choice is the simplest one: DBRL with the Euclidean distance. This is what we will take as a benchmark for comparison with our rank-based record linkage. In practice, DBRL is implemented as follows. Let $X = (x_{ij}; i = 1, 2, \ldots, n; j = 1, 2, \ldots, m)$ represent the original data set and let $Y = (y_{lj}; l = 1, 2, \ldots, n; j = 1, 2, \ldots, m)$ represent the masked data set, where n is the number of records and m is the number of attributes. In DBRL, for a given original record, we compute the Euclidean distance to this record from every masked record. However, since the variance of the individual attributes can be different, the distance is measured using the standardized (to mean 0 and unit variance) attributes, rather than the original ones. Let $x_{i*} = (x_{i1}, x_{i2}, \ldots, x_{im})$ represent the target original record. Then for each masked record y_{l*} we compute the distance measure as

$$d_{il}^2 = \sum_{j=1}^{m} \left\{ \left(\frac{x_{ij} - \mu_{X_j}}{\sigma_{X_j}} \right) - \left(\frac{y_{lj} - \mu_{Y_j}}{\sigma_{Y_j}} \right) \right\}^2.$$

Then x_{i*} is linked to the masked record l with $Min(d_{il}^2)$. The process is then repeated for every record i in the original data. Many variations of the basic DBRL have been proposed. A comprehensive discussion of these variations is beyond the scope of this paper. We refer the interested reader to Winkler [16].

3 Rank-Based Record Linkage

Many masking techniques, such as rank swapping and others, are based on ranks. For these techniques, it makes more sense to consider a rank-based record linkage procedure rather than traditional (distance-based) record linkage procedures based on magnitude. Specialized procedures based on ranks have been developed to assess the disclosure risk characteristics of rank swapping [13] and micro-aggregation [13, 17]. The interval-based procedure described in [11] can also be considered as rank based as well: given the value of a masked attribute, the procedure checks whether the corresponding original value falls within an interval centered on the masked value, where the interval width can be in terms of rank (a rank difference).

The above rank-based procedures above have been designed for application to specific procedures and/or specific data sets. In this study, we offer a new general rank based record linkage procedure that can be applied to data masked using any mechanism. We also provide a general comparisons of the performance of RBRL and DBRL across different (including non-rank based) masking mechanisms.

3.1   Rank Based Record Linkage Description

Let $R = (r_{ij})$ and $S = (s_{lj})$ represent the rank matrices corresponding to X and Y, respectively; that is, r_{ij} represents the rank of record i in X with respect to attribute j; similarly, s_{lj} represents the rank of record l in Y with respect to attribute j. The procedure is implemented as follows:

> *For $i = 1$ to n*
> > *For $k = 1$ to n*
> > > *Compute $d_{ik} = Criterion(Abs(r_{i1} - s_{l1}), \ldots, Abs(r_{im} - s_{lm}))$*
> > > *Next k*
> > > *Linked index of $i = \arg\min_l (d_{il})$*
> > *Next i*

The procedure described above can be considered as the *standardized L_1* metric version of DBRL performed on the rank matrices R and S rather than X and Y. The use of ranks has several advantages. First, no further scaling or standardization is necessary to account for the difference in the magnitudes. Second, while DBRL is almost always implemented using the minimum distance criterion, rank-based procedures lend themselves to different criteria for selecting the match based on the characteristics of the masking procedure.

In this study, we consider two criteria, namely *Sum* and *Maximum*. The rationale for the *Sum* criterion is evident – it is simply the application of the "minimum sum of squared distance" criterion of DBRL to the RBRL approach. The *Maximum* criterion is intended for procedures such as rank swapping where the records are swapped but are subject to an upper bound in the swap. In these cases, using the *Sum* criterion may be misleading. Consider rank swapping with an upper bound of p (that is, only records within a rank difference of p are eligible to be swapped). Even if p is not released to the public, if the adversary knows that rank swapping has been performed, then it is clear that the absolute difference in rank between the target record and a masked record has an upper bound (albeit unknown). Hence, the adversary would use the maximum (across all attributes) as the measure of match and then choose the masked record that has the minimum (across all masked records) of the maximum (across all attributes) as the appropriate match.

In the following section, we describe an experimental investigation to compare the performance of DBRL and RBRL across several masking methods.

4   Experimental Investigation

The objective of this experimental investigation is to evaluate the effectiveness of RBRL and DBRL at linking masked records to original records for different masking mechanisms. The masking mechanisms that we choose are among the commonly used ones, namely: independent additive noise, correlated additive noise, multiplicative noise, rank swapping, and microaggregation. This choice of mechanisms represents well some of the diversity of principles used in microdata masking: noise-based, rank-based and cluster-based. We have used as experimental data sets [1]: (1) the Census data set, consisting of 1080 records and 13 numerical attributes, and (2) the EIA data set consisting of 4092 records and 10 numerical attributes. These data sets have been used many times in the literature to evaluate masking mechanisms in terms of utility preservation and disclosure risk protection.

In each of our experiments, we first generated the masked version of the data using a specific masking mechanism; then DBRL and RBRL (both versions) were used to perform record linkage; finally, the number of records correctly linked by each of the approaches was recorded. For masking methods involving randomness (all except microaggregation), the experiment was replicated by generating 100 masked versions of the original data; the results report the percentage of correctly linked records on average over the 100 replications. Since microaggregation involves no randomness, the corresponding experiment was performed only once, and the result is the percentage of correctly linked records in that single run.

4.1 Additive Noise

First we consider independent additive noise which is implemented as follows. The original observation is modified by adding noise independently for each attribute to result in the masked observation:

$$y_{ij} = x_{ij} + e_{ij}; e_{ij} \sim N(0, a\sigma_j),$$

where a represents the perturbation level and $N(0, a\sigma_j)$ represents a normal distribution with mean 0 and the specified standard deviation, with σ_j being the standard deviation of X_j. Since the noise terms are generated independently for each attribute, the correlation between the noise terms is zero. This can also be viewed as univariate noise addition where the noise added is based only on the standard deviation σ_j of the given attribute (and independent of the correlation between the attributes).

Given that DBRL is almost exactly the reversal of the noise addition procedure, we expected DBRL to perform better than RBRL for masking using independent noise. The results in Table 1 confirm this. For every perturbation level, DBRL outperforms both RBRL-Sum and RBRL-Max (with the *Sum* and *Maximum* criteria, respectively). In some cases (Census data, $a = 0.1$), the difference is large with DBRL correctly identifying 98.3 % versus RBRL-Sum correctly identifying 90.0 % of the records. For higher levels of perturbation, the difference is smaller, but consistent.

The re-identification by all methods is much lower for the EIA data. However, in terms of relative performance, the results for the EIA data are very similar to the ones for the Census data: DBRL consistently outperforms both RBRL methods. Thus, these results show that for independent noise addition, DBRL is the best record linkage approach.

Between the two RBRL approaches, Table 1 shows that RBRL-Sum consistently performs better than RBRL-Max: there is a significant drop in the percentage of correctly linked records when using RBRL-Max compared to RBRL-Sum. This again is

Table 1. Percentage of correctly linked records with distance-based and rank-based record linkage for independent additive noise and several values of the perturbation level a

a	Census data			EIA data		
	DBRL	RBRL-Sum	RBRL-Max	DBRL	RBRL-Sum	RBRL-Max
0.10	98.4 %	90.0 %	57.4 %	20.2 %	12.1 %	7.9 %
0.25	69.4 %	53.7 %	33.7 %	7.6 %	3.7 %	2.3 %
0.50	26.4 %	18.9 %	12.6 %	2.5 %	1.2 %	0.7 %
0.75	11.6 %	7.8 %	5.4 %	1.1 %	0.6 %	0.4 %
1.00	6.2 %	4.2 %	2.9 %	0.6 %	0.3 %	0.2 %

not surprising since the *Maximum* criterion was intended to perform well only in specific scenarios (such as data swapping).

An alternative to independent noise addition is correlated noise, where the noise added has the same correlation structure as the original attributes as follows:

$$y_i = x_i + e_i; e_i \sim MVN(0, a\Sigma_{XX})$$

where x_i is the i-th original record, y_i is the corresponding masked record, e_i is a noise vector and $MVN(0, a\Sigma_{XX})$ represents a multivariate normal distribution with mean vector equal to 0 for all attributes and covariance matrix a times Σ_{XX}, with the latter being the covariance matrix of the original data set. With correlated noise, we observe a smaller percentage of records correctly linked at all perturbation levels. Comparing the different methods of linkage, we find that the results for correlated noise are very similar to those observed for independent noise. Due to space limitations, we have not included these results.

4.2 Rank Swapping

For numerical data, data swapping is usually implemented by swapping values based on the ranks [12]. The swapping parameter p (the maximum allowable distance between the ranks of the swapped values expressed as a percentage of n) is specified. For each attribute j, the rank of a given record is computed and swapped with another record within rank distance $\pm np$ (with respect to attribute j). The process is repeated for all records and (independently) for all attributes. As we mentioned earlier, the primary reason to develop the RBRL approach is to investigate its effectiveness in the context of rank-based masking procedures. In addition, the *Maximum* criterion was developed specifically for rank swapping. Hence, we expect the performance of RBRL-Max to be the best. The results of record linkage for rank swapping are provided in Table 2.

Table 2 indicates that, as expected, RBRL-Max outperforms the other two methods of record linkage. The interesting aspect is that the percentage of records re-identified with RBRL-Max is higher than with the other two methods by a considerable margin. For the Census data, even when the swapping distance is 25 % (records with rank difference as high as 270 are swapped), RBRL-Max correctly links over 37.5 % of the records. This is much higher than the results observed for rank swapping by [4] using

Table 2. Percentage of correctly linked records with distance-based and rank-based record linkage for rank swapping and several values of the swapping parameter p

p	Census data			EIA data		
	DBRL	RBRL-Sum	RBRL-Max	DBRL	RBRL-Sum	RBRL-Max
1 %	98.8 %	100.0 %	100.0 %	75.6 %	85.8 %	93.6 %
5 %	88.8 %	99.4 %	100.0 %	13.8 %	18.2 %	45.1 %
10 %	60.1 %	84.0 %	98.5 %	2.5 %	4.0 %	10.0 %
25 %	7.2 %	10.5 %	37.5 %	0.2 %	0.3 %	0.5 %
50 %	0.6 %	0.7 %	1.0 %	0.1 %	0.1 %	0.1 %
100 %	0.1 %	0.1 %	0.1 %	0.0 %	0.0 %	0.0 %

DBRL. Our results are consistent with those observed by [13]. However, the *ad hoc* re-identification procedure in [13] relied on knowledge of the swapping parameter p. In RBRL, we do not assume that this information is available to the adversary. The performance of RBRL-Max is superior across the board. For the Census data, *rank swapping provides little protection against rank-based record linkage* (re-identification) unless the swapping parameter is $(n/2)$ or higher. But with this high level of swapping, data utility is likely to be very poor.

For the EIA data, fewer records are correctly linked across all methods. Unlike the Census data where a large number of records were identified, with the EIA data, rank swapping performs well for $p > 10$ %. This is consistent with the results in [13]. In terms of the performance of the record linkage procedures however, the results for the EIA data are consistent with that of the Census data. We again find that RBRL-Max performs significantly better than the other two procedures across all levels, followed by RBRL-Sum, followed by DBRL.

Both RBRL procedures outperform DBRL for both data sets across all swapping levels. This indicates that for rank swapping, a rank-based record linkage approach is superior to a distance-based one. While this may not seem a surprising result, what is surprising is that we do not see rank-based record linkage normally used in the literature to evaluate rank-based masking.

4.3 p-Distribution Rank Swapping

One of the key issues with rank swapping is that p imposes a strict upper bound on the swapping distance. When p is known, Nin et al. [13] showed that this could allow the adversary to link swapped records to the original ones. Even when p is unknown, the results above indicate that this is a serious threat. To overcome this problem, [13] proposed a modified rank swapping mechanism, which they call p-Distribution Rank Swapping. In this approach, rather than using p as a constant maximum swapping distance for all records, they use it as a parameter for a normal distribution with $\mu = \sigma = p/2$, from which, for each record, they sample a maximum swapping distance for that record, with the usual restriction that one cannot swap the current record to an index below 0 or greater than n; then the actual swapping distance for the record is randomly chosen between 0 and the record's maximum swapping distance. The advantage of this approach is that there is

Table 3. Percentage of correctly linked records with distance-based and rank-based record linkage for p-distribution rank swapping and several values of parameter p

p	Census data			EIA data		
	DBRL	RBRL-Sum	RBRL-Max	DBRL	RBRL-Sum	RBRL-Max
1 %	98.4 %	100.0 %	100.0 %	66.8 %	77.9 %	85.3 %
5 %	81.3 %	97.4 %	99.1 %	8.5 %	11.8 %	21.8 %
10 %	40.8 %	63.5 %	81.9 %	1.3 %	2.1 %	4.9 %
25 %	2.9 %	3.7 %	10.1 %	0.1 %	0.2 %	0.3 %
50 %	0.2 %	0.2 %	0.5 %	0.0 %	0.0 %	0.0 %
100 %	0.0 %	0.0 %	0.0 %	0.0 %	0.0 %	0.0 %

no strict upper bound on the swapping distance across the records. Theoretically, it is possible for the swap value for a given record to be as high as n. When this modified rank swapping mechanism is implemented, the procedure that [13] used for re-identification performs poorly. In this section, we investigate the performance of the three record linkage procedures for p-distribution rank swapping.

Table 3 provides the percentage of correct linkage resulting from the application of the three linkage procedures for selected parameter values of p-distribution rank swapping for the Census and EIA data. The results indicate that, for both data sets, RBRL-Max performs better than the other two approaches. Not surprisingly, the percentages of correct linkages are somewhat lower than those obtained for simple rank swapping (Table 2), since some records could possibly have been swapped by a distance potentially as high as n. While the average swapping distance for a given value of p is likely to be the same as before, the variability introduced by the use of a random variable does lower the record linkage success. And as observed for the other masking methods, a higher percentage of records are identified in the Census data compared to EIA data. In terms of relative performance among the record linkage procedures, the results are similar to those of simple rank swapping: RBRL-Max correctly links the most records, followed by RBRL-Sum, and DBRL correctly links the fewest records.

4.4 Multiplicative Noise

One of the features of noise addition is that the magnitude of the noise is independent of the magnitude of the original value. When the original values are from a skewed distribution, this leads to a situation where too much noise is added to small values and too little noise is added to large values. To overcome this problem, multiplicative noise has been used to mask data as follows:

$$y_{ij} = x_{ij} \times e_{ij}$$

The selection of the distribution of e_{ij} must be done carefully so as to ensure that the resulting masked values are in the appropriate range. In this study, we use a simple distribution for $e_{ij} \sim Uniform(1 - b, 1 + b)$ where b represents the perturbation level.

Table 4. Percentage of correctly linked records with distance-based and rank-based record linkage for multiplicative noise and several values of the perturbation level b.

b	Census data			EIA data		
	DBRL	RBRL-Sum	RBRL-Max	DBRL	RBRL-Sum	RBRL-Max
0.10	99.0 %	99.7 %	98.9 %	64.2 %	76.5 %	79.4 %
0.25	64.7 %	81.5 %	73.5 %	19.2 %	32.6 %	38.5 %
0.50	18.2 %	31.5 %	22.5 %	4.1 %	11.4 %	13.7 %
0.75	6.1 %	10.7 %	7.9 %	1.5 %	4.6 %	5.6 %
1.00	3.0 %	4.1 %	3.5 %	0.8 %	1.7 %	2.0 %

Multiplicative perturbation is performed independently for each attribute. The results of applying the three record linkage procedures to multiplicative perturbation are provided in Table 4.

We had hypothesized that DBRL would perform better for magnitude-based masking mechanisms and RBRL (Sum or Max) would perform better for rank-based masking mechanisms. Hence, for multiplicative noise (which is magnitude based), we expected DBRL to perform better than both RBRL procedures. However, Table 4 indicates that RBRL performs better than DBRL for multiplicative perturbation. For the Census data, across all perturbation levels, RBRL-Sum results in a higher percentage of correctly linked records, in some cases, by a large margin. For instance, for the Census data, when $b = 0.25$, RBRL-Sum correctly links approximately 81.5 % of the records, while DBRL correctly links only 64.7 % of the records. The performance of RBRL-Max is in the middle, correctly linking 73.5 % of the records. This is observed consistently across all perturbation levels for the Census data (with the exception of the perturbation level of 0.05 where RBRL-Max performs slightly worse than DBRL).

As before, for the EIA data set, fewer records are correctly linked for every perturbation level. RBRL performs better than DBRL for this data set as well. Between the two RBRL methods, RBRL-Max performs consistently better than RBRL-Sum. This is different from the results observed for the Census data and somewhat surprising. But perhaps more surprising is the fact that RBRL methods outperform DBRL in all cases across both data sets. This result was not expected and deserves further investigation.

4.5   Microaggregation

The last masking mechanism that we considered is individual-ranking micro-aggregation. In this mechanism, an aggregation parameter k is specified. For each attribute, the individual original values are replaced by the average of the k closest neighboring values. The procedure is then repeated for each attribute. Table 5 shows the record linkage results using the three approaches for different values of k. Unlike the previous methods where a common masking parameter was used across both data sets, for microaggregation we used different values of k for the two data sets (larger values for the larger data set). As indicated earlier, there is no random component in the implementation of microaggregation. Hence, the results presented correspond to a single run of the mechanism on the Census data.

For the Census data, Table 5 indicates that both RBRL methods outperform DBRL by a significant margin (except when k is small, in which case all methods are equally successful). For instance, with $k = 270$, DBRL correctly links 38.8 % of the records while both RBRL methods correctly link approximately 87.9 % of the records. Even with what would be considered an extremely high value of $k(= 360)$, the two RBRL approaches correctly link 65 % of the records in the Census data. This again confirms the conclusions of [2, 4] that individual-ranking microaggregation is highly susceptible to re-identification (in contrast, multivariate microaggregation satisfies the k-anonymity privacy model provided that all quasi-identifier attributes are microaggregated together, and hence can offer probability of re-identification $1/k$ [6]).

The difference between RBRL and DBRL is even more dramatic for the EIA data. With $k = 341$, DBRL correctly links only a small percentage (6.4 %) of the records, while both RBRL methods correctly link almost ten times as many records (61.4 %). Across all values of k, we note that both RBRL methods provide significantly higher linkage percentages compared to DBRL methods.

Table 5. Percentage of correctly linked records with distance-based and rank-based record linkage for microaggregation and several values of the aggregation parameter k.

Census data				EIA data			
k	DBRL	RBRL-Sum	RBRL-Max	k	DBRL	RBRL-Sum	RBRL-Max
10	99.8 %	100.0 %	100.0 %	341	6.4 %	61.4 %	61.4 %
54	96.5 %	100.0 %	100.0 %	372	5.4 %	56.3 %	56.3 %
108	89.7 %	100.0 %	100.0 %	682	0.9 %	30.7 %	30.7 %
270	38.8 %	87.9 %	87.9 %	1023	0.1 %	15.4 %	15.4 %
360	21.3 %	65.0 %	65.0 %	1364	0.0 %	8.7 %	8.7 %
540	9.6 %	24.5 %	24.5 %	2046	0.0 %	3.3 %	3.3 %

Another result is of particular interest, is that the performance of both RBRL methods is exactly the same; *every correctly linked observation using RBRL-Sum is also correctly linked by RBRL-Max*. The possible explanation for this result is the following. With rank swapping, the masking is performed using only the ranked values. With multiplicative noise, the masking is performed by relative magnitude. Micro-aggregation is a combination of the two – the groups are identified by ranks and then aggregated by magnitude. This combination yields the interesting result that both RBRL methods perform exactly the same.

5   Conclusions and Future Research

We have presented a general rank-based record linkage procedure, as an alternative to distance-based record linkage and probabilistic record linkage. We then have given an empirical performance comparison with distance-based record linkage at re-identifying records masked with several different masking methods. Our results indicate that, whenever a masking method explicitly or implicitly ranks the original data, RBRL outperforms distance-based record linkage. Furthermore, and rather unexpectedly, RBRL also outperforms distance-based record linkage for masking by multiplicative noise. We have considered rank-based record linkage with two different criteria (the sum of rank differences and the maximum rank difference). None of the two criteria outperforms the other for all the methods tried: which one is best depends on the masking method being attacked. Also interesting is the fact that for microaggregation, both RBRL procedures (Sum and Maximum) perform in an identical manner, correctly linking exactly the same records.

Originally, we had hypothesized that distance-based record linkage would perform better for masking procedures based on magnitude (additive and multiplicative noise) and rank based record linkage for masking procedures based on ranks (data swapping). Microaggregation was a special case where the identification of the records to be aggregated is based on ranks but the actual aggregation is performed on the values. Here we expected both record linkage methods to be equally effective. The results however indicate that rank-based record linkage outperforms distance-based record linkage for both multiplicative noise and microaggregation. This suggests an alternative explanation, namely that *rank-based record linkage performs better for masking methods where the level of masking is a function of the magnitude or rank of the actual original value*. The only masking methods in our experiment where the masking is independent of the actual value are independent and correlated noise procedures. Distance-based record linkage performs better only in this case. In all other methods, the masking is a function of either the magnitude (multiplicative noise), or the rank (rank swapping, p-distribution swapping), or both (microaggregation), of the actual original value. Rank-based record linkage performs better in all these cases. We believe that this is an interesting phenomenon that deserves further investigation.

As additional future work, we plan to extend our experimental work by comparing RBRL also against probabilistic record linkage, and increasing the number of masking methods tested. Particularly intriguing is the re-identification effectiveness of RBRL for other masking methods implicitly based on ranks, such as PRAM [8].

Acknowledgments and Disclaimer. The second author is partly supported by the European Commission (projects H2020-644024 "CLARUS" and H2020-700540 "CANVAS"), by the Government of Catalonia (ICREA-Acadèmia prize and grant 2014 SGR 537) and by the Spanish Government (projects TIN2014-57364-C2-1-R "SmartGlacis" and TIN2015-70054-REDC). The second author leads the UNESCO Chair in Data Privacy, but the views expressed in this paper are the authors' own and are not necessarily shared by UNESCO.

References

1. Brand, R., Domingo-Ferrer, J., Mateo-Sanz, J.M.: Reference data sets to test and compare SDC methods for the protection of numerical microdata. Deliverable of the EU IST-2000-25069 "CASC" project (2003). http://neon.vb.cbs.nl/casc/
2. Domingo-Ferrer, J., Oganian, A., Torres, A., Mateo-Sanz, J.M.: On the security of micro-aggregation with individual ranking: analytical attacks. Int. J. Uncertainty Fuzziness Knowl. Based Syst. **10**(5), 477–491 (2002)

3. Domingo-Ferrer, J., Sánchez, D., Rufian-Torrell, G.: Anonymization of nominal data using semantic marginality. Inf. Sci. **242**, 35–48 (2013)
4. Domingo-Ferrer, J., Torra, V.: A quantitative comparison of disclosure control methods for microdata. Confidentiality, Disclosure and Data Access, Theory and Practical Applications for Statistical Agencies, pp. 111–134. North-Holland, Amsterdam (2001)
5. Domingo-Ferrer, J., Torra, V.: Disclosure risk assessment in statistical disclosure control of microdata via advanced record linkage. Stat. Comput. **13**(4), 343–354 (2003)
6. Domingo-Ferrer, J., Torra, V.: Ordinal, continuous and heterogeneous k-anonymity through microaggregation. Data Min. Knowl. Discov. **11**(2), 195–212 (2005)
7. Fellegi, I., Sunter, A.B.: A theory for record linkage. J. Am. Stat. Assoc. **64**(328), 1183–1210 (1969)
8. Gouweleeuw, J.M., Kooiman, P., De Wolf, P.-P.: Post randomisation for statistical disclosure control: theory and implementation. J. Official Stat. **14**(4), 463–478 (1998)
9. Hundepool, A., Domingo-Ferrer, J., Franconi, L., Giessing, S., Schulte Nordholt, E., Spicer, K., De Wolf, P.-P.: Statistical Disclosure Control. Wiley, Hoboken (2012)
10. Jaro, M.A.: Advances in record linkage methodology as applied to matching the 1985 census of Tampa, Florida. J. Am. Stat. Assoc. **84**(406), 414–420 (1989)
11. Mateo-Sanz, J.M., Sebé, F., Domingo-Ferrer, J.: Outlier protection in continuous microdata masking. In: Domingo-Ferrer, J., Torra, V. (eds.) PSD 2004. LNCS, vol. 3050, pp. 201–215. Springer, Heidelberg (2004)
12. Moore, R.A.: Controlled Data Swapping for Masking Public Use Microdata Sets. Research report series (RR96/04), Statistical Research Division, US Census Bureau, Washington, DC (1996)
13. Nin, J., Herranz, J., Torra, V.: Rethinking rank swapping to decrease disclosure risk. Data Knowl. Eng. **64**(1), 346–364 (2008)
14. Soria-Comas, J., Domingo-Ferrer, J., Sánchez, D., Martínez, S.: Enhancing data utility in differential privacy via microaggregation-based k-anonymity. VLDB J. **23**(5), 771–794 (2014)
15. Torra, V., Domingo-Ferrer, J.: Record linkage methods for multidatabase data mining. In: Torra, V. (ed.) Information Fusion in Data Mining. Studies in Fuzziness and Soft Computing, vol. 123, pp. 99–130. Springer, Heidelberg (2003)
16. Winkler, W.E.: Matching and record linkage. In: Business Survey Methods, pp. 355–384. Wiley, Hoboken (1995)
17. Winkler, W.E.: Masking and re-identification methods for public-use microdata: overview and research problems. In: Domingo-Ferrer, J., Torra, V. (eds.) PSD 2004. LNCS, vol. 3050, pp. 231–246. Springer, Heidelberg (2004)

Computational Issues in the Design of Transition Probabilities and Disclosure Risk Estimation for Additive Noise

Sarah Giessing[(✉)]

Federal Statistical Office of Germany,
65180 Wiesbaden, Germany
Sarah.Giessing@destatis.de

Abstract. The Australian Bureau of Statistics has developed an additive noise method for automatically and consistently confidentialising tables of counts 'on the fly'. Statistical properties of the perturbation are defined by a matrix of transition probabilities. The present paper looks at mathematical and computational aspects of an approach mentioned in the literature for how to design those probabilities. In the second part, the paper proposes computation of feasibility intervals as a technique to compare the effects of different variants for post-tabular perturbative protection methods on disclosure risk and provides experimental results.

1 Introduction

In the process of modernizing their output production process statistical institutes seek to move away from a traditional way of disseminating a fixed (in case of some statistics: quite large) set of aggregate tables on one hand while afterwards producing numerous additional tables to answer specific user queries, towards flexible, directly user demand driven online tabulation tools. The Australian Bureau of Statistics (ABS) has set up such an online tool allowing users to define and download tables. For this tool the ABS has developed a disclosure control technique on the basis of additive noise that does not require human intervention but can be applied 'on the fly'.

As pointed out in [14] a key feature of the method is the "perturbation look-up table" implementing transition probabilities which ultimately determine the statistical properties of the noise. [14] also outline a mathematical approach for computing transition probabilities satisfying certain basic properties, originally defined in [6], plus some additional constraints.

Especially in the light of growing interest that additive noise techniques may rise among SDC experts[1], this paper examines the proposed approach (Sect. 2) and presents a few test results computed with the NLopt-package for non-linear optimization [11].

In Sect. 3 a certain type of disclosure risk is addressed. The section outlines how to implement a variant of the shuttle algorithm [2] to evaluate this type of risk. Preliminary

---

[1] At its first meeting in April 2016 the European Working Group on Methodology identified "confidentiality on-the-fly" as an item to be given some priority (c.f. [4]).

© Springer International Publishing Switzerland 2016
J. Domingo-Ferrer and M. Pejić-Bach (Eds.): PSD 2016, LNCS 9867, pp. 237–251, 2016.
DOI: 10.1007/978-3-319-45381-1_18

test results on disclosure risk (or risk potential) associated with several variants of deterministic rounding and additive noise are presented in Sect. 4.

2 Basic Properties of Random Noise for Protection of Counts Data

According to [14], the noise distributions used in the ABS Census TableBuilder are designed to ensure that the perturbations take integer values and that the following criteria hold for the perturbations:

1. the mean of the perturbation values is zero;
2. the perturbations have a fixed variance;
3. the perturbations will not produce negative cell values or positive cell values below a specified threshold j_s; and
4. the absolute value of any perturbation is less than a specified integer value D.

This kind of a discrete perturbation process can obviously be modeled as a simple two-state Markov chain where one state refers to the original data and the other to the data after perturbation. In this terminology, state changes are called transition probabilities, given by a (single) transition matrix. The transition matrix determines the (conditional) probability distribution of the perturbed data given the original data, each row of the matrix referring to a particular original count, stating the probability distribution of the perturbed data for all the table cells with this original count.

Restoring to the denotation of [9] P is the $L \times L$ transition matrix[2] of conditional probabilities: p_{ij} = P(perturbed cell value is j| original cell value is i), where p_i refers to the ith row-vector of matrix P. Let v_i the column vector of the noise which is added, if an original value of i is turned into a value of j. I.e. the jth entry of v_i is $(j - i)$, for example $v_i = (-1, 0, 1, 2, 3, \ldots, L-2)'$. The four requirements for the ABS Census Table Builder noise are then expressed by the following equations which must hold for all rows i of the transition matrix

(1) $p_i v_i = 0$
(2) $p_i (v_i)^2 = V$
(3) $p_{ij} = 0$ for $0 < j \le j_s$
(4) $p_{ij} = 0$; if $j < i - D$ or $j > i + D$.

As each p_i should define a probability distribution we need the two additional constraints to hold for all rows i of the transition matrix, e.g.

(5) $p_i u_i = 1$, where u_i is a column of $1's$.
(6) $0 \le p_{ij} \le 1$.

---

[2] As index j may take a value of zero (when a cell value is changed to zero), in the following we start counting matrix and vector indices at 0, enumerating rows and columns of the $L \times L$ matrix by 0, 1, 2, ..., $L-1$. The number of rows and columns L, which we assume w.l.g. to be the same, differs for different set ups.

In order to define noise distributions satisfying the above listed criteria we have to compute, for each row i of the transition matrix a feasible solution p_i^* of the linear constraint set (1–6).

Note, for all rows after row $D+j_s$, condition (3) is always satisfied, when (4) holds. This means (3) can be dropped from the system of constraints for those rows. Because of (4), for rows i from $D + j_s + 1$ onwards, all entries in v_{i+1} multiplied with non-zero entries of p_{i+1} in (1) and (2) are identical to respective entries of v_i. Thus, for those rows, a solution p_i^* for the constraint set (1–4) corresponding to row i can be transformed into a solution for row $(i + 1)$ by simply shifting the set of non-zero entries one place to the right. We therefore assume here that it is enough to compute solutions p_i^* for each row i up to $D + j_s + 1$.

Moreover, for $i = D + j_s + 1$ the elements v_i corresponding to non-zero entries of p_i are the numbers from $-D$ to $+ D$:$(-D, -D+1, \ldots, -1, 0, 1, \ldots, D-1, D)$. Because of the symmetry around zero in this sequence, it is enough to compute only one side of the probability distribution p_i, f.e. $(p_{ij})_{j=i,\ldots,i+D}$ and then let $(p_{ij})_{j=i-1,\ldots,i-D} :=$ $(p_{ij})_{j=i+1,\ldots,i+D}$. With such a symmetric distribution (1) will always hold, and can thus be dropped from the system. In the following we refer to the instance $i = D+j_s+1$ as to the "*symmetric case*".

Now let $\tilde{A}$ the $(3 \times L)$ matrix with the first row given by v_i', the second row by the squared entries of the first row, i.e. $((v_i)^2)'$, and the 3rd row by u_i. From $\tilde{A}$ we derive matrix A by removing all columns corresponding to entries of p_i which have to be zero according to (3) and (4). As explained in [9] finding a feasible solution p_i^* of the linear constraint set (1–5) is equivalent to finding a solution x^* for the linear equations system

(7) $A x - b = 0$, with coefficient matrix A and $b = (O, V, 1)'$.

The elements of a solution vector x^* correspond to those elements of p_i which are not fixed to zero because of (3) and (4).

Note that in the symmetric case, we can simplify (7): Because, as mentioned above, (1) drops out in that case, we drop the first line of (7), and because we only determine p_{ii} and $(p_{ij})_{j=i+1,\ldots,i+D}$ we also drop the first $D + i + 1$ columns of matrix A. Let A^s the simplified coefficient matrix. The simplified system can be stated as

(8) $A^s x - b^s = 0$, where $b^s = (\frac{V}{2}, (\frac{1-p_{ii}}{2}))$, and $A^s := \begin{pmatrix} 1 & 4 & & D^2 \\ & & \cdots & \\ 1 & 1 & & 1 \end{pmatrix}$.

In all relevant cases, A has more columns than rows. So usually, there is no unique solution for (7)[3] or (8). In [9] a simple ad hoc procedure was suggested to compute feasible probability distributions: A sequence of simple LP problems would be solved, fixing at each step of the sequence at least one of variables (i.e. elements of p_i) by deliberation or by setting it to its upper or lower feasibility bound. However, this can

---

[3] See [9] for a simple illustrative instance (i.e. the case: $i = 1, D = 3$), where (7) has a unique solution, if V is at least 2 and infeasible otherwise (when (6) is considered as well).

turn into a tedious exercise, especially when the number of variables is large (as it is in case of large maximum allowed deviation D).

2.1 Maximum Entropy Based Approach

In this section we explore the suggestion of [14] to determine the probability distributions of the noise by maximizing entropy. Assuming as entropy measure the Shannon entropy, in the above terminology this means we maximize, subject to the equality constraints of (7):

(9) $S(p_i) := -\sum_{j \in J_i} p_{ij} log_2(p_{ij})$.

A strategy to solve this (obviously non-linear) optimization problem is the method of Lagrange multipliers. With Lagrange multipliers λ_1, λ_2 and λ_3 and J_i the set of indices j corresponding to those elements p_{ij} of p_i not fixed to zero because of (3) and (4), the Lagrange function of this problem is: $\mathcal{L}\left((p_{ij})_{j \in J_i}, \lambda\right) = \sum_{j \in J_i} p_{ij} log_2 p_{ij} + \lambda_1 g_1(p_i) + \lambda_2 g_2(p_i) + \lambda_3 g_3(p_i)$. The three constraint functions $g_k(p_i)$ correspond to the rows of the linear equation system (7): $g_k(p_i) := a_k (p_{ij})_{j \in J_i} - b_k$, where a_1, a_2, a_3 refer to the three row-vectors of matrix A, and b_k to the three entries of $b = (0, V, 1)'$.

The first step to find a maximum entropy solution is to compute "critical points", i.e. where the gradient of the Lagrangian is zero, i.e. we solve $\Delta_{(p_{ij})_{j \in J_i}, \lambda} \mathcal{L}\left((p_{ij})_{j \in J_i}, \lambda\right) = 0$ which is equivalent to

$$\begin{cases} (10) \ \Delta_{(p_{ij})_{j \in J_i}} \sum_{j \in J_i} p_{ij} log_2 p_{ij} - \lambda_1 g_1(p_i) - \lambda_2 g_2(p_i) - \lambda_3 g_3(p_i) \\ (11) \ g_1(p_i) = g_2(p_i) = g_3(p_i) = 0 \end{cases}$$

The second step is to compute for those critical points the entropy and identify the point of maximum entropy.

Computing the gradient in (10), $\Delta_{(p_{ij})_{j \in J_i}} \mathcal{L}\left((p_{ij})_{j \in J_i}, \lambda\right)$, i.e. computing for each $j \epsilon J_i$ the first order derivative $\frac{\delta}{\delta p_j} \mathcal{L}\left((p_{ij})_{j \in J_i}, \lambda\right)$, leads to a system of $|J_i|$ equations: $-\left(\frac{1}{ln2} + log_2(p_{ij})\right) + \lambda_1 v_{ij} + \lambda_2 v_{ij}^2 + \lambda_3 = 0$. Those can be transformed into

(12) $p_{ij} = 2^{-\frac{1}{ln2} + \lambda_3 + \lambda_1 v_{ij} + \lambda_2 v_{ij}^2}$.

To complete the first step, we now have to find $\lambda = (\lambda_1, \lambda_2, \lambda_3)$ with the property that (11) holds for the p_i resulting from (12). This shall be demonstrated below for the instance of the comparatively simple "symmetric case".

2.2 Solving a Maximum Entropy Problem Analytically - the Symmetric Case as Illustrative Example

As pointed out above in the symmetric case the constraint system (7) can be replaced by (8). As a consequence, the term $\lambda_1 g_1(p_i)$ can be dropped from (10) and $g_1(p_i) = 0$ can be removed from (11). Then (12) turns into

(12') $p_{ij} = 2^{-\frac{1}{\ln 2} + \lambda_3 + \lambda_2 v_{ij}^2}$ for $j = i + 1, \ldots, i + D$.

According to the definition of b^s and A^s, constraint $g_2(p_i) = 0$ is now $\sum_{j=1}^{D} p_{ij} j^2 - \frac{V}{2} = 0$, and constraint $g_3(p_i) = 0$ is $\sum_{j=1}^{D} p_{ij} = \frac{1-p_{ii}}{2}$.

To make $g_2(p_i) = 0$ hold, we now let $p_{ij}(\lambda_2) := 2^{\lambda_2 v_{ij}^2} \frac{V}{2v(\lambda_2)}$, where $v(\lambda_2) := \sum_{j=i+1}^{i+D} 2^{\lambda_2 v_{ij}^2} v_{ij}^2 = \sum_{j=1}^{D} 2^{\lambda_2 j^2} j^2$. Then, to make $g_3(p_i) = 0$ hold as well, we let $p_{ii} := 1 - 2 \sum_{j=1}^{D} p_{ij}$. Note, this is only well defined as long as $\sum_{j=1}^{D} p_{ij} < \frac{1}{2}$ holds for a choice of λ_2.

If we define $\lambda_3 := log_2 \left(\frac{V}{2v(\lambda_2)} \right) 2^{\frac{1}{\ln 2}}$, then also (12') holds.

So far, the candidate probability distribution p_i actually depends on the choice of λ_2, e.g. $p_i = p_i(\lambda_2)$. The final step is finding the maxima of (9), that is maximize the entropy $S(p_i(\lambda_2))$ with respect to λ_2. This is a straightforward – though a little cumbersome - analytical exercise. Saving the effort here, Fig. 1a presents a graphical solution for the case of D = 2. Obviously, the maximum is around $\lambda_2 = 0.9$. Figure 1b shows the resulting maximum entropy probability distribution.

As final observation: With $\alpha := 2^{-\frac{1}{\ln 2} + \lambda_3}$ and $\beta := e^{(\ln 2)\lambda_2}$, (12') can be re-written as: $p_{ij} = \alpha e^{\beta v_{ij}^2}$, reminding of the functional form of a normal distribution with zero expectation.

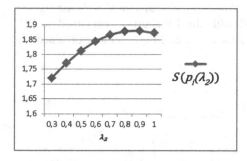
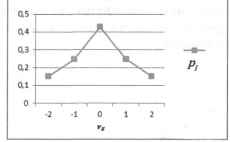

Fig. 1a. Shannon's entropy for the instance $D = 2$; symmetrical case; $0.3 \le \lambda_2 \le 1$

Fig. 1b. Maximum entropy distribution for the instance $D = 2$; symmetrical case

2.3   Software Solution for the Maximum Entropy Problem

While it might be feasible to implement a little procedure to carry out the steps of the approach taken in the illustrative example for the symmetric case, for the general case this would be difficult. It is therefore very advisable to restore to the well-known algorithms for non-linear optimization problems. To obtain the solutions discussed in Sect. 2.4, the Nlopt package [11] has been used via its R-interface routine nloptr [16]. From this package we decided to use a sequential quadratic programming algorithm for nonlinearly constrained gradient-based optimization (supporting both inequality and equality constraints), based on the implementation described in [12] and [13]. In order to execute an instance, seven arguments have to be passed to the interface routine nloptr. See the Appendix for a list and illustration of those arguments.

2.4   Adding Constraints to the Maximum Entropy Based Approach

A simple measure of information loss, easy to communicate to data users is the probability for data to preserve their true value, p_{ii}, or the cumulated probability for the data to change by less than δ (for small δ, like $\delta = 1$): $\sum_{j=-\delta}^{\delta} p_{ij}$. Disregarding the fact that this will reduce the entropy (when the other important parameter, e.g. the variance is fixed), one might be tempted to prefer transition probability distributions with high probabilities in the center (even though this will be at the cost of increasing the tails of the distribution, i.e. the probabilities for larger changes). This can be achieved with the software solution outlined in Sect. 2.3 and in the Appendix by adding constraints to impose lower bounds on the respective probabilities. Especially in combination with such additional constraints, it sometimes happens that in the solution obtained, the left hand side of the probability distribution does not increase monotonously: we get $p_{i(i-k)} < p_{i(i-j)}$ for some $j > k$, which may not be appreciated. To avoid this, more constraints can be added to enforce monotony.

Figures 2a and b compare three transition probability distributions for maximum perturbation $D = 5$ and condition (3) requiring that no cell values 1 and 2 appear in the perturbed data, e.g. $j_s := 2$. Two of the distributions were obtained as maximum entropy solutions for parameter settings of $V = 2$ *vs.* $V = 3$. For $V = 2$, the figures present also another distribution, computed with the LP-heuristic explained in [9]. Figure 2a presents the distributions obtained for $i = 3$, and Fig. 2b those obtained for the symmetric case.

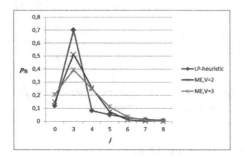
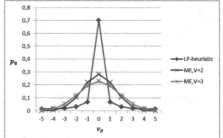

Fig. 2a. Transition probability distributions for $D = 5$; $i = 3$, $j_s = 2$

Fig. 2b. Transition probability distributions for $D = 5$; symmetric case $(i \geq 8)$

Note that a distribution almost identical to the leptokurtic distribution resulting from the *LP*-heuristic of [9] is obtained by the maximum entropy approach, when we add constraints on the elements on the diagonal of the transition matrix, p_{ii}, and on the off-diagonal elements $p_{i(i-1)}$ and $p_{i(i+1)}$, to define the respective probabilities of the 'target' distribution as lower bounds. This is, however, at the "expense" of higher probabilities in the tails of the distribution[4] which are of course rather undesirable.

Another observation from experiments with the approach is that changing the maximum perturbation (to, for example, $D = 4$ or $D = 10$), while keeping the variance $V = 2$ or $V = 3$ constant, reproduces the probability distribution almost exactly, for larger D of course with longer, but very flat tails.

The software solution outlined in Sect. 2.3 facilitates the computation of probability distributions satisfying the basic properties of the random noise outlined in the beginning of Sect. 2, and to experiment with the main parameters (V, D, p_{ii}) which obviously affect the usability of the perturbed data and are easy to interpret. In order to determine these parameters appropriately, however, disclosure risk issues must also be taken into account.

3 Evaluating Disclosure Risk with the Shuttle Algorithm

Disclosure risk measures of the type that measure, for example, the percentage of perturbed data identical with the true data as proposed and discussed in [1, 14] and, lean in some way on a scenario where the users of the data mistake perturbed data for "true", ignoring information on the SDC method offered by the data provider. The present paper assumes a different scenario: we assume a data user (rather "snooper") who is well aware of the perturbation and tries to undo it with certain differencing attacks. On the basis of this scenario the following general requirement for tabular data protection methods makes sense also in the context of a perturbative method *not* preserving additivity, e.g. *not* preserving the "logical" table relations (i.e. interior cells sum up to the respective margins) for the perturbed data.

A method that is in this sense non-additive[5] should ensure that it can be expected that the feasibility intervals that can be computed for protected, sensitive data are not disclosive (c.f. [10], Sect. 4.3.1.) Notably, the idea is to assess this property of a method (*viz.*: choice of parameters) by a rigorous *preparative* study – not in the production phase of table generation and perturbation!

Feasibility intervals are typically computed by means of linear programming (LP) methods (see for example [5]). Solutions can be obtained for each cell of interest by minimizing and maximizing a variable representing its cell value, subject to a set of constraints expressing the "logical" table relations, and some *a priori* upper and lower bounds on unknown cell values. Note, in the setting of a perturbative SDC method, generally all the cells are assumed to be "unknown". Our intruder only assumes to know for each cell a lower and upper bound. This increases of course the size of the problem to be solved and hence computations times, especially when table relations tend to be 'long', with many interior cells contributing to the same marginal.

---

[4] For example, in the symmetric case, with $D = 5$: $p_{i(i-5)} = p_{i(i+5)} \cong 0.0006$ without the additional constraints, $\cong 0.009$ with constraint on the diagonal element, and $\cong 0.01$ with additional constraint on the off-diagonal elements.

[5] Notably, the ABS Census TableBuilder product is additive in this sense, because of a subsequent algorithm implemented in the product that restores additivity to the table.

The objective of the research step envisioned is to compute intervals for all cells of large high-dimensional tables for various settings, regarding for example the assumptions on the *a priori* bounds made by the intruder, different settings for the noise etc. We decided not to rely on a *LP* based tool, but to implement a generalized version of the shuttle algorithm [2] in SAS, which had also certain practical advantages in the processing and management of the instances, even though there might be a theoretical disadvantage: there might be cases, where the LP algorithm produces better solutions. In such cases the shuttle algorithm would underestimate the true risks[6].

The mathematical formulation of the version of the shuttle algorithm we use is briefly presented in the following (for more details and a toy example the reader is referred to [8]):

Let a table be "defined" by a set of cell indices I and a set of table relations J. For cell $c \in I$ let $J_T(c)$ identify the set of table relations in which this cell is the margin, and $J_B(c)$ the set of table relations in which cell c is one of the interior cells. For relation $j \in J$ let $t_j \in I$ denote the index of the margin cell and $B_j \subset I$ the set of indices of the "bottom" cells contributing to this margin. Finally, for $c \in I$ let x_c^U and x_c^L upper and lower bounds for its true cell value which may change in the course of executing the algorithm.

Like the "original" algorithm, the generalized version consists basically of two steps (referred to in the following as L-step and U-step) which are repeated alternatingly. Starting from the original *a priori* bounds (the ones assumed to be known to the intruder) for x_c^U and x_c^L, each step will use the results of the previous step to compute several new candidate bounds for all interior cells and compare those candidate bounds to the "previous" bound: For a given interior cell $b \in \bigcup_{j \in J} B_j$, one candidate bound is computed for every table relation $j \in J_B(b)$ containing this cell.

In the U-step, a candidate bound for cell b is computed by taking the difference between upper bound of the marginal cell $x_{t_j}^U$ and the sum of all lower bounds on the other interior cells in the same table relation $\sum_{i \in B_j, i \neq b} x_i^L$. The upper bound of cell b, x_b^U, will be replaced in the U-step by $x_b^U(j) := x_{t_j}^U - \sum_{i \in B_j, i \neq b} x_i^L$, if for one $j \in J_B(b)$ the candidate bound is tighter, i.e. $x_b^U(j) < x_b^U$.

The L-step works in the same way, but here we take the difference between lower bound of the marginal cell $x_{t_j}^L$ and the sum of all upper bounds x_i^U on the other interior cells in the same table relation. The lower bound of cell b, x_b^L, will be replaced, if for one $j \in J_B(b)$ the candidate bound is tighter, i.e. $x_b^L(j) > x_b^L$. The algorithm stops, when no bound improvement is reached in the previous pair of steps, L or U.

In the original setting of [2], the cell values of the margins are assumed to be known. In our setting they are not. We therefore compute when the algorithm starts and after every U-step (before starting the L-step) for every margin cell $t \in \bigcup_{j \in J}\{t_j\}$ new

---

[6] In the literature [3, 15] there is some discussion as to which extent it could be expected that solutions obtained by the shuttle algorithm coincide with the solutions that would be obtained solving the respective linear programming problems. For some (admittedly: few) test tables we have actually compared the results and found no differences.

candidate upper bounds $x_t^U(j) := \sum_{i \in B_j} x_i^U$, replacing in this additional step x_t^U by $x_t^U(j)$, if for one $j \in J_T(t)$ the candidate bound is tighter, i.e. $x_t^U(j) < x_t^U$. Similarly, at the start of the algorithm and between an L- and the next U-step a lower bound x_t^L will be replaced by $x_t^L(j) := \sum_{i \in B_j} x_i^L$, if $x_t^L(j) > x_t^L$ for one $j \in J_T(t)$.

4 Test Application

Our test data set consists of 31 two-way tables available for about 270 geographical units on different levels of a hierarchy. Taking those relations (like f.e. the district – municipality relations) into account too, the tables are actually 3-dimensional. The 31 two-way tables are given by cross combinations of nine variables. Two with 2 categories, one with 3, two with 4, the others with 7, 9, 19 and 111 categories. The geography relations also tend to be 'long' with often more than 10, or even up to 100 categories. The current implementation does not consider any hierarchical relations: bounds are computed independently for all tables. The test tables are tables of population counts.

The shuttle algorithm is used to compare the effects of different protection methods: we look at a number of variants of deterministic rounding and at variants of additive noise. The rounding methods were included for sake of comparison, in the first place.

For the rounding we assume that data users are aware of the *a priori* bounds which directly result from the rounding rule. Denote R the rounding base and $n_c R$ the rounded count for cell $c \in I$. Then $x_c^L := \max(0; n_c R - R^-)$ and $x_c^U := n_c R + R^+)$ are the known *a priori* bounds. R^+, R^- denote the maximum upper and lower rounding deviation. For example in case of rounding base $R = 3$: $R^+ = R^- = 1$.

For the additive noise we assume two different scenarios: A. The maximum perturbation D is published. B. The maximum perturbation D is not published, but the intruder is able to estimate it at $D \pm 2$. Under scenario $A.x_c^L := \max(0; \tilde{x}_c - D)$ and $x_c^U := \tilde{x}_c + D$ are known *a priori* bounds for a true count x_c given the perturbed count $\tilde{x}_c$. Under scenario B. We assume estimated *a priori* bounds $x_c^L := \max(0; \tilde{x}_c - D - 2)$ and $x_c^U := \tilde{x}_c + D + 2$.

In the test application, we used bases of 3, 5, 7, 9, 10 and 11 for the rounding methods. The additive noise was created with the transition probabilities derived in [9] for maximum perturbations $D = 3, 4$ and 5.

During execution of the shuttle algorithm, we record the cases where one of the bounds $x_b^L, x_b^U, x_t^L, x_t^U$ is replaced by a tighter (candidate) bound, noting in particular the iteration step and the equation j at which a replacement occurs. After the execution the results are investigated. Cases where final upper and lower bounds for a cell coincide with each other, matching the true cell value (i.e. they are *exact* bounds) are categorized according to that original cell value (1; 2; 3–10; >10). Figures 3a, b, 4a, b and c present indicators computed on the basis of the evaluation.

Every data point in those figures refers to one of the 31 structural types of our test tables, resulting from the crossing of pairs of the nine variables mentioned above with the geography dimension. Data points are sorted by table size (#cells = number of table cells, summed across all geographical units) to visualize an obvious general trend for higher values of the indicators in smaller tables.

For the indicators, we count for each cell the number of equations (at most 3, for the 3-dimensional test data) with bound replacements, i.e. where during execution of the algorithm an upper or lower bound was replaced by a tighter one. Results are summed across all cells and across all tables relating to the data point, divided by the respective number of table cells and multiplied by 100. So the theoretical maximum for indicators is 300 %. We refer to the indicator defined this way as "*indicator for disclosure risk potential*".

For a second, more direct indicator, the "*disclosure risk indicator*", we count for the enumerator of the indicator equations with bound replacements only for those cells where the algorithm finally returns exact bounds, i.e. where the algorithm discloses the exact true counts.

Figures 3a and b present results for the *disclosure risk indicator*, grouped by original count size (1; 2; 3–10; >10) of the disclosed cells, for rounding to base 3 (Fig. 3a) and rounding to base 5 (Fig. 3b).

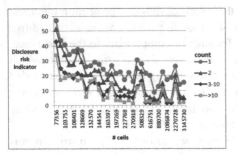

Fig. 3a. Disclosure risk indicator for rounding to base 3, by original count size

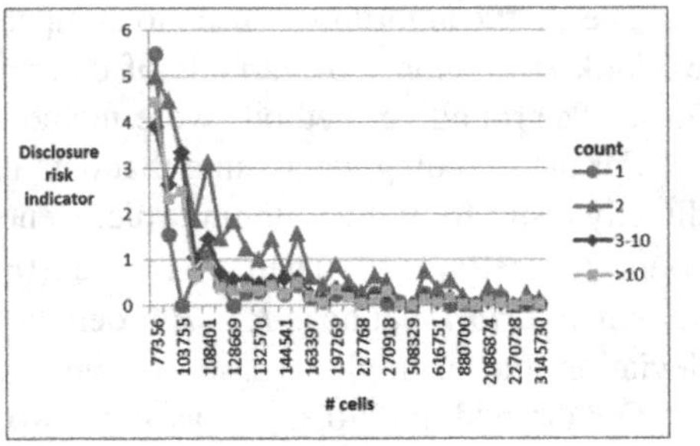

Fig. 3b. Disclosure risk indicator for rounding to base 5, by original count size

An interesting pattern is that many of the peaks in the observations for rounding to base 3 are observed for table types where one of the variables in the respective pair of spanning variables has 2 categories only. For rounding to base 5, on the other hand, most peaks occur at table types involving a 4-categories variable. This can be regarded as empirical confirmation of a theoretical result explained in [7]:

The original values of a table row with l inner cells and a margin cell, protected by rounding to base R, can be retrieved exactly, when $lR^+ + R^-$ or $lR^- + R^+$ is a multiple of R.

For rounding base 3 such a relation exists for the case of $l = 2$ categories ($2R^+ + R^- = 2 + 1 = 3$). For rounding base 5 it exists in the case of $l = 4$ categories ($4R^+ + R^- = 8 + 2 = 10$).

For rounding bases 7 and 10, exact bound retrievals were very rare[7], for rounding bases 9 and 11 no such cases were observed, nor for any of the additive noise variants. Especially for the additive noise this is not so surprising considering another result of [7]:

The original values of a table row with l inner cells and a margin cell, protected by rounding to base R, can be retrieved exactly, when the pattern of the rounding differences is either $(I) R^+ R^+ \ldots R^+ | R^-$, *or* $(II) R^- R^- \ldots R^- | R^+$.

So, all inner cells must have been perturbed by the maximum amount D in one direction, and the total must have been perturbed by D in the other direction. For the stochastic noise method such events occur with approximate probability $\left(p_{i(i+D)} \right)^{l+1}$ which is very small, considering the flat tails of the transition probabilities used to create the noise.

Retrieval of exact bounds and hence the original count is a rather direct risk of disclosure, especially where small cells are concerned. The *disclosure risk indicator* is thus a very informative disclosure risk measure. On the other hand, one should not interpret an indicator value of zero as evidence for zero disclosure risk. After all, such cases might eventually occur in other tables, not taken into account in the testing, or they might have occurred, if the observed data had been a little different, or if additional complexities – like hierarchical relations between some of the variables – had been taken into account.

This risk of underestimating the disclosure risk is reflected in the *indicator for disclosure risk potential*, introduced above. Figures 4a and b compare this indicator for the additive noise variants with maximum perturbations $D = 3$, 4 and 5, for the intruder scenarios A (ABS3, ABS4, ABS5, c.f. Figure 4a) and B (with *a priori* interval bounds increased by +2: ABS3_2, ABS4_2, ABS5_2, c.f. Figure 4b), and for the stronger variants of rounding, i.e. to rounding bases 7, 9, 10 and 11 (Fig. 4c).

For the noise, we observe a strong reduction in the indicators when maximum perturbation is increased. While for ABS3 the indicator values vary between about 15 % and 3 %, for ABS5 the range is on a different scale, between ca. 1.9 % and 0.5 %. The effect is similar (even slightly stronger) when the maximum perturbation is kept, but not published: actually the range for the scenario B variant ABS3_2, for which the shuttle algorithm starts with *a priori* intervals of the same size (i.e.: 10) as for scenario A variant ABS5, varies between 1.8 % and 0.02 %.

Notably, 10 is also the size of the *a priori* intervals for rounding to base 11, but here indicator values are again on a much higher scale (compared to noise), varying between about 60 % and 3 %. For all deterministic rounding variants (including rounding to base 3 and to base 5 not presented in the figure) the indicator values are remarkably similar. Also interesting is the effect that the general trend for higher indicator values in smaller tables, visible for the rounding variants and the least protective noise variant ABS3, vanishes in the variants with the lowest indicator values.

---

[7] Exact bounds were obtained for only 17 of the 31 structural types, for altogether 44 cell/equation cases. Apart from two structural types with disclosure risk indicators of 0.25 % and 0.1 % for original count sizes 3–10, indicators were at most 0.05 %. For rounding base 10, exact bounds were obtained for 20 cell/equation cases, all in the same structural type and only concerning cells with original counts >10.

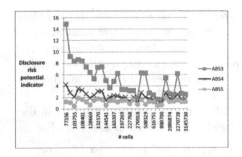

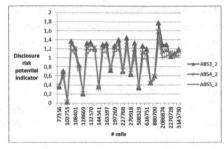

Fig. 4a. Disclosure risk potential indicator for additive noise, scenario A

Fig. 4b. Disclosure risk potential indicator for additive noise, scenario B

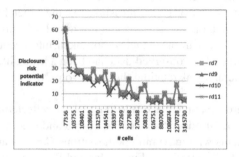

Fig. 4c. Disclosure risk potential indicator for Rounding to bases 7, 9, 10 and 11

A final observation is that most bound improvements occur during the first combination of *U*- and *L*-step. For the additive noise, this is the case in more than 95 % of the cases for almost all structural types of test tables. For deterministic rounding, at least 75 % of bound improvements occur during this first step. The algorithm always came to end after at most 6 steps.

5   Summary and Outlook

In its first part, this paper has explored the suggestion of [14] to determine the transition probability distributions of additive noise by maximizing entropy, looking at the mathematical formulation of the approach. For sake of illustration, the paper has explained how to derive a solution analytically in a relatively simple special case. The first part of the paper concludes with suggestions for how to employ the R-interface routine nloptr for the NLopt nonlinear-optimization package to compute transition probabilities, presenting some examples of distributions produced this way.

The topic addressed in the second part is disclosure risk estimation. The paper has outlined an implementation of a generalized version of the shuttle algorithm [2]. As test application the algorithm has been used on a major set of 3-dimensional tables. We have compared disclosure risks associated to several variants of deterministic rounding and additive noise, using one more and one less direct indicator for disclosure risk. As a side result, this test application has delivered an empirical proof of the high disclosure risks associated to rounding to base 3 as protection method. The paper has also proven empirically that although rounding to larger rounding bases reduces disclosure risks effectively, the disclosure risk potential of additive noise is considerably lower than that of deterministic rounding when we compare variants with the same maximum deviation. For additive noise, looking at only the more direct risk indicator of the two, no cases of disclosure risk were observed, even for rather weak settings of noise parameters.

For future work, it is foreseen to use the algorithm for a rigorous assessment and comparison of disclosure risks and risk potentials for several variants (e.g.: parameter settings) of additive noise, taking into account also more than 3-dimensional instances and, overall, a much larger set of instances. However, it is not at all intended to cover the enormous set of tables which can theoretically be generated by 'on-the-fly' table production. When fixing parameters, this "under-coverage issue" in disclosure risk estimation must be taken into account appropriately.

Acknowledgements. The author gratefully acknowledges the contribution of Jonas Peter who during his time as a student at Destatis has developed the R-code described in Sect. 2.3 and the Appendix.

Appendix

Arguments passed in R to the interface routine nloptr for the package Nlopt.

- **x0:** start-solution for p_i. Use for example a vector of 1's. Vector length must equal the number of columns of the coefficient matrix A.
- **eval_f:** list with two entries representing the objective function of the maximum entropy problem, and its gradient $-\left(\frac{1}{ln2} + \log_2(p_{ij})\right)_{j \in J_i}$ which for nloptr can simply be stated as: "objective" = sum(x * log2(x)), and "gradient" = log2(x)+1)
- **lb, ub:** vectors of upper and lower bounds for p_i. Same vector length as x0. For ub, a vector of 1's will do. For lb, use a vector of $\varepsilon(\varepsilon > 0$, f.i. $\varepsilon := 10^{-7})$. For discussion of more elaborate settings, see Sect. 2.4.
- **eval_g_ineq:** list with two entries representing the inequality constraints, to be stated formally as, e.g. "constraints" =constr, and "jacobian" = grad. The vector constr consists of (at least) two elements. One is the second term in the left hand side of (7), which could be stated as $sum(v^{\wedge}2 * x) - V^8$. The other one relates to condition (6) and could be stated as $x - 1$. Matrix *grad* is composed of (at least) the

---

[8] In our present code the second equality constraint (g_2 (p_i) = 0, i.e. the constraint expressing the requirement that perturbations shall have a fixed variance of V) is handled as inequality constraint, defining V as an upper bound, as in most practical instances the variance of the maximum entropy solution anyway assumes the value of this parameter.

two respective gradient vectors which are actually identical to the last two rows of matrix A. Further inequality constraints could be implemented to enforce monotony[9], c.f. Sect. 2.4.

- **eval_g_eq:** list with two entries representing the equality constraints, to be stated formally as, e.g. "constraints" = constr, and "jacobian" = grad, where constr is a vector listing the first and the last term of the left hand side of (7), like f.e. as $(sum$ $(v * x) - 0, sum(x) - 1)$, and $grad$ is a matrix composed of the two respective gradient vectors, which is obviously identical to the matrix that results when we delete the second row of matrix A.

- **opts:** Argument for the selection of a specific optimization algorithm and respective parameters[10]

References

1. Andersson, K., Jansson, I., Kraft, K.: Protection of frequency tables – current work at Statistics Sweden. In: Joint UNECE/Eurostat Work Session on Statistical Data Confidentiality, Helsinki, Finland, 5 to 7 October 2015. http://www1.unece.org/stat/platform/display/SDCWS15
2. Buzzigoli, L., Giusti, A.: An algorithm to calculate the lower and upperbounds of the elements of an array given its marginals. In: Statistical Data Protection (SDP 1998) Proceedings, Eurostat, Luxembourg, pp. 131–147 (1998)
3. Buzzigoli, L., Giusti, A.: Disclosure control on multi-way tables by means of the shuttle algorithm: extensions and experiences. In: Bethlehem, J.G., van derHejden, P.G.M. (eds.) Computational Statistics 2000, COMPSTAT Proceedings in Computational Statistics 200. Physica-Verlag, Heidelberg
4. EUROSTAT, Methodology for Statistics Unit B1: Minutes of the Working Group on Methodology, 7 April 2016
5. Fischetti, M., Salazar-González, J.J.: Models and algorithms for optimizing cell suppression problem in tabular data with linear constraints. J. Am. Stat. Assoc. **95**, 916–928 (2000)
6. Fraser, B., Wooton, J.: A proposed method for confidentialising tabular output to protect against differencing. In: Monographs of Official Statistics, Work session on Statistical Data Confidentiality, Eurostat-Office for Official Publications of the European Communities, Luxembourg, pp. 299–302 (2006)
7. Giessing, S.: Anonymisierung von Fallzahltabellen durch Rundung. In: Paper presented at the Sitzung des Arbeitskreises für Fragen der mathematischen Methodik am 17.06.1991 in Wiesbaden (in German), Statistisches Bundesamt
8. Giessing, S.: Report on issues in the design of transition probabilities and disclosure risk estimation for additive noise. Statistisches Bundesamt (Unpublished manuscript)

---

[9] Add one element $(x[j] - x[j+1])$ to the vector constr for each index j pointing to the right hand side of the distribution. Extend matrix grad by an additional column vector: The vector entries referring to indices j and j+1 should be 1 and (-1), resp. All other entries are 0.

[10] In our implementation we have defined local_opts as the following list: "algorithm" = "NLOPT_LD_SLSQP", "xtol_rel" = 1.0e−7, "maxeval" = 100000, and "local_opts" = local_opts, where local_opts is another list: ("algorithm" = "NLOPT_LD_MMA","xtol_rel" = 1.0e−7)

9. Giessing, S., Höhne, J.: Eliminating small cells from census counts tables: some considerations on transition probabilities. In: Domingo-Ferrer, J., Magkos, E. (eds.) PSD 2010. LNCS, vol. 6344, pp. 52–65. Springer, Heidelberg (2010)
10. Hundepool, A., Domingo-Ferrer, J., Franconi, L., Giessing, S., Schulte Nordholt, E., Spicer, K., de Wolf, P.P.: Statistical Disclosure Control. Wiley, Chichester
11. Johnson, S.G.: The NLopt nonlinear-optimization package. http://ab-initio.mit.edu/nlopt
12. Kraft, D.: A software package for sequential quadratic programming, Technical Report DFVLR-FB 88-28, Institut für Dynamik der Flugsysteme, Oberpfaffenhofen, July 1988
13. Kraft, D.: Algorithm 733: TOMP–Fortran modules for optimal control calculations. ACM Trans. Math. Softw. **20**(3), 262–281 (1994)
14. Marley, J.K., Leaver, V.L.: A method for confidentialising user-defined tables: statistical properties and a risk-utility analysis. In: Proceedings of 58th World Statistical Congress, pp. 1072–1081 (2011)
15. Roehrig, S.F.: Auditing disclosure in multi-way tables with cell suppression: simplex and shuttle solutions. In: Paper Presented at: Joint Statistical Meeting 1999, Baltimore, 5–12 August 1999
16. Ypma, J.: Introduction to nloptr: an R interface to Nlopt (2014). https://cran.r-project.org/web/packages/nloptr/vignettes/nloptr.pdf

Co-utile Anonymization

Enabling Collaborative Privacy
in User-Generated Emergency Reports

Amna Qureshi[✉], Helena Rifà-Pous, and David Megías

Estudis d'Informàtica, Multimèdia i Telecomunicació, Internet Interdisciplinary
Institute (IN3), Universitat Oberta de Catalunya, Barcelona, Spain
{aqureshi,hrifa,dmegias}@uoc.edu

Abstract. Witnesses are of the utmost importance in emergency systems since they can trigger timely location-based status alerts. However, their collaboration with the authorities can get impaired for the fear of the people of being involved with someone, some place, or even with the same cause of the emergency. Anonymous reporting solutions can encourage the witnesses, but they also pose a threat of system collapse if the authority receives many fake reports. In this paper, we propose an emergency reporting system that ensures the anonymity of honest witnesses but is able to disclose the identity and punish the malicious ones. The system is designed over an online social network that facilitates the indistinguishability of the witness among a group of users. We use a game-theoretic approach based on the co-privacy (co-utility) principles to encourage the users of the network to participate in the protocol. We also use discernible ring signatures to provide the property of conditional anonymity. In addition, the system is designed to provide rewards to a witness and his/her group members in a privacy-preserving manner.

Keywords: Co-utility · Co-privacy · Revocation · Emergency management · Online social network

1 Introduction

An emergency is an unanticipated situation that may lead to the loss of lives (road accidents) or properties (collapsed buildings), to the harm of the physical integrity of human life (robberies), or to the damage of properties (fire) or the environment (wildfire). In such situations, the traditional way to report the incident and ask for help is to call an emergency service that allows the caller to contact local emergency operators for assistance. On average, it takes an emergency operator at least two to three minutes to collect the necessary information in order to respond to the caller [21]. At the time of emergencies, the loss of a few seconds can mean the difference between life and death. Therefore, emergency rescue systems should be fast and efficient in order to ensure a timely response to emergency situations. The recent advances in mobile communication and mobile information systems have made a significant impact on the development of emergency response systems. These systems or platforms allow citizens

J. Domingo-Ferrer and M. Pejić-Bach (Eds.): PSD 2016, LNCS 9867, pp. 255–271, 2016.
DOI: 10.1007/978-3-319-45381-1_19

to communicate location-based emergency information to emergency responders, who, in return, respond quickly to the situation. For example, Alpify [2] is a mobile application that uses a mobile phone's global positioning system (GPS) functionality so that citizens can locate, document and report emergencies (fire or road accidents) to 112/911 emergency services, quickly and effectively.

A main issue faced by the emergency service providers is fake or false emergency calling. A fake call is when a person deliberately calls the emergency service to falsely inform them that there is an emergency when in fact there is not, or when somebody contacts the emergency services for reasons not related to any emergency, or when the situation is not considered an emergency by the emergency services but it is for the caller (e.g. car keys are lost) [5]. The statistics in a recent study show that the emergency services across the United Kingdom (U.K.) receive over 5 million fake calls per year [15]. These fake calls are a misuse of the system and divert emergency services away from people who may be in life-threatening situations and need urgent help. Also, a fake call is an expensive problem because emergency service providers need to multiply their resources to assure that they are not being overloaded by false calls and, therefore, may not be able to respond to true emergencies. Thus, there is a need to figure out mechanisms to prevent people from making fake calls to emergency services so that the true emergencies that require immediate assistance always get a top priority. The communities across Europe and U.K. are trying to face false emergency calls by instituting ordinances and/or special measures by police departments [5]. For example, alternative three-digit numbers for non-emergency calls have been introduced in the recent years in the U.K.

Many systems for emergency management have been envisioned [12,16,18]. In [12], the authors proposed the use of social media in a collaborative effort to inform people about crime events that are not reported to the police. Their wiki website (WikiCrimes) allows users to register criminal events online in a specific geographic location represented by a map; hence, other users can use this information and keep track of the locations to make decisions. However, a limitation of this approach is that each crime registered in WikiCrimes requires confirmation from at least one another person (besides the reporting user) in order to be registered as a true event. In addition, WikiCrimes requires users to log into the system by means of a valid email address, and then tracking the reporting user is possible. In [18], Okolloh proposed Ushahidi, a map-based mash-up tool to visualize crowd-sourced information by allowing citizens to submit information by sending a text message (SMS), a tweet, or an email; or by inputting the information on a form available on Ushahidi's web portal. Though Ushahidi has proven to be a successful online platform for spreading awareness of critical situations worldwide, it faces some limitations, such as the requirement that the reports from incidents have to go through an approval process conducted by a group of volunteers, who publish an online interactive map of reports after successful verification. In [16], a location-aware Smart Phone Emergency and Accident Reporting System (SPEARS) is proposed that allows users of an online social network (Facebook or Twitter) to quickly report emergency situations to the agencies responsible for handling emergency situations. The agencies store their locations via SPEARS, so that users involved in an emergency situation can retrieve the shortest path from the point of alert to the point of care. Though it is an efficient tool for emergency reporting on Android smartphones, it has a few limitations: (1) it can only be used in Thai language, which does not help much for most foreigners living in Thailand, and (2) users must be identified by phone numbers and names before reporting an emergency.

All the systems referred above pose at least one of the following drawbacks: (1) they allow user re-identification, and (2) they require manual sorting of legitimate/fake information. The re-identification of a user by means of his/her email address, phone number or location is a relevant issue, since in most emergency situations, the witnesses are reluctant to report an emergency because they do not want to be identified or reveal their specific location for personal reasons, or because they fear the possibility of being considered as suspects of a crime. Anonymity is, thus, a desired property from the witness point of view. However, total user anonymity is not feasible since it would encourage fake emergencies, which could make the authority collapse. Therefore, the challenge lies in providing anonymity to true reports, but which could be revoked in case of a false emergency report.

Contribution and plan of this paper: The contribution of this paper is to introduce a system for the notification of location-based emergency-related information so that the information is managed by an authorized entity that takes appropriate action. Our proposal stems from a game-theoretic design inspired by the co-privacy (co-utility) approach [6,8,9], which leads to a mechanism of rewards and punishments to encourage legitimate information and discourage false reporting. Unlike existing emergency management systems in which witnesses are required to reveal their identities or personal information to the emergency service provider, our system protects the witness's identity by using the concept of groups, in such a way that the witness is indistinguishable from other members of the group. Here, we propose to use an online social network, whose scope is to facilitate the social interaction among an interconnected trusted network of people, for creating dynamic and location-based user groups. Anonymity is considered in the system by means of ring signatures, i.e. the emergency reports are linked to the groups instead of individual users (witnesses). However, to avoid complete anonymity in our scheme, an accountability property is provided in the sense that a malicious witness who sends false reports can eventually be identified by the collaborative effort of other members of the group. Furthermore, a source routing protocol (similar to onion routing [19]) is used to provide anonymous communication with the authority. The group members would help to run the appropriate protocol for the system to fulfil the requirements.

The rest of this paper is organized as follows. In Sect. 2, we overview the functionality of our system. In Sect. 3, the reward and punishment model of the proposal based on game theory is detailed. Section 4 presents the protocol for sending and managing anonymous emergency reports. In Sect. 5, we discuss the security and privacy aspects of the protocol. Finally, Sect. 6 concludes the paper.

2   Overview of the System

This section describes the architecture of the system proposed for the notification of location-based emergency-related information to a so-called Emergency Management System (EMS) that takes appropriate action to solve the emergency.

A. Requirements of the system: The design requirements of the system are as follows: (1) The system must be efficient to minimize the time taken by the emergency responders to reach the location of the emergency. (2) The amount of fake reports that are considered by the EMS needs to be limited since the management of a false emergency leads to a waste of the resources of the EMS. (3) The system must provide privacy guarantees so that the identities of the users reporting the emergencies remain hidden to everyone, i.e. to the EMS, the OSN and the users of the network. (4) The exact location of the incident must be reported to the EMS so that it can immediately respond to the emergency. (5) The users of the system are organized in groups, which are dynamically formed by the witness of an emergency. Group members must be active users (online contacts), who lie within the vicinity, i.e. within a pre-defined distance of the witness. (6) When the awardees redeem their rewards at City Council, it should not be able to link the recipient with any reward assigned previously.

B. Design assumptions: In our proposed system, the dynamic and location-based user groups are created by assuming users to be registered members of a popular OSN, Facebook [1]. There are mainly two reasons for selecting Facebook as a choice for the OSN: (1) it provides its data to external applications via application programming interfaces (APIs), and (2) it does not require an authorization before using an API.

In the following, the security and general assumptions related to the design of the emergency reporting system are defined: (1) Each user is a registered member of Facebook. Users can log in via a Facebook account to access and use the emergency reporting system on their smartphones. (2) A public key infrastructure is considered for providing cryptographic keys in such a way that each entity of the system has a public and a private key. (3) A group created by a witness can contain up to $n \geq 3$ users (minimum 3 users are required such that one user acts as a witness and, second and third users would be the hop and the reporter, respectively) (4) In case of a false emergency report, a threshold of t users of the reporting group will be able to disclose the identity of the witness. This t is set to 60% of n (i.e. more than half of the users in a group (of $n \geq 3$)). (5) The public keys and the parameters of the ring signature (of each group member) and the public key of the EMS are publicly available. (6) Threshold discernible ring signatures (TDS) [14] provide unforgeability and signer anonymity (details of TDS can be found in Appendix B). (7) The system proposes to leverage GPS and signal triangulation technologies to automatically sense device location. Triangulation is used only if a GPS signal is unavailable. (8) The system provides three user status modes: online (available or busy), idle (away) and offline. In online mode, the actual location is available to the users' friends, showing a person icon, his/her location coordinates and description of a distance on their map-based screens, whereas, in idle and offline modes, the last recorded distance interval of the user along with his/her last online visibility status are provided.

C. System entities: Figure 1 illustrates the model of the proposed emergency reporting system that contains the following basic entities: **(1) The witness:** The user who witnesses an event and reports it. This user wants to safeguard his/her identity. **(2) The social group:** A group in which the witness is a member. **(3) The system manager:** A service provider who is responsible for executing the emergency reporting system via a Facebook API. It also manages the registration of the users and imports a list of users' friends from Facebook (who are already the members of the reporting system). Additionally, the system manager uses location information to calculate the distance between the users and display it on the Google Maps along with a person icon. **(4) EMS:** An entity that receives and manages the emergency reports. On receiving the report, the EMS forwards it to emergency entities for validation. **(5) The reporter:** A friend of the witness (both are members of the same group). The reporter helps the witness to send an emergency report to the EMS. This user can be identified by the EMS. **(6) The intermediate hops:** The users (members of the same group) that serve as report forwarding agents. **(7) The City Council (CC):** A trusted entity from which the witness, the reporter and the group members can redeem their rewards in form of vouchers, one-time discount coupons or tax payments. Also, CC issues punishment to the witness for false reporting. **(8) The emergency entities (EE):** Entities such as police stations, hospitals, rescue units and fire stations. **(9) The Certification Authority (CA):** A trusted entity that has pre-generated key pairs and issues a key pair upon successful authentication. It is an offline process and thus does not affect the performance of the system.

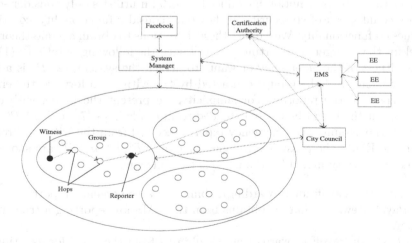

Fig. 1. Overview of the system

It can be seen, in Fig. 1, that the interaction between the witness and the EMS is carried out through multiple intermediary hops and a member of the group (i.e. the reporter), who assumes the responsibility of submitting the witness's report (similar to the multi-hop protocol proposed in [7]).

3   Co-utility Model for the Proposed Solution

The proposed emergency reporting system uses a "reward and punishment" mechanism to reward legitimate reports and punish fake ones. We use a co-utility model based on game theory (see Appendix A) to examine the implications of the witness and the members of his/her social group. We assume that users are interested in two aspects: (1) obtaining rewards, and (2) keeping their anonymity. The co-utility model presented below considers these two aspects. We borrow from [11] the following definition of co-utility:

Definition 1 (Co-utility). *Let Π be a game with self-interested, rational players $P^1, \cdots, P^N$, with $N > 1$. The game Π is said to be* co-utile *with respect to the vector $U = (u_1, \cdots, u_N)$ of utility functions if there exist at least three players P^i, P^j and P^k having strategies s^i, s^j and s^k, such that: (i) s^i involves P^i expecting co-operation from P^j and P^k; (ii) s^j involves P^j co-operating with P^i and P^k; (iii) s^k involves P^k co-operating with P^i and P^j; and (iv) (s^i, s^j, s^k) is an equilibrium for P^i, P^j and P^k in terms of u_i, u_j and u_k, respectively. In other words, there is co-utility between P^i, P^j and P^k, for some $1 \leq i, j, k \leq N$ with $i \neq j \neq k$, if the best strategy for P^i involves expecting co-operation from P^j and P^k, the best strategy for P^j is to co-operate with P^i, and the best strategy for P^k is to co-operate with P^i and P^j.*

If the equilibrium in Definition 1 is a Nash equilibrium, we have *Nash co-utility*. If the utility functions U in Definition 1 only considers privacy, co-utility becomes the co-privacy notion introduced in [6,9]; if utilities only consider security, we could speak of co-security; if they only consider functionality, co-utility becomes co-functionality. We can use these definitions to obtain a game-theoretic model for the emergency reporting protocol with the following notations: (1) P^i is the witness of the emergency or wants to attack the system; (2) P^j is a hop (another member of the group) contacted by the witness to forward the emergency report to the reporter. For simplicity, we present the model with only one hop, but the it can be easily extended to multiple hops [7]; and (3) P^k is a reporter who submits the emergency report (received from the witness through P^j) to the EMS. The possible strategies for player P^i, P^j and P^k are shown in Table 1. The utility model for the game is the following:

- $-c$: Negative payoff for forwarding/submitting an emergency report.
- d_i: Payoff (reward) that P^i obtains from the EMS for reporting a true emergency.
- d_j ($d_j < d_i$): Payoff (reward) that P^j obtains from the EMS for assisting in the submission of a true emergency report.

Table 1. Possible strategies of players

No.	Possible strategies of players		
	P^i	P^j	P^k
1.	S_0^{ii}: Reports a true emergency directly to the EMS.	W_0^{jk}: Forwards the emergency report to P^k.	T_0^k: Submits the emergency report to the EMS.
2.	S_1^{ii}: Reports a false emergency directly to the EMS.	W_1^j: Ignores the emergency report.	T_1^k: Ignores the emergency report.
3.	S_0^{ij}: Forwards a true emergency report to P^j.	W_2^{jl}: Deviates from its pre-defined routing path and does not deliver the report to P^k.	T_2^k: Joins other players that may include P^j to reveal the source P^i to the EMS after being accused of sending a false emergency report.
4.	S_1^{ij}: Forwards a false emergency report to P^j.	W_3^{jk}: Joins other players that may include P^k to reveal the source P^i to the EMS after being accused of sending a false emergency report.	
5.	S_0^{ik}: Forwards a true emergency report to P^k.		
6.	S_1^{ik}: Forwards a false emergency report to P^k.		
7.	S_2^i: Ignores a true emergency and does not report it.		

- d_k $(d_k > d_i > d_j)$: Payoff (reward) that P^k obtains from the EMS for submitting a true emergency report.
- $-v_i$: Negative payoff (punishment) that P^i obtains from the EMS for reporting a false emergency report.
- $-v_j$ $(v_j < v_i)$: Negative payoff (punishment) that P^j, P^k and all the other group members obtain from the EMS for forwarding a false emergency report.
- r_j $(r_j < r_k)$: Reward that P^j and the remaining group members obtain after revealing the source of a false emergency report to the EMS.
- r_k: Reward that P^k obtains after revealing the source of a false emergency report to the EMS.
- $-w_j$: negative payoff that P^j incurs from not following the fixed routing path.
- $-w_k$: negative payoff that P^k obtains due to a loss of privacy w.r.t the EMS.
- $-z_k$: negative payoff that P^k incurs due to a false accusation by the EMS. Typically, $z_k = 0$ if the protocol guarantees that P^k is not the creator of the report.

The values of the utility functions for P^i, P^j and P^k are presented in Table 2.

We can have two possibilities in this situation: P^i either witnesses a true emergency or generates a fake emergency report. In the former case, the witness P^i can decide either to ignore the emergency and obtain a neutral (0) payoff, or to report the emergency an obtain a maximum payoff $d_i - c > 0$ if he/she decides to use the hop P^j and the reporter P^k. In this case, the maximum payoff that

Table 2. Utility functions of P^i, P^j and P^k

Players' strategies	Utilities		
	u_i	u_j	u_k
$S_0^{ii}, \emptyset, \emptyset$	$d_i - c - w_k^{(a)}$	×	×
$S_1^{ii}, \emptyset, \emptyset$	$-c - v_i - w_k < 0$	×	×
$S_2^i, \emptyset, \emptyset$	0	×	×
$S_0^{ij}, W_0^{jk}, T_0^k$	$d_i - c^{(b)}$	$d_j - c^{(c)}$	$d_k - c - w_k^{(d)}$
$S_0^{ij}, W_0^{jk}, T_1^k$	$-c < 0$	$-c < 0$	0
$S_0^{ij}, W_1^j, \emptyset$	$-c < 0$	0	×
$S_0^{ij}, W_2^{jl}, \emptyset$	$-c < 0$	$-c - w_j < 0$	×
$S_0^{ik}, \emptyset, T_0^k$	$d_i - c^{(b)}$	×	$d_k - c - w_k^{(d)}$
$S_0^{ik}, \emptyset, T_1^k$	$-c < 0$	×	0
$S_1^{ik}, \emptyset, T_0^k$	$-c - v_i < 0$	×	$-c - v_j - z_k < 0$
$S_1^{ik}, \emptyset, T_0^k + T_2^k$	$-c - v_i < 0$	×	$-2c - v_j + r_k^{(e)}$
$S_1^{ik}, \emptyset, T_1^k$	$-c < 0$	×	0
$S_1^{ij}, W_0^{jk}, T_0^k$	$-c - v_i < 0$	$-c - v_j < 0$	$-c - v_j - z_k < 0$
$S_1^{ij}, W_0^{jk}, T_1^k$	$-c < 0$	$-c < 0$	0
$S_1^{ij}, W_1^j, \emptyset$	$-c < 0$	0	×
$S_1^{ij}, W_2^{jl}, \emptyset$	$-c < 0$	$-c - w_j < 0$	×
$S_1^{ij}, W_0^{jk} + W_3^j, T_0^k + T_2^k$	$-c - v_i < 0$	$-2c - v_j + r_j^{(f)}$	$-2c - v_j + r_k^{(e)}$

Comments: (a) $c + w_k$ must be smaller than d_i to be positive; (b) c must be smaller than d_i to be positive; (c) c must be smaller than d_j to be positive; (d) $c + w_k$ must be smaller than d_k to be positive; (e) positive if $r_k > v_j + 2c$; and (f) positive if $r_j > v_j + 2c$.

P^j can obtain from the EMS is $d_j - c > 0$ for relaying the emergency report from P^i to P^k. Also, P^k obtains a maximum payoff $d_k - c - w_k > 0$ by reporting the emergency to the EMS. The Nash equilibrium $(S_0^{ij}, W_0^{jk}, T_0^k)$ for P^i is to report the emergency using P^j and P^k, for P^j is to forward the report to P^k and for P^k to submit the report to the EMS. In the latter case, if P^i reports a fake report either directly or through P^j and P^k, group members will obtain positive payoff by revealing the source P^i of the message, who would then be punished by getting a negative payoff $-c - v_i$. P^j will obtain a smaller payoff $r_j - 2c - v_j > 0$, P^k will obtain a major payoff $r_k - 2c - v_j > 0$, and the remaining group members of the group a smaller payoff $r_j - c - v_j > 0$. Hence, there is no profit in generating a fake emergency report, unless some (small) probability may exist that a fake emergency report is taken to be valid by the EMS. In any case, the risk of receiving a punishment should be enough to discourage users from generating false emergency reports.

Note that, in both cases, the best strategy for P^j and P^k is to co-operate with the witness P^i, since they can obtain a positive payoff either by forwarding a true emergency report or by accusing P^i as the source of a fake emergency report. P^k will only succeed in accusing P^i if P^j and other group members collaborate, but since this is also the best strategy for group members, the dominant strategy $(S_1^{ij}, W_0^{jk} + W_3^j, T_0^k + T_2^k)$ for P^k is to forward emergency reports always. Thus, the dominant strategy is, in particular, strictly co-utile [8]. Of course, there are several possible attacks in this scheme to try to obtain a positive payoff. For example, a player P^i may cause an emergency and forward it to P^j for submission to the EMS in order to obtain a positive payoff. This is not exactly an attack to the system, since that would be a real emergency after all (and there is a risk of being traced by the authorities anyway). Another possibility is to try to impersonate another user to generate a fake report, forward it to the EMS as P^k, and obtain a positive payoff by revealing the impersonated source. This is not possible since the signature algorithm of TDS (Appendix B.1) used in the protocol provides unforgeability.

4   Proposed Protocol

In this section, we present the protocol for sending and managing anonymous emergency reports to the EMS. The protocol mainly consists of three phases: witnessing an emergency, managing and processing the emergency report, and the witness distinguisher.

A. Witnessing an emergency: When a user wants to report an emergency, he/she proceeds as follows. **(1)** The witness logins to the system, using his/her Facebook account details, and looks for nearby online contacts in the system. **(2)** The witness creates a dynamic and covert group of $n \geq 3$ nearby users. Since the users share location information with each other, the witness does not require any assistance of the system manager or the users to form a group. **(3)** An online group member (reporter) is selected by the witness to assist him/her in reporting the emergency to the EMS. **(4)** Multi-hop routes are computed at the witness's end to forward the emergency report to the reporter. The report is propagated along a selected route from hop to hop until it reaches the reporter. The hops simply forward the report without checking its content, which is encrypted and unreadable for them. **(5)** The witness prepares a report message r, which is a tuple $r = \{R_{id}, STdata, Content, k_m\}$: R_{id} is a report identifier; $STdata$ is a spatio-temporal tag; the $Content$ is the information of the emergency; and k_m is a random symmetric key that the user generates to establish an anonymous confidential channel between himself and the EMS. **(6)** The witness ciphers the report r with the public key of the EMS: $m = E_{K_{p_{EMS}}}(r)$, where $E()$ is a public-key cipher. **(7)** The witness signs the ciphered report m applying the signing procedure of the TDS scheme (Appendix B.1). With his/her private key x_i and the public keys of the group members, he/she generates the signature: $\sigma = S_{TDS}(g, x_i, y_1, \cdots, y_n, \alpha_1, \cdots, \alpha_n, t, m)$. **(8)** The witness sends the signed and ciphered report request (m, σ) to the reporter through a pre-defined routing path. Assuming that the path consists of two hops (P^{j_1}, P^{j_2}). The first hop P^{j_1} receives the packet: $(Sign_\sigma(ID_{witness}), \{((m, \sigma)_{y_k}, P^k)_{y_{j_2}}, P^{j_2}\}_{y_{j_1}})$. It decrypts the destination field to check whether it is the destination or not. If not, it generates a session key K_{m_1}, encrypts it with the public key of EMS ($K_{p_{EMS}}$), adds it into the packet and sends $(Sign_\sigma(ID_{witness}), E_{K_{p_{EMS}}}(K_{m_1}), \{(m, \sigma)_{y_k}, P^k\}_{y_{j_2}})$ to P^{j_2}. P^{j_2} would do the same thing to execute the similar operation and forward the packet $(Sign_\sigma(ID_{witness}), E_{K_{p_{EMS}}}(K_{m_1}), E_{K_{p_{EMS}}}(K_{m_2}), (m, \sigma)_{y_k})$ to the reporter P^k. **(9)** On receiving the packet from P^{j_2}, P^k checks the destination field of the packet. If no further hop is present, P^k decrypts the payload to obtain (m, σ). Then, P^k verifies whether the signature is discernible, authentic and integral by applying the verifying procedure of the TDS scheme (Appendix B.2). If the signature is verified, he/she submits $(P^k, y_k, (m, \sigma), E_{K_{p_{EMS}}}(K_{m_1}), E_{K_{p_{EMS}}}(K_{m_2}))$ to the EMS in accordance with the strategies explained in Sect. 3.

B. Managing and processing the emergency report: The EMS receives a signed and ciphered report request from a reporter. The EMS obtains the identity data of the reporter; the reporter is responsible for the information in front of the EMS, although the EMS knows that the reporter is not the witness of the event but a proxy chosen by the actual witness. The EMS also receives the session keys K_{m_1} and K_{m_2} of the intermediary hops.

Following are the steps that EMS follows to process the emergency report. **(1)** The EMS verifies the TDS signature generated by the witness. **(2)** The EMS deciphers the report using its private key: $r = D_{KS_{EMS}}(m)$, with $D()$ a public-key decipher; **(3)** The EMS obtains the public keys of n group members from the TDS signature and the report identifier from the report. It signs a group acknowledgement of emergency receipt $Ack = \{R_{id}, Group_{info}\}$, where $Group_{info}$ contains the public keys of n group members. It sends this acknowledgment Ack to the system manager, who sends it to all the group members in such a way that the witness knows about the report reception. If the witness does not receive Ack in a timeout t_0, he/she will try to send the report through another route or reporter; **(4)** Then, after verifying the correctness of the reported information (i.e. the emergency was true), the EMS prepares a reward or a punishment response. This response will be signed using the private key of the EMS. If the report is correct, the EMS first generates a hash value, $H_{EC} = H(ID_{EMS} ||Date||Time||STdata)||R_{id}||nonce_{R_{id}}||y_i)$ (where $H()$ is a collusion-resistant hash function, $nonce_{R_{id}}$ is a fixed value assigned to all the group members that have submitted the emergency report (R_{id}) and y_i is a public key of a group member), signs it and then generates the following rewards: (1) for the reporter P^k, which consists of the payoff ciphered with the reporter's public key y_k and a signed H_{EC}: $Reward_R = \{P^k, E_{y_k}(payoff), Sign_{KS_{EMS}}(H_{EC})\}$, (2) for the intermediary hops with the payoffs ciphered with the received symmetric keys K_{m_1} and K_{m_2} and a signed hash value: $Reward_H = \{Group_{info}, C_{k_{m_1}}(payoff), C_{k_{m_2}}(payoff), Sign_{KS_{EMS}}(H_{EC})\}$, and (3) for the witness, ciphered with the symmetric key received from the witness k_m and signed hash value: $Reward_W = \{Group_{info}, C_{k_m}(payoff), Sign_{KS_{EMS}}(H_{EC})\}$ (with $C()$ a symmetric key cipher). The EMS sends these rewards to the system manager, who forwards the first reward $Reward_R$ to P^k and broadcasts the remaining two rewards $Reward_H$ and $Reward_W$ to all group members. Only P^{j_1}, P^{j_2} and the original witness P^i will be able to decipher $Reward_H$ and $Reward_W$, respectively, in order to redeem them from the CC. Also, the EMS sends a signed H_{EC} to the CC for later use in the reward redemption phase (see Appendix C). If the report is false, the EMS prepares punishments $Punishment_k = \{P^k, R_{id}, E_{y_k}(payoff)\}$ and $Punishment_x = \{y_x, R_{id}, E_{y_x}(payoff)\}$ (where $x = 1, \ldots, n-1$) for P^k and the remaining group members, respectively. Then, EMS sends $Punishment_k$ and $Punishment_x$ to the system manager, who retransmits them among the respective users. Also, the EMS requests the system manager to forward the identities of the group members ($Group_{info}$) to the CC, so that they get punished for reporting a false emergency; and (5) If the EMS repeatedly receives false information from the users of $Group_{info}$, the EMS puts them on a black list and no longer pays attention to the reports coming from them.

C. The witness distinguisher: If the group has been punished for a false emergency report, a subgroup of t users can join to reveal the identity of the malicious witness in order to obtain compensation (in terms of rewards) for the punishments inflicted on them by the EMS. The steps of the process are as follows. (1) A user P_u that participates in the disclosure process deciphers his/her share V_u of the request secret parameter and obtains ρ_u. He/She enciphers this information for the EMS, makes a personal signature, and sends the result to the EMS. (2) The EMS deciphers and verifies the secret shares it receives. It also checks that the secret shares ρ_u received indeed correspond with the encrypted secret shares V_u. (3) When the EMS has the secret shares ρ_u of t users, it triggers the distinguisher algorithm of TDS (see Appendix B.3). It reconstructs the secret f_0 using the public parameters $(\alpha_1, \cdots, \alpha_n)$ and the secret shares $(\rho_1, \cdots, \rho_n)$ of the t participating users. Using f_0, the EMS can recover the identity of the original signer. (4) Then the EMS generates a nominal punishment for the malicious witness $Punishment_W = \{P^i, R_{id}, payoff\}$ and, at least, t rewards (one for each participant in the distinguisher process). It ciphers each payoff using the recipient's public key and sends $Reward_{P_u} = \{y_{p_u}, Sign_{KS_{EMS}}(H_{EC}), E_{y_{p_u}}(payoff)\}$ to the system manager, which distributes it to the respective members. The members can then redeem their rewards from the CC through by executing reward redemption protocol (Appendix C). The punishment for the malicious witness is sent to the witness as well as the CC, who will issue a penalty (fee) to the witness.

5   Discussion

The proposed protocol encourages users to send anonymous reports regarding some witnessed emergency. Anonymity is provided in two ways: (1) in the network layer using multi-hop report retransmissions, and (2) in the application layer using strong cryptography. Regarding multi-hop retransmissions, a witness forwards the emergency report through a fixed routing path (nearby online friends) to another online friend (within his/her vicinity), who in turn sends it to the EMS. This scenario, together with co-privacy, is analogous to the problem of user-private information retrieval [10]. If a witness sent his/her emergency report directly to the EMS, the EMS would know the IP address of this user and get his/her location, so his/her privacy would be surrendered. With this information and the emergency location (this data is always present in the report), the EMS could require more information of the reporter and the intermediary hops and involve them in the investigation of the events. Thus, users are always advocated to select user proxies for sending emergency reports.

When an emergency report is sent to the EMS, all group users are responsible for that report, although the main responsible entity is the reporter. If the report is true, the reporter receives a major payoff and the hops receive nominal payoffs, but if it is false, all group users are punished with the aim that they collaborate to find out the true witness. If the true witness can be discovered, the group members that participated in the witness distinguisher protocol, share some stipulated payoff and the reporter receives a major reward. The witnesses who sent false reports are never rewarded with a payoff even if they participated in the distinguisher protocol. The entire payoff that the EMS pays to the hops, the reporter and the users involved in the distinguisher protocol, is always smaller than the punishment for the malicious witness. This discourages Sybil attacks, where a user generates multiple accounts in order to gain a disproportionately large influence in the group and eventually obtain a global benefit although one of his/her identities (the witness) is severely punished.

In the protocol, the group is created dynamically based on the users' locations to avoid re-identification by strong adversaries. Thus, we propose to use a group consisting of users who are all in the partition where the emergency is located. This reduces the risk of re-identification of the witness even if the system manager and the EMS collude. However, there is a possibility that a witness finds only one user within a pre-defined distance to forward the report to the EMS. This implies that the identification of the witness would be immediate. A possibility to solve this problem is to step-wise increase the distance threshold (in meters). Since the reporting system is proposed for smart cities, it is highly likely that the witness could find at least three users within his/her close vicinity to form a group.

The proposed protocol uses cryptography to provide anonymity and authenticity in the application layer. Our proposal to protect users' identities is to work with TDS that authenticate a group of users (friends) instead of individual users. If a witness sends a report on the group's behalf, it should be impossible to identify which user is the originator. The security of TDS holds in the random oracle model [3], similar to the majority of the ring signature schemes. The security of these signatures has two aspects: unforgeability and signer anonymity. Unforgeability means that an external member of a group cannot create a ring signature with non-negligible advantage in polynomial time. Anonymity entails that at least t ring members of the group are required to discover the original signer of the t-threshold ring signature (with non-negligible advantage in polynomial time). It is worth noting that, in the presented protocol, anonymity is provided to the users without the presence of trusted third parties. The system manager and the EMS do not know the identity nor the IP address of the witness. However, two trusted parties (CA and CC) are required in the reward redemption protocol (Appendix C) so that the users can redeem their rewards in a privacy-preserving manner.

6   Conclusions and Future Work

We have presented an emergency reporting system that ensures the anonymity of honest witnesses but is able to disclose the identity and punish the malicious ones by using TDS. The system is designed using the Facebook API that facilitates the creation of a group of users among which a witness can become indistinguishable. For a group formation or submission of the report, the witness does not need the assistance of the system manager and hence, it could not figure out the group's location. A game-theoretic approach based on the co-privacy principles is used to encourage the users to participate in the protocol.

Future research should be directed: (1) To make the emergency information public and show it on a map (a feature which entails privacy risks that shall be examined and prevented); (2) to extend the co-utility model using multiple hops; and (3) to address the possibility of collusion between ring members such that each member gets a reward for reporting.

Acknowledgment. This work was partly funded by the Spanish Government through grants TIN2011-27076-C03-02 "CO-PRIVACY" and TIN2014-57364-C2-2-R "SMART-GLACIS".

A   Basics of Game Theory

As detailed in [17], a game is a protocol between a set of N *players*, $\{P^1, \cdots, P^N\}$ who must choose among a *set S_i of possible strategies*. Let $s_i \in S_i$ be the strategy played by player P^i and $S = \Pi_i S_i$ the set of all possible strategies for all players.

The vector of strategies $s \in S$ chosen by all players determines the outcome of the game for each player which can be thought of as a payoff or a cost. For all players, a preference ordering of these outcomes should be given in the form of a complete, transitive and reflexive relation on the set S. A simple and effective way of achieving this goal is by defining a scalar value for each outcome and each player. This value may represent a payoff (if positive) or a cost (if negative). A function that assigns a payoff to each outcome and each player is called a utility function: $u_i : S \longrightarrow \mathbb{R}$.

Given a strategy vector $s \in S$, s_i denotes the strategy chosen by P^i, and let s_{-i} denote the $(N - 1)$-dimensional vector of the strategies chosen by all other players. With this notation, the utility $u_i(s)$ can also be expressed as $u_i(s_i, s_{-i})$. A strategy vector $s \in S$ is a *dominant strategy solution* if it yields the maximum utility for a player irrespective of the strategy played by all other players, i.e. for each alternate strategy vector $s' \in S$, maximum utility is $u_i(s_i, s'_{-i}) \geq u_i(s'_i, s'_{-i})$.

In addition, a strategy vector $s \in S$ is said to be a *Nash equilibrium* if it provides the largest utility for all players, larger than any other alternate strategy

$s_i' \in S_i$ or $u_i(s_i, s_{-i}) \geq u_i(s_i', s_{-i})$. This mean that, in a Nash equilibrium, no player will be able to change his/her strategy from s_i and achieve a better payoff when all the other players have chosen their strategies in s. Note that Nash equilibria are self-enforcing if players behave rationally, since it is in all players' best interest to stick to such a strategy. Obviously, if all players are in a dominant strategy solution at the same time, this is a Nash equilibrium. More game theory information can be found in [17].

B   Threshold Discernible Ring Signatures

We base our system on threshold discernible ring signatures (TDS), which were introduced by Kumar et al. [14]. In a t-threshold discernible ring signature, a user in the system can generate a signature using his/her own private key and the public keys of the other n ring members (with $n > t$). A verifier is convinced that someone in the ring is responsible for the signature, but he/she cannot identify the real signer. The identity of the signer can only be revealed if a coalition of at least t members of the group cooperates to open the secret identity. In the following, three TDS operations that were used in the proposed protocol are outlined.

B.1   Signature

The signing algorithm $S_{TDS}(g, x_i, y_1, \cdots, y_n, \alpha_1, \cdots, \alpha_n, t, m)$ generates a ring signature of a message m and a set of verifiable encrypted shares of a secret that allows disclosing the identity of the original signer. The secret, which we call f_0, can only be revealed when a group of t ring members brings together some information. For signing a message m, the user first generates t random numbers $f_j \in Z_q^*$ and computes $F_j = g^{f_j}$ for each of them. The first random number f_0 is used as a trapdoor to hide the real signer of m, and hence, this f_0 is partitioned using the Shamir's secret sharing scheme [20] and verifiably encrypted (VE) in n shares V_k, one for each user of the group, using the public parameters of all the group members $\{(y_1, \alpha_1), (y_2, \alpha_2), \cdots, (y_n, \alpha_n)\}$.

$$s_k \leftarrow f_0 + \sum_{j=1}^{t-1} f_j \alpha_k^j, k = 1, \cdots, n,$$

$$V_k \leftarrow VE_{y_k}(s_k : g^{s_k} = \hat{g} \prod_{j=1}^{t-1} F_j^{\alpha_k^j}), k = 1, \cdots, n, \text{where}, \hat{g} \leftarrow g^l.$$

Then, the user generates another tuple of n random numbers $r_j \in Z_q^*$ and computes $w_j = g^{r_j}$ for each of them. He/She also calculates $\hat{y}_w \leftarrow \hat{g}^{x_i + r_i}$. Finally, he/she computes an equality signature [13] $(EC, ES) \leftarrow S_{SEQDL}(\hat{g}, g, x_i, r_i, \hat{y}_w, Y, W, m)$ and n knowledge signatures $\{(kc_k, ks_k) \leftarrow S_{SKDL}(g, w_k, m), k = 1, \cdots, n\}$ (with $Y \leftarrow y_1, \cdots, y_n, W \leftarrow w_1, \cdots, w_n, KC \leftarrow kc_1, \cdots, kc_n, KS \leftarrow ks_1, \cdots, ks_n$) that allow the signer to prove in zero-knowledge the integrity of the signed report and its group authenticity. The output of the signature algorithm is a threshold discernible ring signature $\sigma = (\sigma_1, \sigma_2)$ where, $\sigma_1 \leftarrow (\hat{g}, \hat{y_w}, Y, W, EC, ES, KC, KS)$ and $\sigma_2 \leftarrow (V, F)$ with $V \leftarrow V_1, \cdots, V_n$, and $F \leftarrow F_1, \cdots, F_t$.

B.2   Verification

The verification algorithm $V_{TDS}(m, \sigma)$ contains two actions: (1) checking the origin discernibility of the signature, i.e. the encrypted shares of the secret f_0 are verifiable and thus, a coalition of t users could reveal the identity of the signer,

$$Verify(VE_{y_k}(s_k : g^{s_k} = \hat{g} \prod_{j=1}^{t-1} F_j^{\alpha_k^j}) = 0, \text{ for any } i = 1, \cdots, n).$$

and (2) verifying the ring signature, i.e. checking that some member of the group with a valid private key has signed m and, thus, that m is authentic and integral. For this, a user first executes a proof of knowledge procedure [4] $V_{SKDL}(g, w_k, m)$ for any $i = 1, \cdots, n$, to check that the signer knows the n random numbers $r_j \in Z_q^*$ used in the signature. Then, it executes the verification algorithm of the signature of knowledge of equality of discrete logarithms $V_{SEQDL}(\hat{g}, g, \hat{y}_w, Y, W, EC, ES, m)$.

B.3   Threshold Distinguisher

The threshold distinguisher algorithm requires that at least t members of the ring decrypt their secret share V_i with their private key x_i to obtain ρ_i. Then, these users have to share their respective ρ_i's to disclose the secret element of the signature f_0. This can be computed using Lagrange's interpolation formula. After obtaining f_0, the users will be able to discover the signer of the message yielding the user P^i that matches the following equation: $(y_i w_i)_0^f = \hat{y}_w$.

C   The Reward Redemption Protocol

The EMS responds the witness, the hops and the reporter (immediately or after some days) with a reward for reporting a true emergency. The witness receives a reward encrypted with k_m, which is only known to the witness. The hops and the reporter receive the rewards encrypted with their corresponding public keys.

In order to redeem the rewards from the CC, the awardees proceed as follows. (1) Each awardee A_i generates a pseudo-identity (PI) with the help of a CA. This PI is used by A_i for redeeming a reward at the CC anonymously. (2) On receiving a request from A_i for generation of PI, the CA selects a secret random number $b \in Z_p^*$, encrypts it with A_i's public key and sends it to A_i. Thus, CA and all the awardees share a secret number b. A_i deciphers b, selects a random number $a \in Z_p^*$ and uses his/her secret key to sign $\{ID_{A_i}, \text{Cert}_{CA}(A_i), b, a\}$. A_i computes his/her PI by using a hash function: $PI_{A_i} = H(ID_{A_i}, \text{Cert}_{CA}(A_i), b, a, \text{Sign}_{A_i}(\text{Cert}_{CA}(A_i), b, a))$. (3) A_i generates a key pair $(y_{A_i}^*, x_{A_i}^*)$, signs the public key with his/her private key, and sends $\text{Sign}_{A_i}(y^*_{A_i}, PI_{A_i})$ to CA. CA verifies the signature using the public key of A_i. If valid, CA generates an anonymous certificate $\text{Cert}_{CA}(PI_{A_i}, y^*_{A_i})$ and sends it to A_i. **(4)** A_i sends a payoff redeem request, $payoff_{Req} = \{PI_{A_i}, \text{Cert}_{CA}(PI_{A_i})\}$, to the CC. **(5)** CC verifies the received certificate from the CA of the system. If verified, CC generates a session key k_{A_i}, encrypts it with A_i's public key and sends it to A_i. Otherwise, CC aborts the redemption process. **(6)** A_i encrypts the received *payoff* and the signed hash using k_{A_i} and sends $payoff_{Req} = \{C_{k_{A_i}}(payoff, Sign_{KS_{EMS}}(H_{EC})), \text{Cert}_{CA}(PI_{A_i}), PI_{A_i}\}$ to CC. **(7)** CC performs decryption with k_{A_i} and obtains the clear text $Sign_{KS_{EMS}}(H_{EC})$ and *payoff*. CC first checks if PI_{A_i} has already redeemed the *payoff* by looking up $\{Sign_{KS_{EMS}}(H_{EC}), payoff, PI_{A_i}\}$ in its database. If no such entry exists, CC sends $Sign_{KS_{EMS}}(H_{EC})$ to the EMS for validation. If the *payoff* has already been redeemed by PI_{A_i}, CC aborts the redemption process. **(8)** If the received H_{EC} is equal to the stored H_{EC}, the EMS sends *accept* notification to the CC. On receiving *accept*, CC sends rewards to A_i. CC then sets a redemption flag to 1 and stores $\{FL = 1, \text{Cert}_{CA}(A_i), PI_{A_i}, payoff, Sign_{KS_{EMS}}(H_{EC})\}$ in its database.

References

1. Facebook (2004). http://www.facebook.com/. Accessed 23 Jun 2016
2. Alpify: An app that can save your life (2014). http://www.alpify.com. Accessed 23 Jun 2016
3. Bellare, M., Rogaway, P.: Random oracles are practical: a paradigm for designing efficient protocols. In: Proceedings of the 1st ACM Conference on Computer and Communications Security, CCS 1993, NY, USA, pp. 62–73. ACM, New York (1993)
4. Camenisch, J.L.: Efficient and generalized group signatures. In: Fumy, W. (ed.) EUROCRYPT 1997. LNCS, vol. 1233, pp. 465–479. Springer, Heidelberg (1997)
5. Committe, E.: False emergency calls. Operations Document 3.1.2, European Emergency Number Association (EENA) (2011)
6. Domingo-Ferrer, J.: Coprivacy: an introduction to the theory and applications of cooperative privacy. SORT Stat. Oper. Res. Trans. **35**, 25–40 (2011)
7. Domingo-Ferrer, J., Gonzàlez-Nicolàs, U.: Rational behavior in peer-to-peer profile obfuscation for anonymous keyword search: the multi-hop scenario. Inf. Sci. **200**, 123–134 (2012)
8. Domingo-Ferrer, J., Sànchez, D., Soria-Comas, J.: Co-utility-self-enforcing collaborative protocols with mutual help. Prog. AI **5**(2), 105–110 (2016)
9. Domingo-Ferrer, J.: Coprivacy: towards a theory of sustainable privacy. In: Domingo-Ferrer, J., Magkos, E. (eds.) PSD 2010. LNCS, vol. 6344, pp. 258–268. Springer, Heidelberg (2010)
10. Domingo-Ferrer, J., Bras-Amorós, M., Wu, Q., Manjón, J.: User-private information retrieval based on a peer-to-peer community. Data Knowl. Eng. **68**(11), 1237–1252 (2009)
11. Domingo-Ferrer, J., Megías, D.: Distributed multicast of fingerprinted content based on a rational peer-to-peer community. Comput. Commun. **36**(5), 542–550 (2013)
12. Furtado, V., Ayres, L., de Oliveira, M., Vasconcelos, E., Caminha, C., D'Orleans, J., Belchior, M.: Collective intelligence in law enforcement - the wikicrimes system. Inf. Sci. **180**, 4–17 (2010)

13. Klonowski, M., Krzywiecki, Ł., Kutyłowski, M., Lauks, A.: Step-out ring signatures. In: Ochmański, E., Tyszkiewicz, J. (eds.) MFCS 2008. LNCS, vol. 5162, pp. 431–442. Springer, Heidelberg (2008)
14. Kumar, S., Agrawal, S., Venkatesan, R., Lokam, S.V., Rangan, C.P.: Threshold discernible ring signatures. In: Obaidat, M.S., Tsihrintzis, G.A., Filipe, J. (eds.) ICETE 2010. CCIS, vol. 222, pp. 259–273. Springer, Heidelberg (2012)
15. Meier, P.: Digital Humanitarians: How Big Data is Changing the Face of Humanitarian Response. CRC Press, Boca Raton (2015). chap. 4
16. Namahoot, C.S., Brückner, M.: SPEARS: smart phone emergency and accident reporting system using social network service and Dijkstra's algorithm on Android. In: Kim, K.J., Wattanapongsakorn, N. (eds.) Mobile and Wireless Technology 2015. LNEE, vol. 310, pp. 173–182. Springer, Heidelberg (2015)
17. Nisan, N., Roughgarden, T., Tardos, E., Vazirani, V.V.: Algorithmic Game Theory. Cambridge University Press, New York (2007)
18. Okolloh, O.: Ushahidi or 'testimony': web 2.0 tools for crowdsourcing crisis information. Participatory Learn. Action 59, 65–70 (2009)
19. Reed, M., Syverson, P., Goldschlag, D.: Anonymous connections and onion routing. IEEE J. Sel. Areas Commun. 16(4), 482–494 (1998)
20. Rivest, R.L., Shamir, A., Tauman, Y.: How to leak a secret. In: Boyd, C. (ed.) ASIACRYPT 2001. LNCS, vol. 2248, pp. 552–565. Springer, Heidelberg (2001)
21. How mobile and cloud technologies are reducing emergency response times. White paper, Tapshield (2014). http://tapshield.com/white-paper-mobile-cloud-technologies-reducing-emergency-response-times. Accessed 23 Jun 2016

Author Index

Printed in the United States
By Bookmasters